DEMOLITION AND REUSE OF CONCRETE AND MASONRY

Publisher's Note

This book has been produced from camera ready copy provided by the individual contributors.

This method of production has allowed us to supply finished copies to the conference delegates in advance of the meeting.

DEMOLITION AND REUSE OF
CONCRETE AND MASONRY

VOLUME ONE

Demolition Methods and Practice

Proceedings of the Second International Symposium held by RILEM (the International Union of Testing and Research Laboratories for Materials and Structures) organized by the Building Research Institute, Ministry of Construction, Japan and co-organized by Nihon University, Japan.

Nihon Daigaku Kaikan
Tokyo, Japan
November 7–11, 1988

EDITED BY
Y. Kasai

Taylor & Francis
Taylor & Francis Group

LONDON AND NEW YORK

First published in 1988 by
Taylor and Francis,
2 Park Square, Milton Park, Abingdon,Oxon, OX14 4RN
52 Vanderbilt Avenue, New York, NY 10017

First issued in paperback 2020

*Taylor & Francis is an imprint of the Taylor & Francis
Group, an informa business*

British Library Cataloguing in Publication Data

Demolition and reuse of concrete and
   masonry
   1. Concrete structures. Demolition
   I. Kasai, Y.
   624'.1834

ISBN 0 412 32110 6
ISBN 0 412 34480 7 V.1
ISBN 0 412 34490 4 V.2

ISBN 13: 978-0-412-32110-8 (set)
ISBN 13: 978-0-367-65945-5 (pbk)
ISBN 13: 978-0-412-34480-0 (hbk)

ORGANIZING COMMITTEE

Dr S. Fujimatsu	*General Director, Building Research Institute, Ministry of Construction, (Chairperson), Japan*
Professor A. Enami	*Nihon University, Japan*
Professor K. Kamimura	*University of Utsunomiya, Japan*
Professor Y. Kasai	*Nihon University, Japan*
Mr K. Kato	*Tokyo Electric Power Company, Japan*
Professor K. Kishitani	*Nihon University, Japan*
Mr S. Koizumi	*Building Research Institute, Ministry of Construction, Japan*
Professor S. Nagataki	*Tokyo Institute of Technology, Japan*
Mr N. Sato	*Ministry of Construction, Japan*
Mr T. Tanimoto	*Public Works Research Institute, Ministry of Construction, Japan*
Professor F. Tomosawa	*University of Tokyo, Japan*
Mr M. Yokota	*Japan Atomic Energy Research Institute, Japan*
Professor T. C. Hansen	*Technical University of Denmark, Denmark*
Mr M. Fickelson	*General Secretariat, RILEM, France*

NATIONAL EXECUTIVE COMMITTEE

Professor K. Kamimura	*University of Utsunomiya (Chairperson), Japan*

INTERNATIONAL SCIENTIFIC COMMITTEE

Professor Y. Kasai	*Nihon University, (Chairperson), Japan*
Mr T. Egashira	*Kaihatu Denki Co., Ltd, Japan*
Mr T. Fujii	*Shimizu Corporation, Japan*
Professor T. Fukuchi	*Nihon University, Japan*
Professor T.C. Hansen	*Technical University of Denmark, Denmark*
Dr C.F. Hendriks	*Road Engineering Division, Rijkswaterstaat, Netherlands*

Dr H. Kaga	*Taisei Corporation, Japan*
Dr T. Kemi	*Institute of Technology, Toda Construction Co., Ltd, Japan*
Dr K. Kleiser	*University of Karlsruhe, West Germany*
Mr S. Kobayashi	*Public Works Research Institute, Ministry of Construction, Japan*
Mr E.K. Lauritzen	*Demex, Denmark*
Dr P. Lindsell	*University of Oxford, England*
Professor Y. Malier	*Laboratoire Central des Ponts et Chaussees, France*
Dr Y. Masuda	*Building Research Institute, Ministry of Construction, Japan*
Mr M.T. Mills	*Institute of Demolition Engineering, Griffiths-McGee Demolition Co., Ltd, England*
Mr P. Mohr	*A/S Skaninavisk Spendbeton, Denmark*
Dr C. Molin	*Statens Provningsanstalt (National Testing Institute), Sweden*
Professor T. Mukai	*Meiji University, Japan*
Dr M. Mulheron	*University of Surrey, England*
Professor S. Nagataki	*Tokyo Institute of Technology, Japan*
Professor N. Nishizawa	*Chuou University, Japan*
Mr C. de Pauw	*Centre Scientifique et Technique de la Construction, Belgium*
Dr G. Ray	*Concrete Paving Consultant, USA*
Mr E. Rousseau	*Centre Scientifique et Technique de la Construction, Belgium*
Professor M. Sakuta	*Nihon University, Japan*
Dr R. Schulz	*Institute fur Baustoffprufung, West Germany*
Professor T. Soshiroda	*Shibaura Institute of Technology, Japan*
Mr Y. Takahashi	*Building Research Institute, Ministry of Construction, Japan*
Professor F. Tomosawa	*University of Tokyo, Japan*
Professor K. Torigai	*Science University of Tokyo, Japan*

Contents

VOLUME ONE   DEMOLITION METHODS AND PRACTICE

VOLUME TWO REUSE OF DEMOLITION WASTE

xiv

Preface

RILEM Technical Committee 37 DRC on Demolition and Reuse of Concrete was formed in 1976 and held its first meeting at the Building Research Station in Garston (UK) in June of 1977 under the chairmanship of Dr L.H. Everett. In 1978 the first RILEM TC 37–DRC state-of-the-art report was published on recycled concrete as an aggregate for concrete [1].

After the committee was reorganized in 1981 and the author of this preface became chairman, a second committee meeting was held in Copenhagen (DK) in December 1982. Since then the committee has held yearly meetings in the Netherlands, England, Belgium, France and Japan.

The following general terms of reference of the committee were agreed on at the meeting in Copenhagen in 1982.

1. To study the demolition techniques used for plain, reinforced, and prestressed concrete and to consider developments in techniques.
2. To study technical aspects associated with reuse of concrete and to consider economical, social and environmental aspects of demolition techniques and reuse of concrete.

Three task forces were formed, each with its own specific terms of reference.

Task Force 1 surveyed, on the basis of the existing literature, methods of demolition and fragmentation including economic, social and environmental aspects. It published its findings in two state-of-the-art reports, one on demolition techniques in general [2] and another on blasting of concrete [3].

Task Force 2 collected and surveyed codes and regulations concerning demolition in various countries. It did not issue a separate state-of-the-art report. Instead its findings were included in [2,3,4,5].

Task Force 3 studied technical aspects associated with reuse of concrete and considered economic, social and environmental factors. It published its findings in two state-of-the-art reports, one on the reuse of concrete as concrete aggregates [4] and another report on the reuse of mixed concrete and masonry rubble as aggregate for concrete [5].

The committee arranged the first international symposium on demolition and recycling of concrete in Rotterdam in 1985 in co-operation with the European Demolition Association (EDA). The symposium proceedings were published in [6] and [7]. The symposium gave valuable input to the work of the committee from an industrial point of view. Developments were fast, and it was soon decided to hold a second international RILEM symposium on demolition and reuse of concrete already in 1988 in Tokyo in order once more to make it possible for people from science and practice and from all over the world to communicate and exchange experience before the committee is dissolved at a final meeting in Tokyo in 1988. As chairman of RILEM TC 37–DRC I sincerely hope the symposium will be successful and that the proceedings will serve as a source of inspiration for further research and development in the fields of demolition and reuse of concrete.

Moreover, I want to thank the following persons who have served as members and corresponding members of the committee over the years. Members Mr R.C. Basart (NL), Dr Ch.F. Hendriks (NL), Professor P. Lindsell (GB), Professor Y. Kasai (Japan), Dr K. Kleiser (D), Dr R.R. Schulz (D), Professor Y. Malier (F), Mr R. Hartland (GB), Mr T.R. Mills (GB), Mr P. Mohr (DK), Dr C. Molin (S), Mr G. Ray (USA), Mr C. de Pauw (B), Mr E. Rousseau (B), Mr E.K. Lauritzen (DK), Secretary from 1982–1985, and Dr M. Mulheron (GB), Secretary from 1985–1988. Corresponding members: Mr F.D. Beresford (AUS), Mr M. Whelan (AUS), Mr A.D. Buck (USA), Dr S. Frondistou-Yannas (USA), Mr J.M. Loizeaux (USA), Mr J.F. Lamond (USA). Our very special thanks go to the European Demolition Association for its loyal co-operation in the work of the committee, and to the Japanese Building Research Institute for organizing this symposium.

Torben C. Hansen

LIST OF REPORTS ISSUED BY RILEM TECHNICAL COMMITTEE 37–DRC

[1] Nixon, P.J. (1978) Recycled concrete as an aggregate for concrete – a review. *Materials and Structures*, **11** (65) September–October, pp. 371–8.

[2] Task Force 1 – RILEM Technical Committee 37–DRC (1985) *Demolition Techniques*. European Demolition Association, Wassenaarseweg 80, 2596 CZ Den Haag, The Netherlands, Special Technical Publication, May.

[3] Molin C. and Lauritzen, E. (1988) *Blasting of Concrete. Localized Cutting and Partial Demolition of Concrete*. Special Report 09, National Testing Institute, Box 5608, 11486 Stockholm, Sweden.

[4] Hansen, T.C. (1986) Recycled aggregates and recycled concrete aggregate. Second state-of-the-art report, developments 1945–1985. *Materials and Structures*, **19** (111), May–June, pp. 201–46.

[5] Hendriks, Ch.F. and Schulz, R.R. Recycled masonry rubble as an aggregate for concrete. State-of-the-art report, developments 1945–1985. *Materials and Structures* (in press).

[6] EDA–RILEM (1985) *Demolition Techniques. Proc. First International EDA–RILEM Conference on Demolition and Reuse of Concrete*, Rotterdam, 1–3 June, **1**, European Demolition Association, Wassenaarseweg 80, 2596 CZ Den Haag, The Netherlands.

[7] EDA–RILEM (1985) *Reuse of Concrete and Brick Materials. Proc. First International EDA–RILEM Conference on Demolition and Reuse of Concrete*, Rotterdam, 1–3 June, **2**, European Demolition Association, Wassenaarseweg 80, 2596 CZ Den Haag, The Netherlands.

Foreword

The buildings and the urban areas where we live have had to change in harmony with the needs of the age. In these circumstances, it is natural to consider demolition work as being as important as construction work in the sound development of urban areas. Demolition work is inevitable for the revitalization of urban areas. In congested cities, in particular, the quality of the demolition technique becomes an essential element which determines the success of the revitalization of the city. In addition to efficiency in demolition, strategies must be adopted to avoid noise, vibration and dust which affect the surrounding environment and there must be efficient disposal of the waste products. The effective reuse of these waste products can be of great importance socially as well as economically.

As far as Japan is concerned, concrete and masonry construction has only a one hundred year history. Since the great earthquake in the Kanto region, reinforced concrete has been used for the main, large buildings. Soon after the second world war, many buildings of poor quality were constructed because of the need for low-cost buildings and they remain even now. To date, many buildings with reinforced concrete have been demolished and new buildings which satisfy modern requirements have been constructed in their place.

In this symposium, various demolition techniques which have been accumulated and tested in recent years will be presented. They will be helpful for demolition work all over the world. Some presentations will report research carried out on demolition work at atomic power stations or at buildings where asbestos fibres were used as building materials. I believe that they will meet our expectations.

I would like to thank all the members of RILEM TC37 for their honest efforts to undertake research on demolition techniques over a long period. I would also like to extend my thanks to the members of the International Scientific Committee who have worked hard to hold this symposium. I hope it will be useful in efficiently establishing the cities of tomorrow.

Dr Susumu Fukimatsu

Introduction

At this time when the second RILEM Symposium on the Demolition and Reuse of Concrete and Masonry is to be held in Japan, I would briefly like to describe the history of research and development in this field in Japan. In the near future the vibrationless and noiseless methods of demolition developed for the congested urban areas found in Japan will be utilized for demolition work in cities around the world.

As far as concrete structures are concerned the mechanisation of demolition work started in the late 1950s with the introduction of pneumatic hand hammer breakers and steel balls. Using these tools the slabs of multi-storey buildings were first broken up by 'balling' and then the beam ends were crushed by hand hammer breakers. Finally the remaining large walls, often of several spans and storeys, were then felled in a single operation. However, this method was dangerous and resulted in a number of accidents.

Due to the high economic growth policy and changes in peoples' attitudes, there were frequent complaints and claims made against demolition activities in congested cities. As a result the construction industry was obliged to develop safer, quieter and less intrusive methods of demolition.

The research committee on 'Demolition and Removal Methods' was started in these days and the first technical book in this field *Demolition Method for Concrete Structures* was published in 1970. In 1971 the Building Contractors Society (BCS) started the 'Committee on Demolition of Reinforced Concrete Structures' which drew its members from the research and engineering staff of major contractors along with academics from various universities. This committee conducted research and development work into a variety of demolition techniques such as jacking, explosives, and rebar heating methods by either direct or induced current. In 1978 the committee published the *Standard of Public Nuisanceless Demolition Method of Reinforced Concrete Structures* and in 1987 also published a *Recommended Proposal of Demolition Method of Underground Reinforced Concrete Structures*.

During this period, general contractors and manufacturers of construction plant joined together in developing hydraulic 'C-shaped' concrete crushers, diamond cutters and flame jetting methods. A most important development was the introduction of concrete crushers from England in about 1975. Stimulated by this, and using the experience gained from the production of 'C-shaped' crushers a number of highly efficient concrete crushers were developed and these can be seen in use to this day. The development of chemical expansive demolition agents, started in 1967, was a solely Japanese initiative. They were available as commercial products by 1978. Developments have continued in this area and today it is possible to purchase materials capable of crushing concrete within one hour of application.

The study of the demolition of the Japan Power Demonstration Reactor (JPDR) by the Japan Atomic Energy Research Institute (JAERI), which started in about 1979, resulted in many useful developments for the demolition of reinforced concrete structures by explosives, large diamond cutters, core boring machines and abrasive water jetting. In addition methods for stripping surface concrete by the use of microwaves were investigated. The removal of surface concrete by the rebar heating method using alternating current was another Japanese development and after the initial experimental trials were complete in 1968, it was later used for the demolition of special structures. In 1981, diamond wire saw for cutting reinforced concrete was introduced and it is expected that this method will be the subject of further developments over the coming year.

Study of the reuse of concrete waste in Japan started in about 1971. In 1974 the BCS formed the 'Committee on Disposal and Reuse of Construction Waste'. This committee conducted many successful experiments on the production of recycled concrete aggregate, the study of recycled concrete, chippings from waste wood and the production of wood chip concrete. In 1977 this body published the *Proposed Standard for the Use of Recycled Aggregate and Recycled Aggregate Concrete*. Later on, during the period 1981–1985, the Ministry of Construction conducted a study to encourage the reuse of construction waste for new construction work and introduced a standard for the reuse of demolished concrete and waste timber.

In this way the methods of demolishing concrete structures have developed rapidly in Japan in order to meet the strict requirements for demolition methods imposed by its citizens. The successful development of these methods is the result of the efforts of the demolition related industries, academic institutions and public authorities who have joined forces to ensure that these requirements are met. Indeed the reason for holding the Symposium in Tokyo is that the RILEM Committee 37–DRC considered that the development of demolition methods and use of modern demolition techniques was well advanced in Japan.

We sincerely expect that this Symposium will contribute to the development of Demolition and Reuse of Concrete and Masonry Structures.

Yoshio Kasai
Chairman of Scientific Committee

OUTLINE OF VARIOUS DEMOLITION
METHODS AND THEIR EVALUATION

Y. KASAI
College of Industrial Technology, Nihon University, Japan

E. ROUSSEAU
Belgian Building Research Institute, Belgium

P. LINDSELL
Department of Eng. Science, University of Oxford, England

Abstract
This paper describes the general application of demolition methods.
Firstly, they are classified into 11 categories based on the breaking
mechanisms, and according to demand some of them are stated in 2 ~ 7
items. Secondly, the evaluation of main demolition methods is
mentioned and two large tables on evaluation are presented.
Key words: Demolition, Mechanism, Machine, Evaluation, Nuisance.

1. Introduction

The evaluation of demolition methods for some construction is affected
greatly by ① "structural form" e.g. a reinforced concrete and steel
frame reinforced concrete structure and plain concrete and masonry, ②
"scale of construction" e.g. small or large, ③ "location of building"
e.g. placed in urban centre or suburb, ④ "range of demolition" e.g.
partial or whole demolition, ⑤ "use of construction" e.g. ordinary or
special purpose construction, ⑥ "grade of safety in work", ⑦
"permitted noise, vibration, dust or so", ⑧ "period of demolition work"
and the like.
    This paper deals with two viewpoints, one is the outline of current
demolition methods and the other is the evaluation of main demolition
methods based on the various factors mentioned above.
At first, the demolition methods are classified by breaking principles
or mechanisms and potential applications.
    The second, the main demolition methods are evaluated by "oper-
ations", "application fields", "nuisance", "safety measures" and
"performance". The evaluation of demolition methods is a very difficult and complicated job, because these parameters interact with each
other and have many aspects. The evaluation check-list is composed by
considering many factors to provide a representative description.

2. Classification of demolition methods

The demolition methods are classified into principles and mechanisms
of breaking and only demolition methods which are printed in capital
letters are evaluated in Table 1 a) and b).

2.1 Simple demolition (machine)

1

The demolition methods, e.g. bucket, rock shovel bucket, rubble shovel bucket, heavy duty bucket and demolition boom and elephant tusk are classified into the simple demolition method, because these are applicable to demolish masonry in very simple way. These methods are described by the Report of RILEM Technical Committee 37-DRC (1985).

2.2 Hammering
Repeated hammering can be used to crack and break concrete.
(a) Stone chisel; A stone chisel is operated by manpower but its use is limited to demolishing a light brick house.
    (b) HAND HAMMER; There are two types of hammer, one is operated by pneumatic force, the other by hydraulic force. The former is lighter in weight, smaller in size and causes more noise than the latter.
    (c) LARGE SIZE HAMMER; This large and powerful hammer is mounted on a crawler type or wheel type machine. There are two types of energy transmission to a hammering rod, one is pneumatic force and the other is hydraulic force. The pneumatic hammer makes more noise than the hydraulic one, and therefor hydraulic hammer is widely used in urban areas.

2.3 Hitting
Repeated hitting by a steel ball can demolish concrete and masonry.
(a) STEEL BALL; A steel ball of weight 0.5 ~ 2.0 ton is suspended by a crawler type crane and used to demolish concrete and masonry by repeated impact. It is widely used in European countries, but is rarely used in urban areas of Japan, because it poduces too much noise, vibration and dust.
    (b) Balling machine; This machine is built up to demolish a concrete pavement by dropping a steel ball automatically.

2.4 Jack up
The demolition machine using a "HYDRAULIC JACK" was developed to demolish reinforced concrete buildings in Japan. The floor slab is designed to support dead load and live load downwards, so when a pointed load is applied upwards from the bottom face of slab,it is demolished easily. It is rarely used nowadays.

2.5 Breaking by hydraulic pressure
(a) HYDRAULIC BREAKER; This demolition machine can break concrete using hydraulic pressure to hold and crush the section in a solid C-shaped frame or strong jaws. The C-shaped breaker is suspended by a crawler crane. The jaw-type breakers are mounted on top of the arm of the crawler machine. Four or five different types are available.
    (b) Pile crusher; The pile crusher can break a concrete pile having a logitudinal hole in the centre. A strong band combined with a hydraulic jack is set up to hold the pile and the jack-force breaks the concrete pile by crushing across the diameter in various directions.

2.6 Abrasion
There are different demolition machines based on abrasion, e.g. a hammer drill, core boring machine, diamond disc cutter, wire saw, water jet and abrasive water jet.

2

(a) HAND HAMMER DRILL; This drill is used for boring into concrete. The drill bit hits concrete over 1000 times and turns about 300 times in one minute, then concrete is reduced to powder by the abrasion.

(b) LARGE SIZE HAMMER DRILL; This drill is mounted on a self-propelled machine and is used to bore massive or mat concrete.

(c) DIAMOND BORING MACHINE; The diamond core boring machine is mounted on a self-propelled machine and is used to bore holes of 100 ∼ 150 mm in diameter. When the holes are drilled side by side, concrete can be cut. This demolition method is applied to concrete shields in nuclear power plants.

(d) DIAMOND DISC CUTTER; This disc cutter is capable of cutting reinforced concrete. A disc over 1000 mm in diameter is available. It was applied to demolish the telegraph station buildings in Japan.

(e) DIAMOND WIRE SAW; A loop wire saw with a diamond bit can cut around the circumference of the concrete section. This cutting method is effective in demolishing foundations and massive concrete.

(f) Water jet; A high pressure water jet about 250 ∼ 300 MPa from a nozzle of about 0.3 ∼ 0.5 mm in diameter can cut only plain concrete by abrasion.

(g) ABRASIVE WATER JET; An abrasive water jet which contains garnet or steel particles can cut reinforced concrete. This method is now being developed for use in demolishing nuclear power plants.

2.7 Bursting

Concrete and rock can be split by bursting force in a hole. This method is classified into static bursting e.g. burster with wedge, and chemical expansive demolition agent , and dynamic bursting e.g. water gun, gas cylinder, CARDOX, milde explosive and explosive. The safety and environmental control in practice is most important because of spouting out of expansive material or gas.

(a) BURSTER WITH WEDGES; A set of mechanical wedges is forced into a bored hole and breaks the concrete by expanding. This method is used to split plain concrete and masonry.

(b) CHEMICAL EXPANSIVE DEMOLITION AGENT; Unslaked lime composite named "chemical expansive demolition agent or chemical splitting agent" is mixed or absorbed with water and injected or charged into a hole. The expansion of the mixture by hydration causes splitting the concrete.

(c) Water gun; Water having a pressure of 40 MPa shoots into the drilled hole in which is filled with water and splits concrete.

(d) Gas cylinder; A flexible gas pipe which is connected to a inorganic gas vessel is fitted to a bored hole with a special rubber stopper. The gas valve is then opened to allow pressure into the hole to break the concrete.

(e) CARDOX; Liquid carbon dioxide (CO_2) which is enclosed in a metallic tube is inserted into the concrete hole and heated by an electrical filament. It expands immediately and breaks the concrete.

(f) MILD EXPLOSIVES; The blasting speed of these explosives is about 30m/sec or so. These explosives break concrete with little noise and vibration, compared with the ordinary explosives.

(g) EXPLOSIVES; The blasting speed is 4000 ∼ 7000 m/sec. These explosives are very powerful.

2.8 Melting
Concrete and rebars are melted by the heat of combustion of metal or organic fuel, plasma and laser beam.

(a) THERMIC LANCE; The aluminum alloy or iron alloy wires are enclosed in a same metal pipe of about 14 mm or 18 mm in diameter. At first, the metal lance is ignited using acetylene gas flowing between the wires in the pipe. The acetylene gas is turned to oxygen gas and the metal lance continues to burn. The heat of combustion can melt concrete and rebars. Another lance is available using a metal powder which is called a "powder lance".

(b) FUEL OIL FLAME; The heat of combustion of a mixture of a high pressure kerosene and oxygen gas jet melts concrete and rebars. Another gas flame using organic gas is available which is called the "organic gas flame method".

(c) Plasma; Concrete can be melted by Argon-Hydrogen or Argon-Nitrogen plasma. This method is under development.

(d) Laser beam; A carbon dioxide laser beam can melt and cut concrete including the rebars. This is also under development.

2.9 Heating and peeling concrete
Concrete is peeled by heating rebars electrically or heating concrete by a microwave.

(a) HEATING REBARS BY ELECTRIC CURRENT; The rebars are heated one by one by an electric current up to 400 ~ 450 ℃. Then the concrete cracks away from the rebars and as a result the cover concrete to the rebar can be peeled off. This method was developed for the demolition of concrete structures in the nuclear power plant industry. Recently, it has been applied to make a large opening for an under-ground concrete structures.

(b) Heating rebars by induction current; An induction current can heat the rebars from outside the concrete. This method is still under development.

(c) MICROWAVE; Concrete can be peeled by microwave heating the outer 10 ~ 15 mm layer of concrete. This method is also under development.

2.10 Shearing
Shears can only cut rebars and shaped steels.

(a) Hand shears; There are two types of hand shears. One is worked by manpower and the other is worked by a small hydraulic apparatus. They can cut slender rebars.

(b) HYDRAULIC LARGE SIZE SHEARS; These shears are worked by hydraulic power and mounted on a self-propelled machine. This device is capable of cutting small H-shaped steel sections.

2.11 FELLING
(a) Felling of columns and smokestacks; The foot of a column or a smokestack is cut into a V-shape to the intended felling side. It is felled by the imbalance of weight and pulling with wireropes.

(b) Felling by blasting of explosive; The column foot and beam ends are broken by millisecond delayed ignition of explosives. The building should fall down in a predetermined direction.

4

3. Evaluation of main demolition methods

The evaluation of demolition methods is described briefly under the classification of the operations, application fields, nuisance, safety measures, performance.

3.1 Operations
The operations of machines are classified by the "need of pre-work", "size of demolished materials" and "need of second demolition".
(a) Need of pre-work; Most demolition methods do not need pre-work, but the bursting methods e.g. the burster with wedges, chemical expansive demolition agent, explosives and the like need bored holes to insert the wedgas, or to inject or charge bursting materials.

(b) Size of demolished materials; The size of demolished materials depends on the demolition method, e.g. a hammer breaker and a hydraulic breaker can break concrete into small rubble, a steel ball, chemical expansive demolition agent, explosives, generally into medium sizes, and a diamond cutter and wire saw into large sections.

(c) Second demolition; When the size of demolished materials is middle or large, they need to be broken into smaller pieces again because of the permission for use in land reclamation,

3.2 Application fields
The application field of demolition methods is classified by the "type of structure","structural member","demolition range", "application in urban centre" and "frequency".
(a) Type of structure; The application of demolition methods to structures has a peculiar aspect, because of the principle of demolition or breaking. The type of structure is classified broadly into four types by materials e.g. "reinforced concrete", "steel frame reinforced concrete", "plain concrete" and "masonry". A chemical expansive demolition agent can demolish the plain concrete and masonry, but can not demolish the heavily reinforced concrete structures.

(b) Structural member; The application of demolition methods to structural members depends on the principle of breaking, size and power of the machine, e.g. a large size hammer is applicable to all members, but the chemical expansive demolition agent and explosives can not usually be applied to thin walls and floor slabs. Large size shears can cut rebars and medium size steel sections.

(c) Demolition range; The demolition range varies according to the demolition methods, e.g. a large size hammer and hydraulic breaker can demolish selected parts or a whole structure, but a steel ball and instantaneous felling demolition by explosives are only applicable to total demolition. The diamond disc cutter and abrasive water jet can cut members, so these methods are suitable to partial demolition.

(d) Application in urban centre; If a demolition method produces a lot of noise, vibration and dust, a lot of care is needed to decide whether it should be used in an urban centre. The hydraulic breaker is an excellent demolition machine, but instantaneous felling demolition is very difficult to use in crowded urban centres such as Tokyo.

Table-1 a) Evaluation of main demolition methods
-Operations and application fields-

| Demolition methods | Principle of breaking | Mounted or accompanied machine | Operations | | | Application fields(2) | | | | | | Frequency |
			Need of prework	Size of the demolished materials	Type of construction (1)	Column Beam	Wall slab Floor slab	Foundation	Partial demolition	Whole demolition	Application in urban centres	
Hand hammer	Chopping of material by repeated shock at a point	An air compresser or hydraulic pump	No	Small	Rc. C.M.	O	O	O	O	—	High	Special use
Large size hammer		Crawler or wheel type machine	No	Small	Rc. C.M.	O	O	O	O	O	High	Quite Common
Steel ball	Repeated hitting by a steel ball	Crawler type machine	No.	Med-ium	Rc. C.M.	O	O	X	X	O	Low nui-sance	Rare(Japan) Quite Common (EC.)
Hydraulic jack	Bending and shearing break of floor slab and beam	Crawler type machine	No	Med-ium	Rc.	C:X B:△	W:X F:O	X	O	X	High	Rare
Hydraulic breaker	Breaking by hydraulic pressure of the jaws	Crawler type machine	No	Small	Rc. C.M.	O	O	—	O	O	High	Quite Common
Hand hammer drill	Drilling by repeated and rotational percussion	An air compresser or hydraulic pump	No	—	Rc. C.M.	O	O	O	O	X	High	Quite Common
Large size hammer drill	Drilling by a bit	Crawler or wheel type machine	No	—	Rc. C.M.	O	O	O	O	X	Med-ium	Special use
Diamond boring machine	Drilling by rotation and abrasion of a diamond drill	Crawler or wheel type machine	No	Large	Rc. C.M.	O	O	O	O	△	High	Special use
Diamond disc cutter	Cutting by abrasion with a diamond disc	Crawler or wheel type machine	No	Large	Rc. C.M.	O	O	—	O	△	High	Special use

Method	Principle	Equipment	Need a hole	Size	Applicable structure						Nuisance	Use
Diamond wire saw	Cutting by abrasion with a diamond wire saw	Rotating machine of wire saw	No	Large	Rc. C.M.	O	△	O	O	△	High	Special use
Abrasive water jet	Water jet containing abrasive particles	Super high pressure pump	No	Large	Rc. C.M.	O	O	O	O	△	Med-ium	Special use
Burster with wedges	Splitting by hydraulic wedges	High pressure pump	Yes	Large	C.M.	△	△	O	O	△	High	Special use
Chemical expansive demolition agent	Expansion of quicklime	None (Need a hole)	Yes	Large	C.M.	△	△	O	O	△	High	Special use
CARDOX	Expansion of CO2 gas	None (Need a hole)	Yes	Large	Rc. C.M.	O	△	O	△	O	Low nui-sance	Rare
Mild explosives	Blasting by mild explosives	None (Need a hole)	Yes	Med-ium	Rc. C.M.	O	△	O	△	O	Low nui-sance	Special use
Explosives	Blasting by explosives	None (Need a hole)	Yes	Med-ium	Rc. C.M.	O	△	O	△	O	Low nui-sance	Special use
Thermic lance	Melting by heat of oxidation flame of metal	Oxygen tank and metal lance	Yes	Large	Rc. C.M.	O	O	-	O	-	High	Special use
Fuel oil flame	Melting by flame heat of fuel	Kerosene and jet oxygen tank	No	Large	Rc. C.M.	O	O	△	O	△	Low sound	Special use
Electric heating of rebar	Cracking of concrete by heating rebar	Transformer and frequency amplifier	Yes	Large	Rc.	O	O	△	O	X	High	Special use
Microwave	heating concrete by microwave	Magnetron and wave guide	No	Small	Rc. C.M.	O	O	-	O	X	Under development	
Large size shears	Shearing rebar and shape steel	Crawler type machine	No	-	S.	-	-	-	-	-	High	Quite Common
Felling	Breaking structure by felling	V cutting column feet or blasting column and beam	Yes	Large	Rc. C.M.	O	W:O F:X	X	X	O	Med-ium	Common

(1) Rc.: Reinforced concrete structure and steel frame reinforced concrete structure
    C: Plain concrete structure M: masonry
(2) O: Generally applicable, △: Specially applicable, X: Not applicable

Table-1 b) Evaluation of main demolition methods
-Nuisance, security measures and performance-

| Demolition methods | Nuisance(1) | | | | Security measures(1) | | Performances(1) | | Remarks |
	Noise	Vibration	Heat Fire Water	Dust Fume Projection	Environmental protection	Hazards to the worker	Execution rapidity	Cost	
Hand hammer	4	1	3 Dust	0	3	5	5	5	o Dust-proof mask,glasses,ear defenders,vibration-proof gloves and safety band are needed. o Need a working stage.
Large size hammer	5	4	4 Dust	1	4	3	3	3	o Noise and dust insulation fence are required according to demand. o Rigid working floor is required.
Steel ball	4	5	5 Dust	4 Pro.	5	5	1	1	o Prohibited to entry into the working area. o Taking care of hitting wrongly. o Not to over-turn the mounted machine.
Hydraulic jack	2	1	3 Dust	2 Pro.	1	3	3	4	o Taking care of falling materials. o Rigid working floor slab is required. o Scarcely used nowaday.
Hydraulic breaker	2	2	3 Dust	3 Pro.	1	3	3	4	o Taking care of falling materials. o Rigid working floor slab is required. o Widely used in urban centre.
Hand hammer drill	5	2	5 Dust	1	4	5	2	2	o Same as mentioned in "Hand hammer". o Bore a hole to apply a bursting demolition.
Large hammer drill	5	3	5 Dust	0	4	3	2	3	o Same as mentioned in "Large size hammer". o Bore a hole to apply a bursting demolition.
Diamond boring machine	2	0	2 Water	0	0	1	5	5	o Bore a series of holes along a cutting line. o Used for the demolition of nuclear power plants.
Diamond disk cutter	4	1	3 Water	0	3	3	5	5	o Crane is needed to suspend cut materials. o Rigid working floor is required.

Method									Remarks
Diamond wire saw	4	2	3 Water	0	3	3	5	5	o Same as mentioned in "Diamond disc cutter". o Protection is needed in case the wire saw snaps.
Abrasive water jet	4	1	4 Water	3 Pro.	4	4	2	5	o Protection is needed for water jet. o Wear ear defenders o No crossing a water jet.
Burster with wedges Boring	2	0	2 Dust	0	2	1	4	4	o Noise and dust are generated while boring, otherwise noiseless and vibrationless.
Chemical expensive demolition agent Boring	2	0	2 Dust	0	2	2	5	4	o Same as mentioned in "Burster with wedges". o Wear protective glasses, never watch a charged hole from upright position
CARDOX	4	3	4 dust	4 Pro.	4	4	2	4	o Noise and dust occur while boring. o Taking refuge is needed at blasting time. o Blasting finishes immediately.
Mild explosives	4	4	5 Dust	5 Pro.	5	3	1	2	o Same as to mentioned in "CARDOX". o Licensed specialist is required. o A written consent from neighbourhood is required.
Explosives	5	5	5 Dust	5 Pro.	5	3	1	1	o A blast fence is needed at blasting time.
Thermic lance	1	0	4 Fume	4 Fire	2	3	5	5	o Effect of fumes and fire prevention are needed.
Fuel oil flame jet	5	0	3 Fume	5 Fire	4	4	4	4	o Same as mentioned in "Thermic lance". o Can not be used in an urban centre because of noise.
Electrical heating of rebar	2	1	1	3 Heat	2	2	5	5	o Noise and dust or spray are generated during exposure of rebar and removal work, otherwise no noise and no vibration. o Developing for use in demolition of atomic power plants.
Microwave	3	0	1	1 Heat	4	4	4	5	o Anti-leakage device of microwave is required. o Prevent interference for TV and communication facilities.
Large size shears	2	2	0	1	1	3	1	2	o Rigid floor slab is required. o These Shears can cut a medium sized steel secton.
Felling	4	5	5 Dust	5 Pro.	5	4	1	2	o Protect the felling in opposite direction. o Protect the underground services

(1)In increasing order of number, it shows from favourable to very unfavourable for this item.

3.3 Nuisance

The nuisance occurring in demolition work generally comes from the individual breaking mechanisms. It is an important factor for evaluation, e.g. a hammer creates vibration and noise, and though a steel ball and explosives also makes a lot of noise, vibration, dust and even flying debris, the hammer is a persistent nuisance while the explosives finishes in a short time.

3.4 Performance

The performance of demolition methods is classified into the "execution speed", "cost" and "frequency".

(a) Execution speed; The execution speed is very important. If the working speed of demolition is slow, the period for demolition becomes longer and the demoliton cost increases greatly. The execution speed of the handy hammer, diamond disc cutter and similar tools is slow, but the steel ball and explosives are quick.

   (b) Cost; The cost is a very important factor in evaluating a demolition method, e.g. the cost of a handy hammer, an abrasive water jet, a microwave and the like is expensive, but that of a hydraulic breaker and chemical expansive agent is reasonable.

   (c) Frequency; The frequency of application is a result of the evaluation from the suitability, performance and nuisance.
The large size hammer, hand hammer, hydraulic breaker and hand drill are currently in use, and the large size drill, diamond disc cutter, wire saw are used in special circumstances.

3.5 Safety measures

Safety measures are classified into "environmental protection" and "safety of the worker".

(a) Environmental protection; Environmental protection should be examined according to the demand of each demolition method, e.g. the large size hydraulic breaker does not need protection for neighbours, but the large size hammer needs a noise shielding fence in urban areas.

   (b) Safety of the workers; The health and safety of workers are essential on site, and in general the need for this comes from the principle and mechanisms of breaking concrete, so the effective safety preparation, equipment and manual should be prepared.

4. Conclusion

The outline of demolition methods and evaluation are mentioned, but in general the demolition work is carried out by the combination of two or three methods. The feasability of combining different methods should be investigated in the future.

References

Report of RILEM Committee 37 DRC, Task Force 1 (1985), Demolition Techniques, European Demolition Association (EDA).

   Report of the Committee on Demolition of Reinforced Concrete Structures (1972-1978), Building Contractors Society of Japan (BCS).

   Research Group of Demolition Methods (1979), Demolition Methods and Estimation of RC structures, Economic Investigation Inc. (in Japanese).

DEMOLITION OF CONCRETE STRUCTURES IN CZECHOSLOVAKIA

A. HÖNIG   Brno Technical University
Z. KRAJČA   Industrial Constructions Enterprise

Abstract
An overview of employed demolition methods for concrete
structures and buildings in Czechoslovakia is demonstrated
on an enumeration of certain demolition activities within
the last twenty years. A processing line for the production of material from demolished concrete is described.
Key words: Demolition, Crushing and Separating Techniques,
Recycled Aggregate.

1. Demolition of concrete structures

The Special Operations Works of the national enterprise
Průmyslové stavby /Idustrial Constructions/ Brno is carrying out demolition work since its foundation, i.e. from
the year 1969.
    During the starting period such work depended upon the
availability of suitable mechanization and blasting technique means. Even at that time demolitions were implemented which were unique, e.g. the blasting of blocks of flats
in Prague and in Brno as well as of chemical plants and
other industrial constructions, the blasting of chimneys
in dense built-up spaces and the blasting of a television
tower. In the years 1974 - 79 was demolished the entire
town of Most to clear the ground for open-cast mines,
further was demolished one metallurgical plant including
blast furnaces, two deep mines and many other structures.
    Abroad the Works have blasted head frames in a Belgian
mine and demolished an integrated iron and steelworks,
the LEUNA Chemical works and a sports hall in Berlin in the
German Democratic Republic.
    Successively,in parallel with the development of blas-
ting technique means and mechanization were planned and
implemented also complicated demolition jobs in closely
built-up areas of chemical plants as well as municipal
agglomerations. This concerns heavy both reinforced concrete and steel structures, reinforced concrete cooling
towers, blast furnaces, reinforced concrete as well as
steel bridge structures, and others (Fig. 1., Fig. 2.).

Fig. 1. First timed explosion during the demolition of a
block of houses for preparing the building site
of the Intercontinental Hotel in Prague.

Fig. 2. Third timed explosion during the demolition of a
block of houses in a dense built-up area for preparing the building site of the Intercontinental
Hotel.

1.1 Employed demolition methods
The Special Operations Works of Průmyslové stavby in Brno
are employing a number of well known technologies:

- Successive dismantling - especially in situations where
  no large mechanisms can be used.
- Demolition with the aid of mechanisms - when heavy
  mechanization is being utilized, e.g. salvage tanks,
  heavy cranes, excavators, loaders, hydraulic demolition
  hammers, hydraulic jaws, etc.
- Aluminothermic separation of steel structures of roofs,
  bridges, masts, cable support structures, etc.
- Demolition of structures from plain concrete, stone or
  bricks with the aid of hydraulic cylinders and expanding
  cement, respectively. This technology is employed in
  situations where for reasons of undesirable vibrations,
  environmental protection or for other reasons blasting
  operations cannot be utilized
- Blasting which form in actual fact part of the technological procedures of demolition jobs, the result of
  which is the fastest possible liquidation of the
  buildings.

1.2 Larger demolition jobs
To give a more perfect idea of the scope of the demolition
work, certain activities will be described which are this
field interesting and unique, be it from the point of view
of complexity, extent, originality of the solution, etc.

1.2.1 Demolition of power plant
One of the largest demolition jobs carried out by our Works
in the years 1980 - 86. There were demolished and removed
buildings with a total space of 1,1 million m^3, in physical
volumes of 248 thousand m^3 of brick masonry, 200 thousand
m^3 of concrete masonry, out of this 140 thousand m^3 of
reinforced concrete and 40 thousand tons of steel structures. The consumption of explosives amounted to 73 863 kg
and 189 374 pieces of primers.
    Demolished were all overground as well as underground
parts of the buildings. This concerned above all heavy
reinforced concrete structures with brick filler masonry,
a series of heavy steel structures and extensive overground
and underground structures.
    The surrounding built-up areas allowed to a maximum
extent the utilization of blasting operations and also the
testing of technologies which have prior to this been
worked out only theoretically. The obtained experience
was utilized successfully in further demolitions of similar
structures, but already under operational conditions of

plants and continuously built-up areas.

With success was tested a new demolition technology for reinforced concrete cooling towers in the shape of a rotary hyperboloid. By the creation of horizontal ring-shaped destruction sections and partial weakening of the tower shell there occures an insertion of the individual shell parts "into one another". Thus is formed a sufficient fall height which leads to the destruction of the entire tower shell, and that on its own ground plan. This method allows also the demolition of cooling towers in built-up areas of plants. In the course of this activity were successfully demolished 4 cooling towers with a height of 60 m and a diameter of 40 m, the concrete wall thickness being variable from 330 mm to 120 mm. For blasting were employed for each tower approx. 100 kg explosives and 2 300 pieces primers. This cooling tower demolition technology is protected by a patent.

1.2.2 Demolition of industrial constructions of a chemical plant

Within the framework of modernizing the operations, demolition work is being carried out. The most exacting job was the demolition of a coal bunker with a total built-around space 73 thousand m^3; 19 thousand m^3 of reinforced concrete were demolished.

The demolition of the heavily reinforced concrete building with the ground plan dimensions 27 x 84 m and height 31 m was carried out with the aid of 4 sectional blastings with a total consumption of 350 kg of explosives and 3 810 pieces primers. Before these main blastings it was necessary to weaken to the smallest possible measure the whole construction. During these jobs were consumed 2 298 kg explosives and 15 020 pieces primers. The complexity of the demolition was above all in the massive structure of the building which was within a dense and complicated built-up area of the chemical plant. Along the entire building and technologically within a distance of approx. 8 m was a pipeline bridge, the slightest damage of which would have caused great damage and endangered the entire operation.

1.2.3 Demolition of a granulation tower

Cylindrical reinforced concrete tower ,diameter 15 m,height 56 m and wall thickness 300 mm. The total weight of the tower, including the technology located in its upper part, was 2 200 t. Here it was necessary to accurately direct the fall of the building and to eliminate seismic effects /from the falling debris/ onto the surrounding built-up area. For blasting 27 kg explosives and 540 pieces primers were employed. During blasting occured a complete dis-

integration of the concrete shell. The hydrogen compressor
station /approx. 15 m from the building in parallel with
the axis of fall/, the main electric power distribution
plant /25 m from the place of fall/, the underground power
supply duct /located at the location of the fall/ and also
the steel tanks and railway siding in a close vicinity behind the tower, have in no way been affected by the demolition.

1.2.4 Demolition of a railway overbridge
Reinforced concrete arch bridge with two fields with a
span of 2 x 32 m and a weight of 2 x 720 t. The selection
of the technology was essentially determined by the requirement of the Czechoslovak State Railways ČSD not to
limit the traffic on the highly utilized line under the
bridge. Liquidation procedure:
- With the aid of a provisional steel framed structure
  which was inserted into the arch, the bridge was lifted
  hydraulically above the support sleepers. The bridge
  arches themselves were carried by eleven cross beams
  and tie members. The cross beams were mounted on a
  series of carts which travelled along the steel structure.
- Shifting of the bridge structure by approx. 4,5 m above
  the bridge head with the aid of a salvage tank.
- Execution of the blasting of the exposed part of the
  bridge and its further separation into pieces suitable
  for loading and removal.

These operations were repeated on both sides of the bridge.
    After the execution of the bridge structure demolition
were liquidated also the approaches and bridge abutments.
    In the course of the entire demolition work the opera-
tions on the railway line were not interrupted. All the
work was carried out in the periods between the passing
of trains. The planned stoppages have not been utilized.
Beside the railway line it was necessary to protect also
the operational buildings of the surrounding built-up area
against the effects of blasting work. Not even in those
buildings occured a curtailment of operations and not even
the slightest damage was done to them.
    During the liquidation was carried out a total of 86
sectional blastings with a consumption of 700 kg explosives
and 2 760 pieces primers.

1.2.5 Demolition of a blast furnace
In 1983 was stopped the operation of the blast furnace
and its liquidation, including auxiliary operations, began.
From the technical point of view was highly interesting
the blasting of the air preheaters, the blast furnace

and the blast furnace structure. Also the dismantling of
the lattice girder bridge for the charging of raw materials
was very demanding. The overall length of the structure
was 67 m and its weight 82 t. Underneath the bridge was
located a pipeline which was not allowed to be affected
by the demolition work.
 For dismantling was utilized the COLES 80 heavy crane,
the bridge was divided along its length into 3 sections,
and these were then removed. The pipeline underneath the
bridge was damaged in no way.

1.3 Mechanization employed during demolition jobs
The majority of liquidation jobs carried out is executed
with the aid of explosives. Through the fall of the
demolished structures there occurs a primary separation of
the material. The secondary separation of the material
after the fall of the structure is carried out with the
aid of small explosive charges or mechanical with the aid
of hydraulic presses, jaws and heavy breakers is employed.
For disengaging the structures are frequently used salvage
tanks. During the secondary separation are sometimes
utilized also concrete saws and drilling machines. An ana-
logical destruction procedure is used also for steel
structures, but the material is separated by flame cutting.
 For loading the demolition material are employed current building machines which have been modified for these
special activities. These include loading shovels with
a shovel capacity of 0,5 - 3,0 m^3, excavators with a shovel capacity of 0,4 - 2,5 m^3 which serve at the same time
as carriers for special hydraulic implements, and various
types of clamshells and electromagnets for the loading of
steel scrap. For the loading of structure parts and the
handling of larger material blocks are used mobile cranes
of various types with load-carrying capacities from 5 to
130 tm.
 For work in inaccessible localities are generally
employed erection platforms of various types and reaches
and suspension platforms as accessory equipment to the
mobile cranes.
 The drilling technique employed in the execution of
blasting work includes current pneumatic and hydraulic
drilling machines - portable for holes up to a diameter
of 42 mm and transportable for larger diameter bore holes,
including core bore holes for concrete separation blasting
wedges.

2. Crushing of reinforced concrete

The stockyards of coarsely separated reinforced concrete
in Czechoslovakia are growing. On them are stored ma-

terials from demolished civil engineering constructions,
from scrapped prefabricated element out of the manufacturing facilities and from damaged structural elements in
the course of shipment and erection. Experimentally is
being introduced a processing line for the manufacture of
crushed and screened materials from the demolished concrete.

The overwhelming majority of equipment developed for
the crushing of concrete wastes operates on the mechanical
principle of material disintegration. After the separation of large lumps are utilized crushers designed for
hard rocks and natural aggregates.

Experiments were carried out with a mobile dressing
plant for building wastes and a semi-mobile line, both
with a capacity of 15 t/h of crushed concrete waste
(Fig. 3.) and a project was worked out for the liquidation of waste from large-size elements and lumps down to
fractured demolished concrete. A stationary line /with a
capacity of 50 t/h/ is equipped with hydraulic demolition
shears for the longitudinal separation of panels. The
separated material is fed into a jaw crusher which is til-
ted away from the horizontal plane by 15°- 20°, so that
even long elements can continuously find room between the
jaws (Fig. 4.).

3. Conclusion

Concurrently with the introduction of processing lines it
is essential to solve as quickly as possible:

a/ the rationalization of the actual demolition jobs,
   especially in areas where forceful demolition methods
   cannot be used /housing and civic constructions in a
   dense built-up areas and in industrial operations/, i.e.
   the actual demolition process of concrete structures
   or parts thereof, right up to the obtainment of dimensions necessary for the entry into the processing line,
   including transport facilities.

b/ Investigation of the properties of concrete manufactu-
   red with the full or partial utilization of the crushed
   materials from the processing lines. In this field is
   necessary an intensive cooperation of international
   organizations such as EDA /European Demolition Associ-
   ation/ [2] and RILEM /Réunion internationale des la-
   boratoires d`essais et de recherches sur les matériaux
   et les constructions/ [3]. The result of the international cooperation should be the issueing of recommen-
   dations, guidelines and as a follow-up of national
   standards for the design of structures from demolished
   concrete which exhibit a number of significantly different properties from concretes made with natural ag-

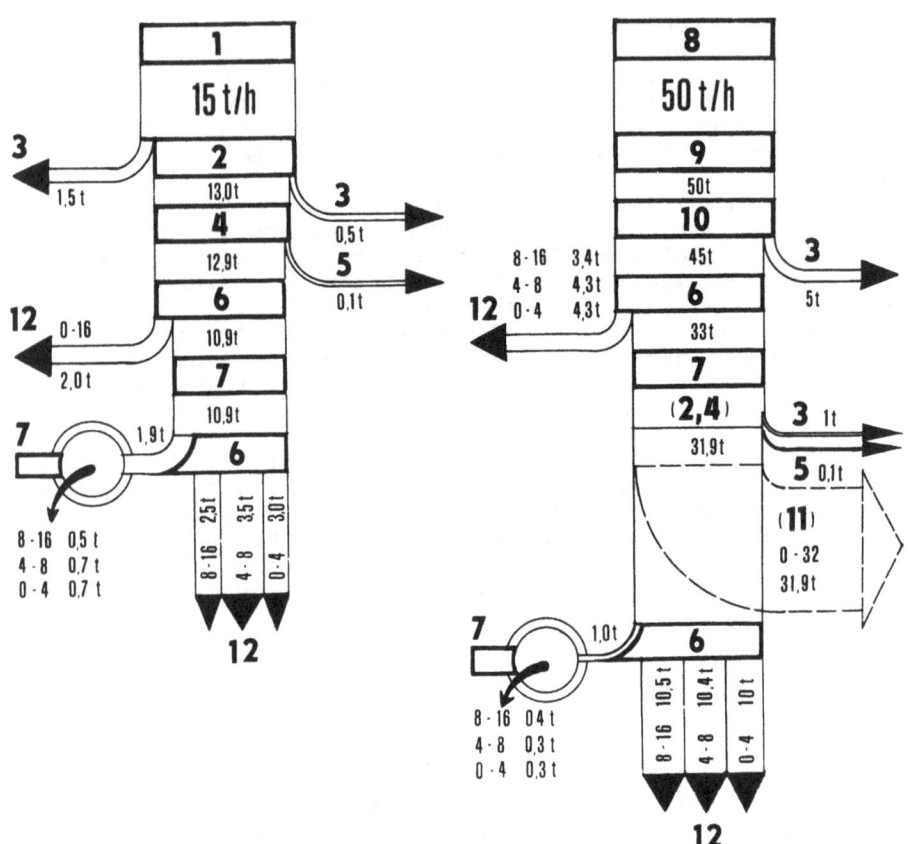

Fig.3. Flow chart of a mobil
plant for production
of recycled aggregate
from concrete debris.
Capacity 15 t/h [1].

Fig.4. Flow chart of a stabil
plant for production
of recycled aggregate
from concrete debris.
Capacity 50 t/h [1] .

1 - primary selection, 2 - magnetic separation, 3 - steel
reinforcement, 4 - polysryren separation, 5 - polystyren,
6 - primary and secondary screening, 7 - primary and secondary crushers, 8 - selective demolition /hydraulic shears/,
9 - loading machine, 10 - impact crusher, 11 - monofraction,
12 - fractions.

In both flow charts (Fig.3. and Fig.4.) the units t read t/h.

gregates /larger batch water consumption, more pronounced volume changes, especially shrinkage, different
strength and deformation characteristics as static and
dynamic modullus of elasticity/.

4. Acknowledgement

The authors would like to thank Jan Bradáč from Hutní projekt Ostrava and to the Research and Development Institute
for Mechanization of the Building Industry Bratislava for
the supplied information, as well as to Torben C. Hansen
for the enormous work he has done as Chairman of the
RILEM TC-37-DRC /Demolition and Recycling of Concrete/
Commission. The results of the work of this Technical Commission, published in their First State-of-the-Art Report
[4] and in the Second State-of-the-Art Report [3] served
as very important fundamental data for the investigation
of recycled aggregates concrete in Czechoslovakia.

References

1 Príprava armovaného betónu pred drvením /1987/. Tech-
      nicko-ekonomická studie, Výskumno-vývojový ústav
      pre mechanizáciu stavebníctva, Bratislava.
2 Re-use of concrete and brick materials /1985/. Demo-
      recycling, Proceedings II, EDA/RILEM Conference,
      European Demolition Association.
3 Hansen, T.C. /1986/ Recycled aggregates and recycled
      aggregate concrete, Second state-of-the-art report,
      Developments 1945-1985, RILEM Technical Committee-
      37-DRC, Demolition and recycling of concrete,
      Materials and Structures /RILEM/, 111, 201-246.

4 Nixon, P.J. /1978/ Recycled concrete as an aggregate
      for conrete - a review. First state-of-the-art
      report RILEM TC-37-DRC, Materials and Structures
      /RILEM/, 65, 371-378.

DEVELOPMENT AND CURRENT SITUATION OF BREAKERS IN JAPAN

SHIGEMASA YOSHINO, Consulting Engineer
HIROSHI KAWASAKI, Research Institute of Architecture, Foundation,
and Disasters Protection

Abstract
Demolition of concrete building structures was performed mainly by the
breaker such as steel-ball and impact type around 1970. However, these
type breakers caused an environmental pollution including noise,
vibration and dust, so they became unsuitable for the social environment. Since then, a variety of non-pollution breakers such as hydraulic type and thermal energy type have been developed. Nethertheless,
none of them was satisfactory for practical use. On the other hand,
the needs for demolition have increased more and more because of urban
redevelopment or reconstruction of old buildings. Around 1978, a
hydraulic nipper type breaker installed on hydraulic excavator as a
base machine was introduced from England. With this introduction,
hydraulic breakers to meet domestic requirements were originally developed in Japan and then widely spreaded all over the country as attachment for hydraulic excavator. The number of monthly production units
is now approximately 60.
Key Words: Breaker, Noise, Vibration, Dust, Hydraulic, Excavator,
Reinforced concrete.

1 Introduction

Demolition of concrete building structures had been mechanically performed through impact destruction by a steel-ball or continuous blow
destruction by a giant breaker. However, these destructive methods
have caused hard pollution of noise, vibration and dust, and the use in
town has harmed the life environment and become a social problem. To
solve this problem, some measure societies for study have started by
related organizations. One of them, RC destructive method committee by
Building Contructor Society (Foundation) started in 1971. Destruction
& demolition work method study committee started also in 1973. These
committees investigated and evaluated the less noise, less vibration
and less dust breakers which were developed in various places for the
social need at that time. The subjected machines and the number of
their manufacturer were many as follows: hydraulic jack type breaker -
7 manufacturers, disc saw cutter - 1 manufacturer (hereafter ditto),
magnetic induction heating breaker - 1, microwave heating breaker - 1,
jet lance boring machine - 1, flame jet cutter - 1 and lime expansion
breaker device -1 university. In 1974, Central district engineering

office, the Ministry of Construction, conducted an evaluation test on
the various new developed machines for noise, vibration and performance. In addition to above subjected machines, each manufacturer
released the machines around that time which had been developed for the
same purpose. As the result, many machines or work methods were developed and released in total. Though these machines achieved the target
to reduce noise, vibration and dust, they were not yet practical
machines which should have provided good workability, safety and
economy. After that, each manufacturer have continually developed such
machine. About that time, in 1977 a hydraulic road destruction &
demolition machine, Nibbler was introduced which Hymac Inc. in England
had developed as a low noise and vibration machine. Taking this opportunity, such the hydraulic breaker has been widely spreaded. The
reason is that the machine was an attachment to the hydraulic excavator
as base machine. In Japan, 50 to 60 thousand hydraulic excavators are
being produced in a year and widely spreaded. This fact expedites the
wide spread of hydraulic breaker and now many manufacturers participate
in this business to develop and produce. When viewing the historical
development and progress in Japan on the demolition machines for the
concrete building structures, the term can be divided into two stages.
The first stage is that the development aiming at only no-pollution is
concentrated. The later stage is that the development aiming at
practical use is concentrated by spreading as an attachment to the
hydraulic excavator. These two stages are explained below. The age
of the first stage is 1971 to around 1977. The later is after that.

2. Various machines during first development stage

2.1 Outline

The feature is that all sorts of energy were used for destruction
machines. Those energy sources were hydraulic force, electric motor,
magnetic induction heat, lime expansion, high temperature of thermit
method and super high pressure water. Followings are typical machines
among them.

2.2 Con-destrer

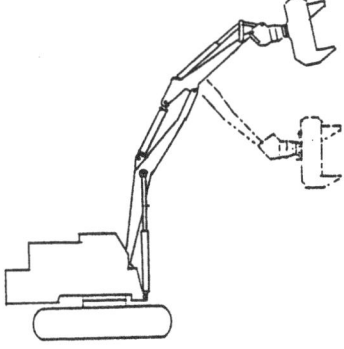

Fig. 1. Con-destrer

21

2.2.1 Major specifications

Model(Manufacturer): III type (Sanwa Kizai Co., Ltd.), Max. opening width: 1000 mm, Max. press force: 300 tf, Weight: 4.7 tf, Base machine: 33 tf class crawler crane, Joint developer: Japan National Railway.

2.2.2 Performance

1) Breaking power - At the test of breaking RC concrete with 29 mm dia reinforcement, this machine applied chisels to the reinforcement at a right angle and out 4 reinforcements with 265 tf force resulting in entire breakage of the blocks.
2) Sound test - 73 phon at 10 m, 68 phon at 20 m, and 63 phon at 30 m each from sound source.

2.2.3 Features

1) The machine was too heavy but the first breaker in Japan to be installed on hydraulic excavator as a base machine.
2) Swing and slue of machine is mechanically drived and the most suitable angle can be selected for breaking.

2.3 Takenaka type silent breaker

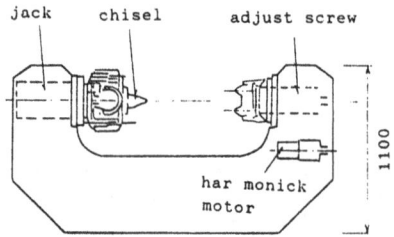

Fig. 2. Takenaka type silent breaker

2.3.1 Major specifications

Model(Manufacturer): No.5 machine (Takenaka Kohmuten Co., Ltd.), Max. opening width: 850 mm, Cylinder stroke: 300 mm, Max. press force: 300 tf, Hooking crane: Tower crane.

2.3.2 Features

1) Elevation work can be performed because of wire rope hooking by tower crane. It does not cause an excessive reaction on the tower crane in comparison with attachment type machine directly mounted on the base machine.
2) Since the chisel inclines freely 10 degree, it moves forward along the crack in lexx than 10 degree direction. Therefore the breaking resistance is small and the bending stress at the chisel mounting section is also small.

22

3) Hooking change only makes easy of accompanying works such as removal of reinforcements or scraps, or carrying in the gas cutters. This can accelerate the whole destructive work.
    4) Sound level - 63 ∿ 66 phon (A) at sound source.

2.4 Asahi Jacker

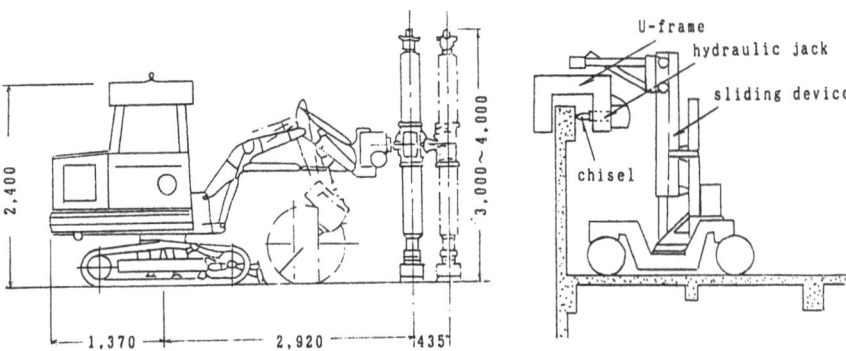

Fig. 3. Asahi Jacker   Fig. 4. Ohbayashi type

                                              silent breaker

2.4.1 Major specifications
    Model(Manufacturer): 160 type (Asahi Chemical Industry Co., Ltd.), Max. load: 160 tf, Jack stroke: 40 cm (upper), 40 cm (lower) total 80 cm, Weight including base machine: 4.4 tf, Joint developer: Building Contructor Society.

2.4.2 Features
    1) The concrete building structure is reinforced so as to resist downward load, so the jack is placed on the floor to break ceiling panels.
    2) The unit weighs light. It can be easily transported and hooked up to the tall building by a crane.

2.5 Ohbayashi type silent breaker

2.5.1 Major specifications
    Model(Manufacturer): COW-C (Ohbayashi CORP.), Size: 2.3 m x 3.5 m x 6.4 m, Weight: 12.7 tf, Press force: 150 tf, Jack stroke: 400 mm, Slide stroke: 4.4 m, Travel speed: 6.2 m/min

23

2.5.2 Features

1) Since it provides free movements of travel, lift, slue and vertical swing, a flexible work can be performed inside the building.
2) A possible remote control provides safety for the indoor work.

2.6 Toda type cutter

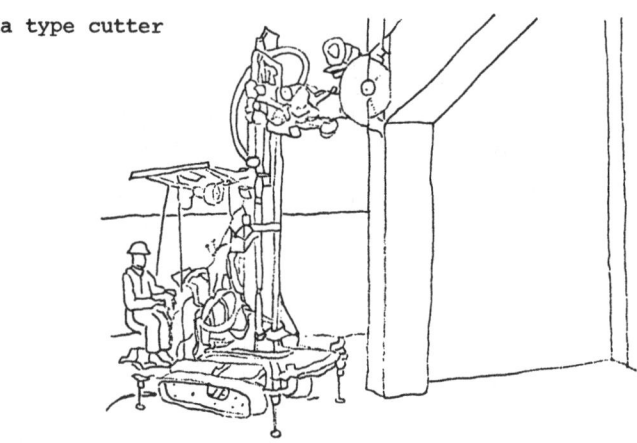

Fig. 5. Toda type cutter slab work

2.6.1 Major specification

Model(Manufacturer): Universal type (Toda Construction Co., Ltd.), Cutter: Diamond blade disc saw type, Max. cutting depth: 17 cm, Cutting speed: 25 ∿ 40 m/min

2.6.2 Features

1) The demolition work is neat because of less scraps.
2) Demolished parts can be reusable.
3) Since this machine was developed with a concept that the demolition order is adverse to the build order, the work can be planned and orderly.
4) Sound level - With sound protective cover, 60 phon at 30 m from sound source.
5) In spite of diamond blade cutter, it is economical.

2.7 Magnetic induction heating breaker

2.7.1 Major specifications

Electric source: 37 kW, 30 kVA, 3 phase 400 Hz Generator, Condenser: 300 µF ∿ 800 µF with selecting device, 150 kg, Magnet: M-80, 40 cm x 60 cm x 35 cm, Manufacturer: Tokyo Electric Machine Co., Ltd., Joint developer: Building Contructor Society.

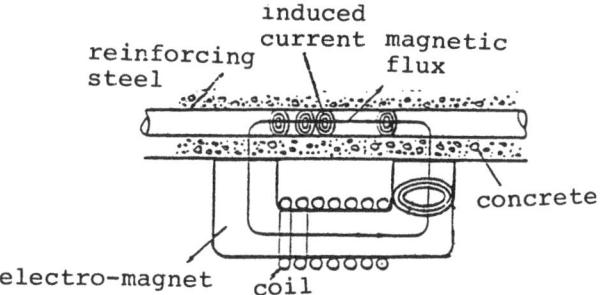

Fig. 6. Principle of magnetic induction heating

2.7.2 Features

1) The external electromagnet simply energized with AC heats the reinforcement and weaken the reinforced concrete.

2) Big concrete blocks after reinforcements exposure are similar to non-reinforced concrete, and makes easy of following works.

3. Various machines during later development stage

3.1 Outline

The machines were developed and spreaded as attachments to hydraulic excavators. There are approximately 10 manufacturers. Among them 5 manufacturers are selected and presented here. These have common features as follows:

1) Both sound and vibration levels are less than the local government's regulation. The dust generated is a little amount and eliminated by the water spray as required.

2) An automatic rotation device is provided between the top end of machine and the arm end of base machine. This device prevents a bad influence caused by the strong reaction to the base machine during breaking. This has also a performance to be always directed to the destructive material at a right angle, resulting in an effective increase of breaking power.

3) The reinforcement cutter is provided as a standard with each proper installation.

4) Besides the above basic technique for countermeasure, each manufacturer is making efforts to upgrade its own technique. This is the current situation in Japan.

The followings are typical 5 manufacturers and their typical machines, and descriptions concentrated in the features.

3.2 Yutani Nibbler

3.2.1 Major specifications (Typical model)

Model(Manufacturer): RC 750 W (Shinko Kobelco Construction Machine Co., Ltd.), Opening width: 735 mm, Breaking force: 56 tf, Weight: 1.7 tf

25

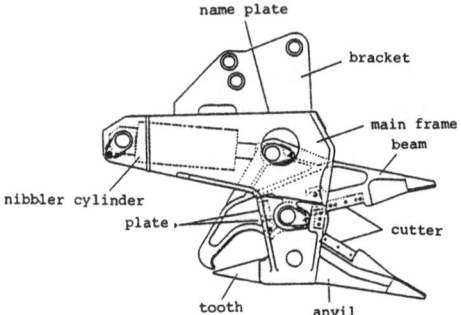

Fig. 7. Yutani Nibbler

3.2.2 Features

1) This has an excellent workability on the close point such as the reinforcement draw, and the consolidated operation including the base machine operation is highly efficient.

2) Breaking claws provided in front and in the rear can be used properly. These are actuated by one hydraulic cylinder, so the work applicability is excellent.

3.3 TS crusher

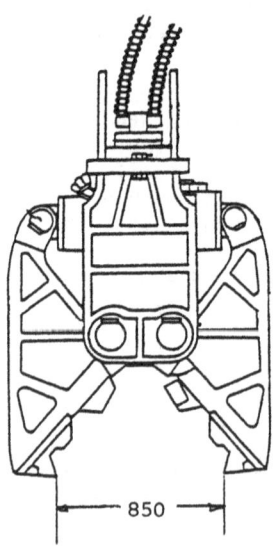

Fig. 8. TS crusher

3.3.1 Major specifications (Typical model)

Model(Manufacturer): TS-800 (Okada-aiyon Co., Ltd.), Max. opening width: 800 mm, Breaking force at tip: 100 tf, Weight: 2.0 tf

26

3.3.2 Features

1) The hydraulic thrust force is generated by the single cylinder with two rods, and the cylinder is mounted on the sliding center trunion to prevent the cylinder oscillation.

2) Each section material has high strength in total and presents an excellent durability.

3.4 Pacler

Fig. 9. Pacler

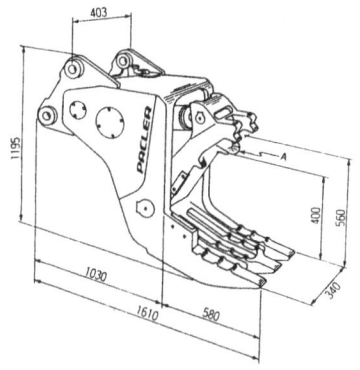

3.4.1 Major specifications (Typical model)

Model(Manufacturer): SPAC-80 (Sakato Kosakusho Co., Ltd.), Max. opening width: 850 mm, Max. breaking force: 93 tf, Weight: 1.7 tf

3.4.2 Features

1) The breaking cutter is a part of disc so that the breaking resistance is small.

2) The machine is lightened by using high tension steel or specific steel casting for the section material.

3) The breaking cutters are arranged properly and efficient for small piece crush.

3.4.3 Breaking resistance test on the breaking cutter

This test was conducted at the Construction Machinery Institute, August 1978.

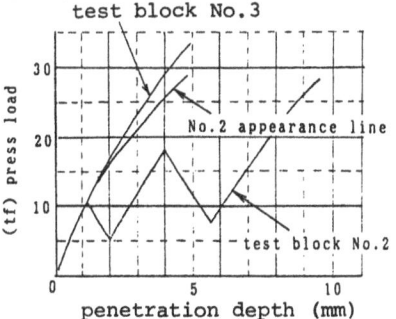

Fig. 10. Penetration amount vs press load

Fig. 11. When the test completed

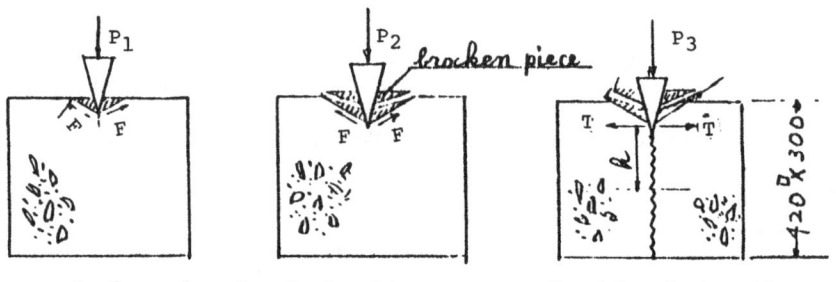

Surface shearing destruction   Cracking destruction

Fig. 12. Breaking mechanism in the concrete block using disc type cutter

$$T = \frac{P_3}{2\tan\alpha} , \qquad \sigma_t = \frac{Th}{2} + \frac{T}{A}$$

Destructive condition: $\sigma_t > 43.8$ kgf/cm^2, σ_t: Compressive crack
P$_3$ test value   : 28.4 tf   test strength

T: Destructive thrust force, P: Hydraulic cylinder force,
α: 1/2 Cutter nose angle + friction angle, h: Bending moment arm by T
Z: Section modulus at destruction plane, A: Area of destruction plane

3.5 Ohsumi

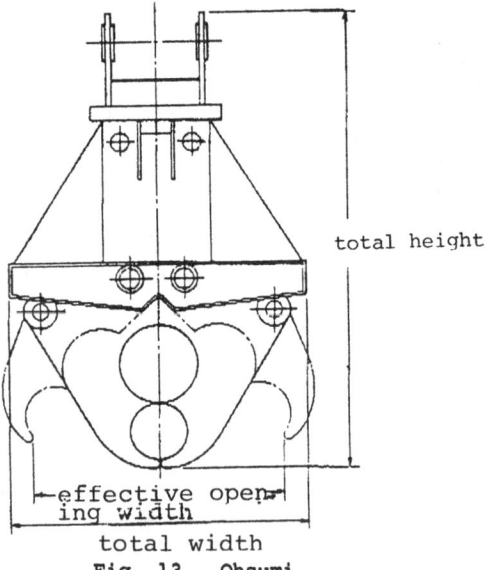

Fig. 13. Ohsumi

3.5.1 Major specifications (Typical model)

Model (Manufacturer): MR 1000-2 (Takachiho Industry Ltd.), Effective opening width: 1000 mm, Max. breaking force: 137 tf, Weight: 2.1 tf

3.5.2 Features

1) Two cylinders are installed almost vertically, and the overall machine length is small when the claws open fully. Therefore, the workability at narrow spot is excellent.

2) Since the cylinders always act symmetrically, the machine own balance is maintained to have good workability.

3) Max. opening width is large because of both ways opening with two cylinders.

4) The light weight is advantageous to the tall building demolition.

3.6 Nippon pneumatic pressure breaker

3.6.1 Major specifications (Typical model)

Model(Manufacturer): S-22R (Nippon Pneumatic Industry Co., Ltd.), Max. opening width: 800 mm, Tip breaking force: 100 tf, Weight: 2.2 tf

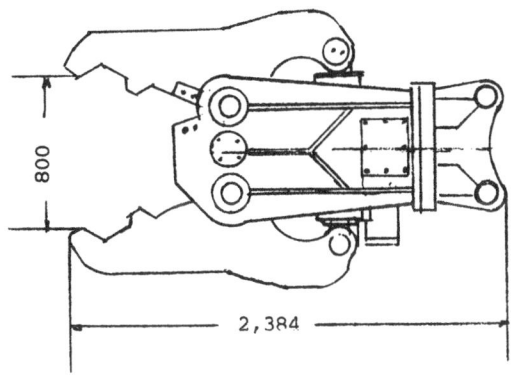

Fig. 14.   Nippon pneumatic pressure breaker

3.6.2 Features

1) Adoption of the booster cylinder shortens the work cycle time and generates a big breaking power.

2) The flat tip claw is advantageous to deal with small pieces such as reinforcement. This figure is effective in total.

3) Both claws move symmetrically through the link mechanism actuated by the single cylinder, so the machine is easy to operate with a good balance.

4. Concluding remarks

The development of Japanese breaking technology for concrete buildings is historically overviewed. The particularly important requirement on the breaking machines in Japan is no environmental pollution and flexibility of installation since such machines have to be often used in more crowded urbans or more dense residence areas than in Western countries. This particular requirement has been well overcome by the long year effort of many Japanese manufacturers. The Japanese breakers have entered the era of practical use and is now rapidly spreading especially as attachments of hydraulic excavators which are already widely spread. This tendency will be more accelerated in future. The future subject required is to enable their use in taller buildings, inside rooms and in other complicated structures.

A STUDY ON DEMOLITION METHOD OF CONCRETE STRUCTURES BY CUTTING

T.KEMI and T.HIRAGA
Institute of Construction Technology, Toda Construction Co.,Ltd.

Abstract
This paper describes the systematic demolition method of concrete
structures by cutting.
　In this method the cutting machines equipped with sawing blade
containing diamond abrasives are used. The dismantling work is
executed by separating reinforced concrete structure into members by
use of the cutting machine and then removing the cut-off members one
by one by cranes as if they were toy building blocks, in the reverse
order to the setting up work of pre-casting concrete plates.
Key words: Demolition, Dismantling method, Harmless, Cutting machine,
Diamond blade, Dismantled member, Taking down, Carrying out.

1. Introduction

Three quarters of a century have passed since the first reinforced
concrete structure made its debut in Japan. However, no earlier
reinforced concrete structure was virtully built in anticipation of
possible future demolition. After 1970, the nation's high pitched
economic growth and the incidence of demolition rose steeply, and the
demand for demolition inducing harm such as noise, vibration, dust,
etc., increased reflecting the social background of the transition to
newer life-styles, urban development and more intensive land use, the
obsolete facilities, etc. This intensified the need for the research
and development of new harmless, economical and urban-oriented
demolition methods.

This paper describes the results of our research and development
work of a new dismantling method - this method differs from
"demolition".

2. Outline of the dismantling method making use of cutting techniques

2.1 Characteristics of the dismantling method

① This dismantling method is almost completely free of vibration and
dust which are usually accompanied with demolition. Noise levels can
be maintained within 70 dB(A) by using relatively simple
noise-suppressing equipment.

31

② This dismantling system is executed orderly and safely according to well-planned procedures in reverse order to pre-casting structure construction.

③ Pieces obtained from dismantled members by crushing at outside disposal facilities may be recycled as they are, in an aim to preserve natural aggregate for concrete.

2.2 Dismantling method

In our dismantling method which uses cutting techniques, a reinforced concrete structure is cut into a number of blocks for piecemeal removal, using a crane as if they were toy building blocks. Dismantling structures is executed by means of cutting machines. A cutting machine equipped with a special diamond blade is used for cutting only the reinforced areas not entire non-reinforced areas under dismantling. During cutting, adequate attention must be paid so that concrete portions, which are subject to shearing and tention, may remain intact as much as possible without being damaged by cuts. Fig.1 and 2 show dismantling procedures using a crane for the removal of such cut members.

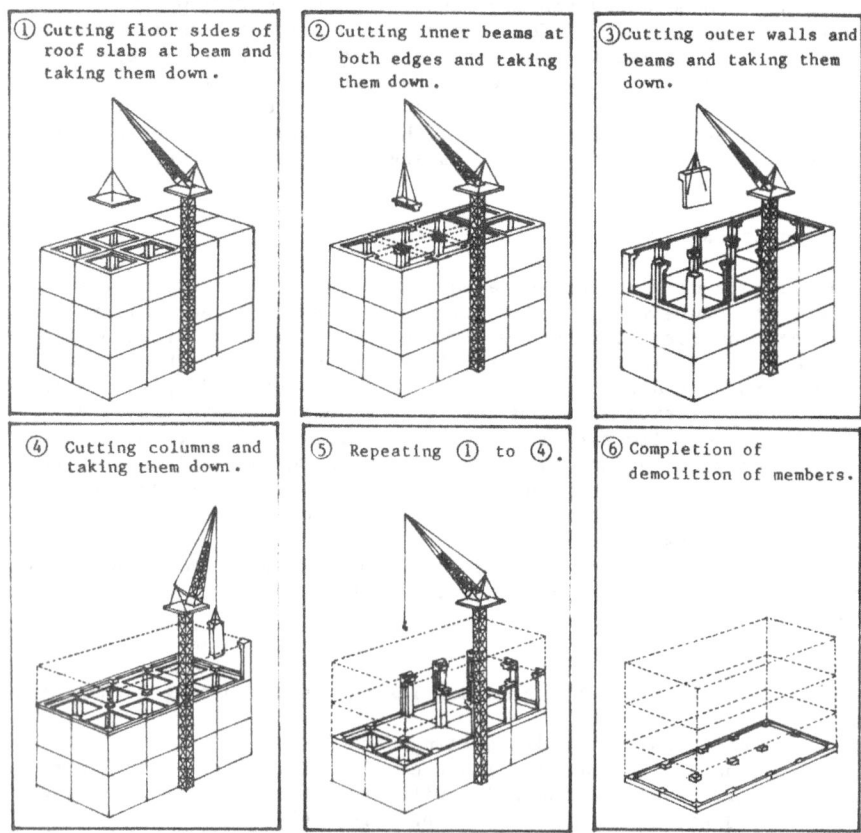

① Cutting floor sides of roof slabs at beam and taking them down.

② Cutting inner beams at both edges and taking them down.

③ Cutting outer walls and beams and taking them down.

④ Cutting columns and taking them down.

⑤ Repeating ① to ④.

⑥ Completion of demolition of members.

Fig. 1. Order of dismantling by cutting.

32

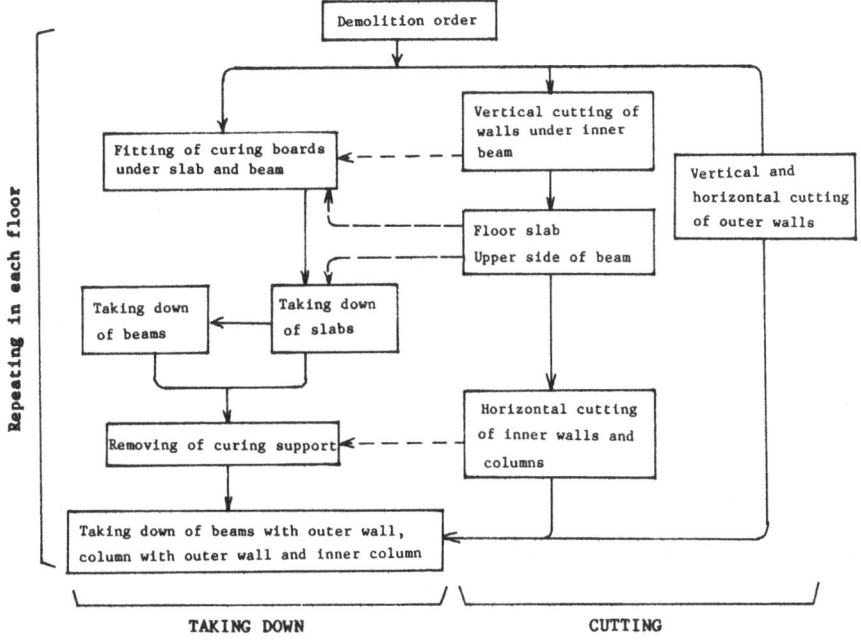

Fig. 2. Demolition order of cutting and taking down members.

3. Dismantling system

3.1 Cutting of structure

1) Shapes and weights of dismantled members

The bigger the dismantled members are, the higher the dismantling efficiency gets in the works of cutting, taking down and carrying out. However, the weights of the members should be 5 to 6 tons per piece on an average in consideration of the ability, working ratio of a crane, the condition of site and the limitation of transportation. As shown in Fig.3, the shapes of dismantled members are the standard types in each member.

2) Cutting position and volume

Cutting of structures should be executed at the position where the cutting length is the shortest. In result the energy of cutting reinforcing bars is at a minimum. Especially, in case of walls the parts near the openings should be cut, so that the cutting length gets shorter. Cutting length and number of dismantled members can be estimated by use of Fig.4.

3) Execution of cutting

In execution of cutting members, cut positions are marked according to the cutting planning and cut by use of a cutting machine as shown in Fig.5. The cutting efficiency of each member varies a bit in compliance with the volume of reinforcing bars in the cut section, the cutting depth, the ability of a cutting machine and the difficulty in execution. The total days required for cutting can be calculated in consideration of the cutting length, cutting efficiency and the real cutting time in one day. The species and numbers of cutting machines are decided by dismantling order and duration of works.

3.2 Taking down of dimantled members
The members which were cut in accordance with the cutting planning are removed safely by use of a jack as an auxiliary tool and taken down by a crane as shown in Fig.6.

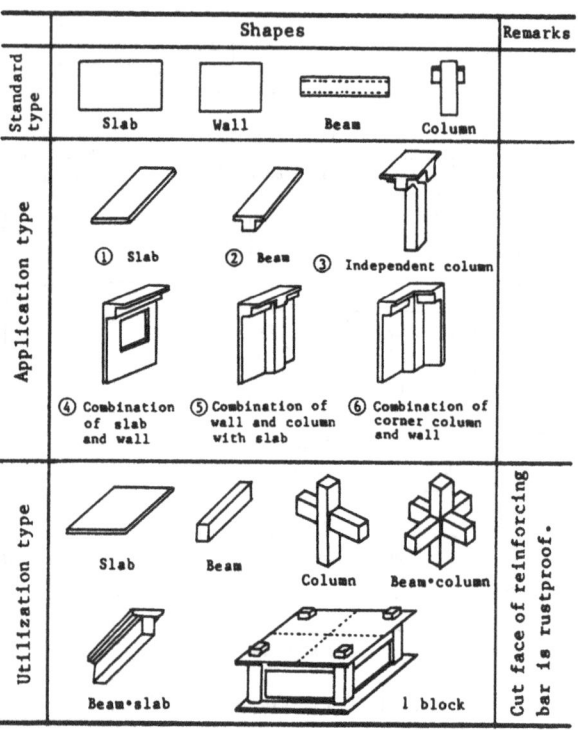

Fig. 3. Shapes of cut members.

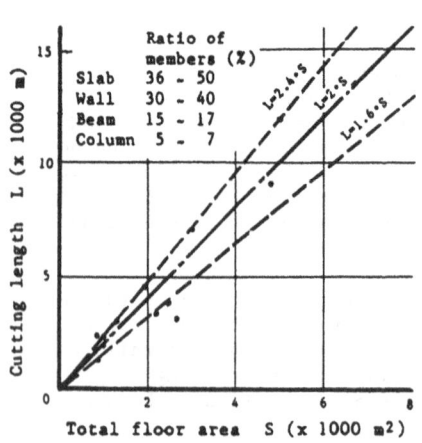

Fig. 4. Estimating method of cutting length by using total floor area.

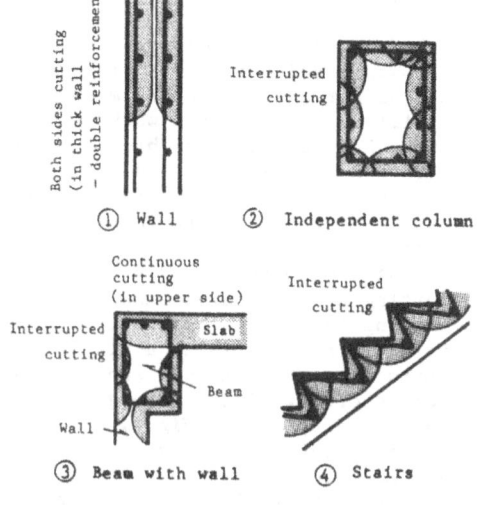

Fig. 5. Cutting method of members.

34

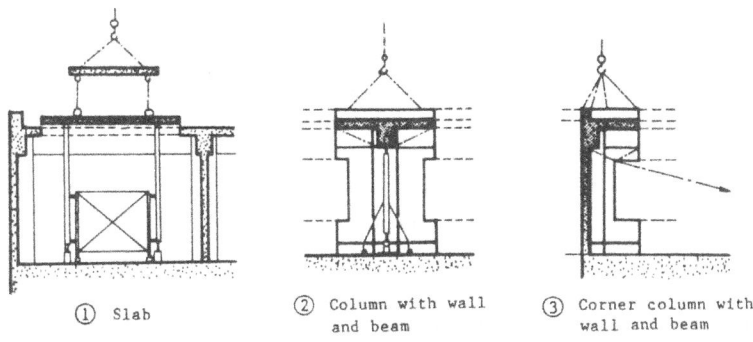

① Slab   ② Column with wall and beam   ③ Corner column with wall and beam

Fig. 6. Taking down method of dismantled members.

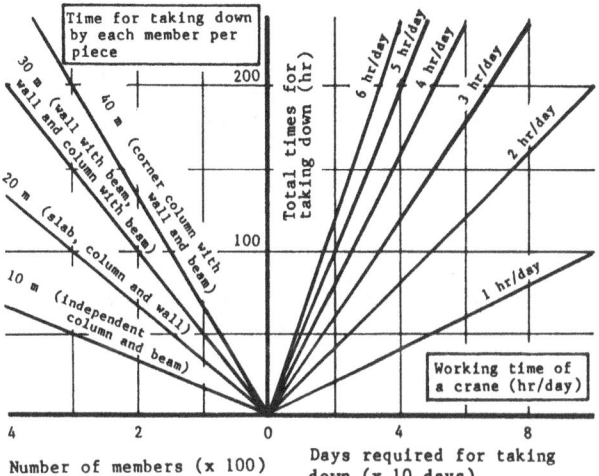

Fig. 7. Estimating method of time for taking down members.

The total days required for taking down members are decided in consideration of the cutting ability, time required for taking down of a crane, working ratio of a crane, breadth of a stockyard, time required for loading members on a truck, holidays, weather, etc. They can be estimated by use of Fig.7. The time required for taking down members is generally in the range of 10 to 40 minutes per piece and the average time is 30 minutes per piece.

3.3 Carrying out and disposal of dismantled members
1) Carrying out
The number of trucks for carrying out dismantled members is calculated in the manner that the total days are divided by the time required of loading on a truck for disposal.
① Loading volume of members is 2.5 to 3.5 m³ per 11 ton damp truck (2 m width x 5 m length), and the average volume is 3 m³ per truck.
② Loading time of a couple of members is 20 to 25 minutes on an average.

35

2) Disposal
In disposal of members, it is then decided whether they should be
reused or disused. If they could not be disposed as they were, they
would be crushed into small pieces (about 30 cm cubic). This should
be done in a proper place near a site where there is little affect to
the surroundings.

4. Cutting machines and their performance

4.1 Species of cutting machine
Some different types of cutting machines for general purpose are used
in compliance of the types of dismantled members. These include a
floor cutting machine, a vertical cutting machine, a horizontal
cutting machine, a stair cutting machine, a wall cutting machine, an
all-round cutting machine, etc. Specifications of them are presented
in Table 1. Photo 1 shows some examples of these cutting machines.

Table 1. Specification of cutting machine.

Species of machine	Use	Species and power of engine	Maximum cutting depth(mm)	Weight (kg)	Cutting speed* (min/m)
TAC-S	Slab upper edge of beam	Electromotor 3 phases,AC-200V,11kw	200	270	2~10
TAC-O	Vertical cutting of beam and wall	Electromotor 3 phases,AC-200V 175+1.5kw	160	1500	5~25
TAC-H	Horizontal cutting of column and wall	Ditto 7.5kw	160	220	5~25
TAC-R	Wall	Ditto 2.2kw	100	70	20~30
TC-Z	Stairs	Ditto 11kw	230	260	20~30

* Cutting speed at 10cm cutting depth.

(a) TAC-S Type (slab use only)

(b) TAC-O Type (all-round use)

Photo 1. Species of cutting machine.

36

The cutting speed of these cutting machines is determined by various factors such as the quality, the peripheral speed, the cutting torque and the rigidity of a sawing diamond blade. However, their cutting speed varies greatly with conditions such as the position of members to be dismantled (portion to be cut out), the configuration of reinforced area, cutting load, driving electromotor, cooling water, etc.

4.2 Sawing diamond blade

Many diamond blades are already on the market, but it is very difficult for them to cut reinforced concrete, because the diamond blades made in Japan are only used for non-reinforced area, building stone, gravestone, asphalt concrete road, etc. So, the specific diamond saws (blades) which cut reinforced concrete structure have been newly developed.

The quality of a diamond blade for cutting reinforced concrete is determined by the degree of concentration of diamond particles contained in their segment (tip) as well as the property of metal bonds binding these diamond particles. The cutting performance of our blades themselves is shown in Table 2. As to it, the optimum peripheral speed for cutting reinforced concrete is 2700 to 3000 m/min, under this condition the cutting torque showed the minimum as shown in Fig.8 and the life of a diamond blade is

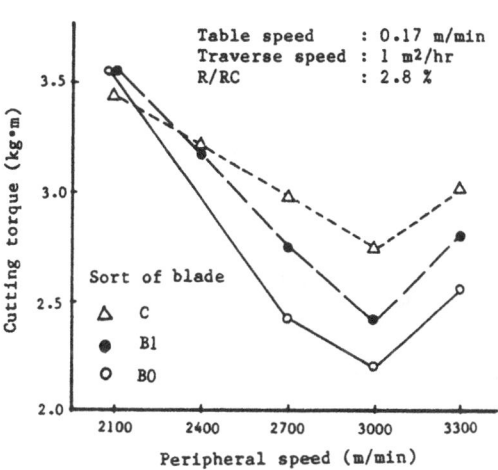

Fig. 8. Relation between peripheral speed and cutting torque.

Table 2. Cutting ability of blades.

Cutting ability	Member	Slab-wall		Beam-column	
	Cutting bar ratio	0.6 ~ 1.5		2 ~ 4	
	Blade	A	B	A	B
Life of blade (m)		36~40	24~31	22~33	17~26
Cutting speed (m/hr)		1.5~2	1.5~2.5	0.6~1.6	0.4~0.8

Table 3. Material mixture and manufacture of new improved
segment for reinforced concrete.

Species	New blade	Usual blade
Cutting object	Reinforced concrete	Stone/concrete
Concentration of diamond*	30 ~ 40 %	20 ~ 30 %
Metal bond	Bronze based or steel based	Tungsten based or super-hard alloy based
Hardness of metal bond	5 ~ 15	30 ~ 45

* 100% concentration indicates 4.4ct/cm3.

lengthened. The torque is usually 3 to 4 kg•m on an average in most
cutting machines in addition that it needs cooling water of 5 to 10
l/min to remove the frictional heat of the diamond blade. The manufacturing method of a newly developed segment is shown in Table 3.

5. Application to demolition work

5.1 Example of execution
This method which has developed from the demolition by crushing so
far in use is harmless to the surroundings, and it has been applied
to some sites shown in Table 4 after the first trial of an experiment
by use of the real sized model. The method has widely been used
today, through the analysis, improvement and development of the
method and the cutting machine. Photo 2 shows examples of executing
demolition works by use of this method.

Table 4. Main examples of demolition work by use of cutting machines.

Name of building	Size of demolition	Duration	Method of demolition
"K" bank	3 floors of RC 950 m2	Apr.1974	The part for carrying things out and into was cut first and taken off, and then cut members were carried out in each block.
"K" school-house	1 floor of RC 816 m2	Mar.~ May 1976	Walls were demolished by use of newly developed vertical and horizontal cutting machines.
"S" transformer substation	3rd floor of RC 280 m2	Nov.~ Dec.1986	The 3rd floor of the transformer substation, where 1st and 2nd floor were in use, were cut for demolition.

(a) Site of under demolition

(b) Taking down of slab.

(c) Cutting of wall (by wall saw).

(d) Taking down of beam
with wall.

Photo 2.   Examples of actual work.

5.2 Work analysis of demolition work of "K" bank
In this cutting work of 1200 m length required 110 cutting machines,
and the cutting speed which showed differences between species of
machines and members was 0.3 to 1.2 m2/hr in each member as shown in
Table 5.   It took 20 days to take down 132 pieces of cut members.
The taking down time which showed differences in each member was 5 to
50 minutes per piece on an average.

Table 5. Cutting time in each member.

Cutting machine	Member	Total cutting area (m2)	(%)	Total cutting time (hr)	(%)	Total working time (hr)	(%)
TAC-S	Slab	85.9	66.9	187	55.3	281	40.2
TAC-H	Wall	11.3	8.8	38	11.2	113	16.2
	Column	1.7	1.3	14	4.2		
TAC-O	Wall-beam	18.0	14.1	41	12.3	122	17.3
TAC-R	Wall	11.2	8.9	57	17.0	184	26.3
Total		128.1		337		700	

6. Conclusion

The dismantling method using cutting techniques was developed in 1972 by Toda Construction Co., Ltd. as a new harmless demolition one. The method has been used for numbers of demolition cases as well as modification cases including repair and reinforcement work. In particular, research and development efforts have currently been made on systematization, robotization and automatic control technologies for the method, so it can be applied to decommissioning at nuclear power plants.

Reference

HIRAGA, T. (1983) A study on the cutting technology of reinforced concrete member and its application in construction sites. Toda Technical Research Report.

DEMOLITION OF CONCRETE STRUCTURES
(SILENT DRILL SYSTEM)

TAKEDA   Technical Planning Dept.
Okada Aiyon Corp.

Abstract
The environment in which we live as we move into the 21st
century is changing at an accelerated rate according to the
needs of the times. Concrete structures are increasing
daily responding to this advanced lifestyle as a result of
demolition and construction. The production of noise and
vibration during demolition work on these concrete
structures has a major negative influence on the present
environment in which we live. As a result, controls on this
form of pollution are being strengthened.
    Along with performing demolition of concrete structures
both safely and with a low level of environmental pollution,
the "Silent Drill System" performs all other related work
including handling of concrete waste materials after
demolition work has been completed.
Key words: Demolition, Noise, Vibration, Silent Drill,
Reinforced Concrete Structures.

1. Introduction

As a method of demolition of reinforced concrete structures
at a low level of environmental pollution, typically, there
are "crushers" which have crushing arms in opposition to
each other that are opened and closed with hydraulic
cylinders. These "crushers" are shown in Photos 1 and 2.
They are mounted on the ends of the arms of an excavator.
The "crushers" compress and crush the concrete structure by
the movement of the arms. Since the demolition efficiency
of these "crushers" decreases remarkably when the size of
the structure demolished is greater than the width of the
opening between the opposing arms, other demolition methods
are necessary.

Photo 1   TS Crusher   Photo 2   Kowari-Kun

Here, ⌀40-50 holes are drilled in the structure beforehand. "Static Crushing Agent" causes cracks by expanding inside the holes, and "splitters" which insert wedges into the holes using a hydraulic cylinder, are then used. When carrying out this work, a rock drill is used to drill the holes to reduce noise and vibrations during drilling work.

Although the rock drill performs drilling by striking and rotating the surface, the accompanying striking action which causes noise and vibration is primary. Cutting by rotation is not the objective. Because of this, in the case of drilling reinforced concrete structures, the reinforcing bars are deformed by the impact of the rock drill. It is therefore tremendously difficult to cut the reinforcing marks.

Therefore, when drilling work is obstructed by the reinforcing bars, drilling unavoidably has to be stopped and attempted at a different location. Furthermore, when drilling with hand-held rock drills, if the drill bit makes contact with a reinforcing bar, it is difficult to maintain control which may cause injury to the operator.

The "Silent Drill System" has been developed to overcome these problems. The purpose of this system is to recycle wasted materials of RC performing demolition at a low level of environmental pollution by organization of demolition work on comparatively large concrete structures.

2.   Silent Drill

The major problem in organizing demolition work is the drilling work. In order to cut through both reinforcing bars and concrete, which have different characteristics, without having too much effect on environmental pollution, we have independently developed a bit for RC which has a W-shaped cross-section. "Silent Drill" was developed using method to cut reinforced concrete by rotation only. Since it also enables the operator to drill from the operator's cabin, mounted on the same excavator as "Silent Drill", it also improves the working environment of the operator.

   In order to satisfy this prerequisite, "Silent Drill" was divided into the rotary unit, which turns the bit using hydraulic pressure, and the feed unit, which moves the rod and the bit connected to the rotary unit back and forth along the guide-shell using hydraulic pressure. "Silent Drill" is mounted onto the end of the arms of the excavator with brackets attached midway on the guide-shell. Therefore, the rotary unit and the feed unit are driven by hydraulic pressure supplied from the excavator. Control is accomplished with a hydraulic control unit mounted at the rear of the guide-shell.

Photo 3   Silent Drill   Photo 4   (Cutting of
                                 Reinforcing Bars)

   On cutting reinforcing bars when drilling into reinforced concrete, a signal is output from the sensor of the hydraulic control unit that identifies the reinforcing bar. This output signal is input into the controller at the operator's seat in the excavator, and the hydraulic control unit is controlled by the control output signal from the controller.

As a result of this electronic control, reinforcing bar and concrete cutting conditions are selected automatically, making possible rotary cutting of two different materials, which was previously quite difficult.

The noise level has been reduced to a level which allows normal conversation 2m from the drilling site. Table 1 shows a summary of the specifications of "Silent Drill".

Table 1. Summary of Specifications [SD50E].

Drilling Diameter ϕmm	Drilling Depth mm	Total Length mm	Weight kgf	Hydraulic Pressure kgf/cm	Oil Volume l/min	No. of Rev. RPM	Power Supply VDC Al
40-50	1000	2360	600	210	120	170	24

With regard to net drilling speed in this case, there is a difference in net drilling speed during cutting of reinforcing bar and cutting of concrete depending on when the aggregate inside the concrete is cut. Therefore, net drilling speed is influenced by the shape and the number of reinforcing bars contained in the concrete.

The amount of SiO_2 contained in the aggregate in the concrete is a factor affecting the wear of the bit. The rate of wear of the bit is increased with amounts of SiO_2. As a result, the number of times of bit polishing per unit of drilling depth is determined mainly according to the aggregate.

3. Experimental Data

Following factory testing of "Silent Drill", averaged data based on those for ones obtained in the field tests indicated in Table 2 are given in Table 3.

Table 2. Field Tests

Silent Drill	SD50E
	Bit Gauge ϕ40 & 48
Excavator Used	UH045 0.45m³ (HITACHI)
	PC200 0.7 m³ (KOMATSU)
	MS110 0.4 m³ (MITSUBISHI)
	HD400 0.4 m3 (KATO), etc.
Local Demolition Structure	Building Foundation
	High Voltage Iron Tower Foundation
	Railway bridge pier, etc.

Table 3. Rough Data.

Avg. Drilling Speed	Repolishing Period	Bit Lifetime	Noise Level (at a distance of 15m)
25cm/min.	Every 5-6m	Approx. 40m	75dB (A)

In addition, this data varies according to the condition of the reinforced concrete.

4. Organization of Demolition Work

The demolition system, from drilling work to handling of waste materials, the first objective achieved by "Silent Drill", handles these tasks following a sequence from the first step to the fourth step. An explanation of these steps is given in Table 4 in sequential order.

Table 4. Sequence of Demolition System

First Step Drilling	Silent Drill "Drilling into demolition structure with low level of environmental pollution"
Second Step Crack Formation	Static Crushing Agent Daruda Cracker $CaO-SiO_2$-based inorganic compound, hydraulic wedge-insertion splitter, hydrostatic splitter "Formation of cracks from drilling sites"
Third Step Fine Crushing	Hydraulic Crusher TS Crusher, Silent Kowari-Kun "Fine crushing of materials from cracked sites"
Fourth Step Disposal of Waste Materials	PCP Portable jaw cracker "Crushing into uniform size for recycling into aggregate"

5. Formation of Cracks

In this method, either static crushing agent or splitters are inserted into the holes that were drilled continuously at intervals in the drilling work with "Silent Drill". The holes are then expanded internally resulting in the formation of cracks. This method is advantageous because of the low level of environmental pollution.

This static crushing agent is filled inside the holes after mixing with water. It is a powder that demonstrates expansion properties due to a hydration reaction. In general, although it expands 10-20 hours after addition of water causing formation of cracks in the concrete, there are also those that expand in a comparatively short period of time.

In this case, since the rate of expansion varies according to the time it is used, there are individual types available in accordance to the particular season (temperature).

The most common method when using the splitter is inserting wedges between the opposing feathers using considerable hydraulic pressure. After inserting the wedges in the holes, the wedges are pressed in the holes are expanded with the feathers resulting in the formation of cracks. Since wider cracks can be formed by replacing the feathers with thicker ones, these thicker types are more suitable for reinforced concrete.

6. Fine Crushing Work

As indicated in Photos 1 and 2, there are two types of systems for the crusher. In one system, both of the opposing arms rotate, and in the other system, one of the arms rotates while moving towards the other arm which is stationary. There is also a great deal of variation and diversity with respect to the shape of the arms, weight, cylinder output, crushing efficiency, etc. In addition, the size of the excavator on which the crushers are mounted varies according to the weight of the crushers. The cylinders which rotate the arms are driven by hydraulic pressure supplied from the excavator.

In the performing of fine crushing work, this system has merit in that even a comparatively small crusher can perform fine crushing work easily as a result of the formation of cracks that was performed in the second step of the process.

In the case of demolition using these methods, there is no production of noise from the main unit of the machine in the same manner as the splitter. The primary sound that is produced is the sound of the excavator engine and that sound produced during crushing of the structure. The majority of the sound at a distance of 30 meters from the worksite is at a level of 70dB (A) or less.

7. Disposal of Waste Materials

The purpose of this operation is to use the waste materials of the reinforced concrete that has been crushed in fine crushing work as recycled aggregate, crushing the material into the size of a clenched fish by a PCP (Portable Concrete Crushing Plant). However, even in the case when the material is inappropriate for use as recycled aggregate, there may be cases when it can be used as temporary materials at the worksite. Even if disposed of, not only will the space factor be improved when loading onto trucks, but since separation of reinforcing bar from concrete is possible, effective processing can be achieved including recycling of the reinforcing bars that have been removed.

Photo 5   PCP

The PCP is a single-toggle type of jaw crusher that has been specially developed for concrete. It is driven by an electric motor. Along with maticulate safety and noise abatement measures having been taken, it has also been designed for ease of transport and setup in order to be used after transporting to the worksite.

The waste materials are placed in a hopper in about 30cm squares. The waste materials in the desired size are then discharged from the outlet in which it is possible to adjust the size of the materials after crushing within a range of 40-100mm.

8. Discussion of the Demolition System

Not only has development of "Silent Drill" made possible
low-noise cutting and drilling, it has also resulted in the
designing of an entire low-noise system that also consists
of crack formation, fine crushing and waste disposal. With
this system, it is possible to suppress the noise level to
roughly 70dB (A) at a distance of 30 meters from the
worksite. This makes it to possible carry out demolition of
comparatively large reinforced concrete structures at a low
level of environmental pollution, which was previously quite
difficult.

In Japan, where the demolition of large structures is
proceeding at a rapid pace based on the current efforts of
the government to increase internal demand, there have been
legal controls placed on demolition work using hydraulic
breakers mounted on excavators for reasons of generation of
excess noise and vibrations. At present, in which
demolition work is predominantly carried out using the
crushers mentioned previously, the "Silent Drill System" is
presented as a demolition method with a low level of
environmental pollution that surpasses the limits of these
crushers.

DEVELOPMENT OF EXPLOSIVES AND BLASTING TECHNOLOGY FOR THE DEMOLITION OF CONCRETE

Erik K. Lauritzen MSc CivEng
DEMEX Consulting Engineers Ltd., Copenhagen, Denmark

Abstract
The conventional use of explosives in the demolition in-dustry has often been associated with magic and danger. This paper gives a general introduction to the research and development of explosives and blasting methods carried out to provide the demolition industry with a safe and competitive tool for concrete fragmentation.

The paper contains a brief outline of modern high and low detonating explosives, their use, and fields of applica-tion. Matters concerning safety, legislation, and the en-vironment are emphasized.

The paper is based on the State-of-the-Art Report on Conc-rete Blasting prepared for RILEM 37-DRC.
Key words: Demolition, Concrete re-use, Blasting, Fragmen-tation, Explosives, Miniblasting.

1. Introduction

1.1. Background, history and purpose.
During the past few years, it has been proved that there is a beginning international acknowledgement of the future need for demolition of plants and buildings. There is also evidence of an increasing interest in demolition techni-ques and the re-use of building materials.

An investigation conducted by the Environmental Resources Ltd (1979) for the EEC Commission envisages an increase in the amount of concrete scrap in the EEC as a whole from 55 million tons in 1980 to 302 million tons in the year 2020.

As safety and environment regulations are becoming increa-singly stringent, demands for improvement and higher effi-ciency of past demolition methods are getting pronounced. Special rules and regulations concerning demolition have been instituted in several countries including Great Bri-tain (UK), Holland, Sweden, and Japan.

There is a great number of different demolition techniques for the removal of concrete in connection with repairs. Each technique has its advantages and its limitations de-pending on the following conditions:

- Geometry and dimensions of the concrete construction.
- Requirements of limited inconveniences regarding working, living, and enviromental atmospheres.
- Restrictions concerning the concrete, and the structures, including requirements for clean surfaces.

Further points to be taken into consideration are: Transportation, time available, and total accumulated financial estimate.

A complete description of all existing techniques of concrete demolition and the promotion involved can be found in a report by the RILEM Technical Commitee 37-DRC (1984).

An overall valuation of the various techniques proves the use of explosives and expansion resources to be among those that are the most adventageous involving the greatest potential.

1.2. Demolition of buildings and tall structures, etc.
The controlled blasting techniques are most often applied for turning over tall structures, towers, chimneys, tower blocks, etc.

Blasting of blocks of flats in cities take place by making the buildings or blocks collapse successively. During recent years, blastings have been carried out in European cities such as Gothenburg (Sweden), Liverpool (England), and La Courneuve (Paris).

Fig. 1. Blasting of a ten-storey residental building at Kortedala, Gothenburg, Sweden. (NITRO CONSULT 1986). Approximately 80 kg of explosives in 3,000 boreholes.

In the United States, blasting of tall buildings with steel frames have been carried out for several years. For this purpose, specially made charges for cutting steel have been used with great success.

With this type of buildings it is necessary to control the collapse and prevent materials from falling outside the building site. To control a vertical collapse, the entire building is often constricted and supported by wires thus avoiding unintended disintegration.

Fig. 2. Blasting of 40 m high, and 1700 t water tower, Frederikshavn, Denmark 1976.
Approximately 90 kg of explosives in 1,000 boreholes.

1.3. Scope
This paper concentrates on blasting as a method of demolishing concrete. Consequently, the only types of blasting described in detail are: Partial demolition, localized cutting, "Miniblasting", and fragmentation of concrete by the use of explosives.

2. Explosives

2.1. Definitions
According to the current NATO definitions, the word **EXPLO-SION** expresses the mechanical and thermal effect of the chemical reaction of an explosive during detonation or deflagration in confinement.

DETONATION is a violent and entirely chemical reaction which proceeds at supersonic velocity within the explosive. It generates gases at extremely high pressures and temperatures. The sudden and powerful pressure of hot gases violently disrupts the surroundings which causes a shock wave to propagate at supersonic velocity.

DEFLAGRATION is a chemical reaction which proceeds at subsonic velocity through - or along the surface of - an explosive substance. It produces hot gases, at high pressures. Deflagration in confinement increases the rate of reaction, and the temperature which may cause a transitior from deflagtion into detonation.

2.2. Detonating explosives.
Blastings are normally performed by the use of detonation explosives which can be separated into the following types:

A. Military high explosives with a velocity of detonation of between 6,000 and 9,000 m/s, e.g. Trinitrotoluene (TNT), Hexogen RDX), Pentaerythritol tetrannitrate (PETN), and composites, such as Comp. B (60% RDX/ 40% TNT), Pentolite (10-50% PETN/ 90%-50% TNT), ect.

B. Commercial explosives, typically the various kinds of dynamite with a detonation velocity of between 3,000 and 7,000 m/s or other explosives such as ANFO (a mixture of fuel oil and ammonium nitrate) water-gel explosives, and special explosives including the Swedish "GURIT". The latter is a low detonating explosive especially suited for careful blasting.

Today, mainly the dynamite explosives are used for the demolition of major concrete structures. They require boreholes, drillings of a diameter of at least 25 mm. The dynamites are normally covered in paper or plastic tube packings. Owing to their content of nitroglycerine, handling often causes headaches.

Where demolition of minor structures is concerned, i.e. "Miniblasting" or fragmentation (see other papers), it is preferable to use high detonating explosives, mainly PETN mixed with a certain percentage of wax. The handy PETN explosives are fit for use in small charges, e.g. 10 g charges, which can easily be placed in boreholes of small diameters, i.e. 10 - 18 mm.

2.3 Deflagrating explosives.
During the latest decades, much research has been carried out in the field of explosives, and extensive studies have been made to find solutions to the problems attached to the use of detonating explosives and their high energy release.

According to Ito, Sassa, and Tanimoto (1972), the so-called "low explosives" or deflagrating explosives were already developed in JAPAN in the early 1970ies.

Explosives of a deflagration velocity of about 100 m/s are developed for urban blasting. The abovementioned Japanese paper gives a detailed description of the breakage mechanism.

	LOW ◄——————— EXPLOSIVES ———————►HIGH					
	CRC	URBANITE	GURIT	DYNAMITE	TNT	PETN
Manufactor	ASAHI	Nippon Oil & Fats	NITRO NOBEL	(General types)		
Country	Japan	Japan	Sweden	-	-	-
Density, g/cm^3	-	1,30	1,30	1,50	1,60	1,65
Velocity of detonation, m/s	60	2,000	3,000	6,000	6,500	7,500
Energy of explosion, kJ/g	-	-	3.8	4.6	4.1	6.1
Volumen of gases, l/kg	50	-	400	750	690	780

Fig. 3 Table giving datas on low and high explosives.

3. Blasting methods.

3.1. Controlled blasting. Controlled or careful blasting of concrete is a technique which is based on the same principles as modern rock blasting, namely the use of a minimum of explosives at the same time as fulfilling great requirements as to control of the effect of the explosives and the prevention of damage to the surroundings.

Modern explosives and blasting accessories enable the performance of controlled blasting anywhere.

The control of blasting operations is made possible by u-
sing divided charges in densely drilled holes ignited at
short-time intervals - normally 20-30 ms. The desired de-
finitive effect can hereby be obtained and unintended effects reduced.

This permits blastings to be carried out indoors as well
as in densely built-up areas without disregarding safety
regulations and environmental requirements. Blastings can
furthermore be effectuated inside buildings without damaging the remaining part of the structure or installations,
etc.

However, the conversion and the transmission of energy to
the surroundings, which are caused by blastings, constitute a very complicated process. That is the reason why it
has, for the present, proved impossible to make practicable calculations of the consequences of a blasting operation. The performance of blastings have therefore always
been based on experience and the use of emperical formulas.

There is a considerable lack of the needed documentation
of blasting techniques. This implies a natural scepticism
by "uninitiated" technicians and others who must evaluate
an explosion or take the responsibility for it.

3.2. General principles of blasting methods. Usually,
concrete blasting is done by the use of small confined
explosive charges which are placed in drilling holes.
Single, double, or triple rows are used in the case of linear cuts of a structure or the demolition of columns,
beams, walls, etc. If the demolition purpose is to create
total fragmentation in the concrete or part of the structure, for instance when blasting foundations, abutments,
openings, etc, the procedure is to arrange a pattern of
holes in a grid with equal distances between the holes.

The explosive charges are normally ignited by millisecond-
-delay detonators (time of delay is a mulitple of 20-30
ms) and special consideration is paid to the safety of the
surroundings and remaining structure just as the optimal
use of the explosive energy is ensured.

The mechanisms of breakage, propagation of cracks, and the
throwing of debris more or less depend on a number of parameters, see tabel fig. 4 . Some of the parameters have
been studied closely. Most blasting parameters are in fact
determined with much uncertainty and it is difficult to e-
stablish theoretical models giving a realistic picture of
the coherence between each parameter.

Some of the most important parameters are discussed
in detail in other papers.

GENERAL PARAMETER:	SINGLE PARAMETER:
Concrete	Strength of concrete Quality of contrete Reinforcement Microstructure Crack propagation
Explosive charge	Strength of explosive Loadingding density Confinement Interaction and cooperation of charges
Geometry	Geometry of object Burden Hole spacing Crater volume Constriction of boreholes Depth of boreholes
Environment	Vibrations Air blast Noise Dust Gases Remaining structure
Practical application	Drilling Covering Fragmentation Handling of rubble

Fig. 4. Table of important blasting parameters

Calculations of blasting charges have always been based on
emperic formulas and the personal experience of the individual blaster. A basic starting point for charge calculation is to determine the average specific charge for the
object. The table in fig 5 shows an example of emperic
specific charges.

Object	Specific charge (kg/m^3)	Hole spacing (m)
Concrete and masonry, poor quality	0.15–0.40	0.70–0.80
Concrete and masonry, good quality	0.30–0.40	0.60–0.70
Reinforced concrete, normal	0.40–0.60	0.40–0.50
Reinforced concrete, heavy	0.60–1.50	0.30–0.50
Reinforced concrete, heavy, high concrete strength	1.50–2.00	0.25–0.50

Fig. 5. A charging table for concrete blasting based on
experience of blasting in drill holes.
Estimated consumption of high explosives per m^3 concrete.
(Lauritzen, 1986)

55

4. Examples of practical applications. Concurrently with the increasing need for concrete repairs, renewal of structure elements, and the demolishing of concrete structures, there is scope for the use of explosives for the dismounting of major elements. See principles in fig. 6.

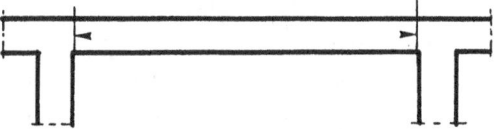

Fig. A. A reinforced concrete beam has to be removed.

Fig. B. Preparing of blasting and supporting.

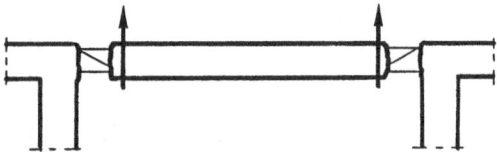

Fig. C. Blasting, cutting of reinforcement bars, and lifting.

Fig. D. Concrete structure after removal of the beam.

Fig. 6. Outline of principles of the dismounting of a reinforced concrete beam by the use of explosives.

Fig. 7. & 8. Concrete pipe before and after blasting by the use of a detonating cord 2x24g/m. (DEMEX 1986)

The common procedure for cutting concrete pipes in sewer conduits, etc. is by the use of diamondtipped tools. The cutting is often slow and laborious. By the use of a detonating cord, a pipe can be cut within less than 10 minutes as shown in fig. 7 and 8.

With reference to the many other contributions providing detailed descriptions of the applicability of blasting techniques within the field of demolition and the reuse of concrete, fig. 9 and 10 constitute our last example.

Fig. 9. The beginning of blasting out the 90 MPa concrete with steel fibre reinforcement cast into a concrete canon.

Fig. 10. Blasting work inside the concrete canon.
       (DEMEX 1986)

5.   Concluding Remarks.

There is no doubt that blasting techniques are among the most competitive methods of demolition. By means of effective covering devices it is possible to carry out blasting under almost any circumstances. But unfortunately, the use of explosives is connected with a certain risk of different kinds of abuse including criminal abuse which has caused the authorities to be under exaggerated apprehension. Added to this apprehension, the traditional view that blastings should only be carried out in remote places creates considerable barriers restraining future development. Mention may be made of the fact that the number of builders injured permanently by the use of traditional equipment, specially concrete hand hammers, by far exceeds the number of injured in the field of blasting. This fact has in Denmark been submit to consideration and has in many cases led the authorities for working environments to request that contractors effectuate blastings instead of manual cutting.

References:
1. Environmental Resources Ltd.: Demolition in Denmark. Consultative paper, October 1978.
2. NATO Working Group AC/258, 1984.
3. RILEM Technical Committee 37-DRC: Report on Demolition Techniques, 1985.
4. Ito I., Sassa K., Tanimoto C.: Blasting by Specially made Explosives for Urban Works. Proc. of JSCE, No. 199, March 1972.
5. Lauritzen E.K.: Mini-blasting for Repair Work. Building Research & Practice, Sept./Oct. 1986.
6. Nippon Oil & Fats, Japan, Company information
7. ASAHI, Japan, Company Information
8. NITRO CONSULT AB, Sweden, Company Information
9. NITRO NOBEL AB, Sweden, Company Information
10. DEMEX Consulting Engineers Ltd., Denmark, Company Information
11. Molin C., & Lauritzen E.K.: State-of-the-art-report of concrete blasting prepared for RILEM TC DRC-37, 1988

DEMOLITION OF CONCRETE STRUCTURE BY BLASTING AND ASSUMPTION OF THE
RANGE OF DESTRUCTION

TAKASHI SHINDO   Engineering Department, Blasting Division
                   Chugoku Kako Co., Ltd.
MASANAO MAEDA   Planning and Development Department
                   Chugoku Kako Co., Ltd.

Abstract
In demolition of concrete structure by means of explosives, more often
than not, it is required to eliminate only certain part of the structure with the remainder to be left as it is. The extent of impact
such remaining part incurs, is therefore an essential point in execution of a job. Results of vibration measurement taken during blasting
for partial demolition in the several operations we have carried out,
were analized for the study of the characteristics.
    Initially, in order to make an assumption of a range of destruction
by comparing such results of analyses with the observation of destuc-
tion, samples were taken with core boring to check their compressive
strength as well as existence of any crack. While the data of stress
in the vicinity were obtained through measurement with strain gauge,
the quantity of such data and correlations among them did not appear
to have been reasonably sufficient to make a quantitative assumption.
As an example for investigation of range of destruction however, such
data will be presented hereunder.
Key words: Blasting, Acceleration, Frequency, Remaining part.

1. Introduction

In recent years in Japan also the demolition of reinforced concrete
structure is calling keen attention in the rush of urban renewal projects because of changing industrial structure or exorbitant increase
of land cost. As a method of demolition, although the mechanical one
wherein hydraulic giant breaker is employed or manual one using conventional portable pneumatic breakers still remain to be the main-
stream, due to issues of noise and vibration particularly in urban
area lately, other methods are frequently employed including those
that use hydraulic concrete crusher, concrete cutter or non-explosive
demolition agent. With any of these method however, required period
of work tends to be longer and cost to be higher, making it rather
unfavorable depending on the shape of structure. Therefore blasting
method that uses explosives is being adopted in parallel.
    Among reinforced concrete demolition work in use of explosives we
have executed to date, the object of demolition was often base concrete in foundation or large footing, with somewhat special cases
having been a retaining or cut-off wall. Normally, the degree of de-

struction by blasting is not such that the reinforced concrete is totally removed but that it is slackened or loosened to be followed up by final destruction with mechanical force of breaker or the like. In other cases, it is requested to remove only certain area of reinforced concrete locally by blasting. Satisfactory investigation as well as level of technic is essential in such occasion as certain dimensional accuracy is required on resultant finish surface of remaining part.

We generally classified the processes of partial demolition into following three categories on the basis of destructing part vs. remaining part. The first is the case wherein a part of continuous single concrete mass is blasted and the vibration wave thereof is propagated in the same concrete. The second is the case wherein only a projected part of the structure is blasted and in respect of the shape, it has corners. The third case consists of two structurally separated parts and the part to remain is located closely to the part to be demolished. In this case the vibration wave is propagated through the medium of different quality.

Hereunder, the blasting procedure, result of measurement and our study will be discussed:

2. Typical demolition works

2.1 Blasting procedure
In accordance with the three categories as classified in the Introduction hereof, Cases-1, -2 and -3 will be presented. The relationship between the part demolished by blasting and the remaining part will be shown in Fig. 1.

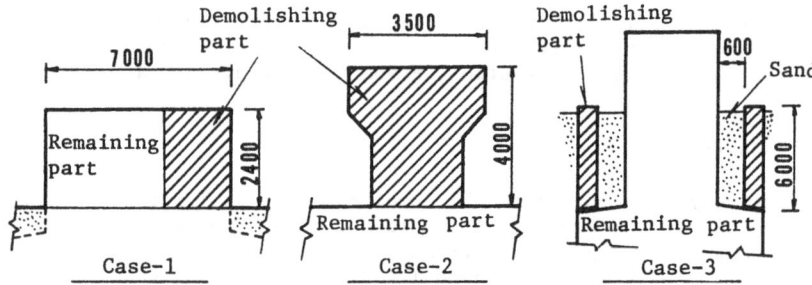

Fig. 1. Relationship between demolishing part and remaining part

Case-1 represents a machine foundation concrete of continuous single concrete mass. The distance of blast hole nearest from the remaining part was 250mm. The minimum burden at the time was 500mm. Drill bit diameter and hole depth were 34mm and 1000mm respectively and no prior cutting of reinforcement was carried out.

Case-2 is a substantially massive machine foundation concrete which is larger than that of Case-1, and the top projection of which was blasted as shown in Fig. 1. The border between base concrete and projected part having been the concrete placement surface of the base concrete, the vibration wave was suspected to be reduced to some extent. However, since it was also assumed that at the corners stress concentration might develop, the distance from blast hole bottom to base concrete surface was arranged to be greater than 2.6m to avoid possible adverse downward effect. The charge for single simultaneous blasting was 800g in maximum and prior cutting of reinforcement was not conducted here either.

Case-3 is an example of cut-off wall blasting. After filling back the space of 0.6m in width between bridge pier and cut-off wall with earth, deck loading was made in blast holes prepared beforehand by dividing the circumference quarterly. The bottom of cut-off wall was separated from pier using rubber plate and the reinforcement was not continuous either. This method allows to remove a cut-off wall that has become useless, quickly and inexpensively and in addition it gives no adverse effect to bridge pier itself or its surroundings.

Table 1. Data of blasting.

Job	Bit dia. (mm)	Minimum burden (mm)	Hole depth (mm)	Specific charge (g/m^3)	Explosive used	Remark
Case-1	34	500	1000	400	#3 Kiri*	
Case-2	34	500	900	740	"	
Case-3	32	400	6000	320	"	Deck loading

*#3 Kiri dinamite: Detonation velocity   5500 $\sim$ 6500 m/sec
Specific gravity   1.3 $\sim$ 1.45

All of above three blasting works were carried out either within plant premise or urban area. Therefore they were executed not for total destruction but under well controlled manner so that flying debris, vibration and noise level were all confined to applicable regulative value. After blasting final destruction with mechanical means was performed, but improvement in overall efficiency was apparent. We have experienced such method that combines the blasting with mechanical force in many other instances. While we have executed approximately 100 concrete blasting operations, we have always successfully controlled flying debris, vibration and noise level and completed the works without incurring an accident.

2.2 Measuring procedure
In any of above cases, measurement was taken on the concrete of the remaining part, wherein pick-ups were installed within the range of 0.25m $\sim$ 18m away from demolishing part. And among vibration waves on concrete surface, mainly the component in vertical direction was caught in the form of acceleration waveform. The reason for this was that firstly it was necessary to check the distance attenuation within relatively immediate distance and secondly to simplify the propagation route of the vibration wave. Also in case of blasting vibration, the waveform of vertical direction is more likely to attenuate in very short distance than that of horizontal direction and because in the past examples too, more vertical ones have been investigated, it seemed easier to make the comparison.

Actual works were executed by delay blasting, therefore practically

it was impossible to use the instantaneous electric detonator only
during the time when measurement was to be taken. However in the
analysis, only the waveforms of instantaneous blasting were identified
to compile the result. In Fig. 2, the location of measuring points
and blast holes are shown.

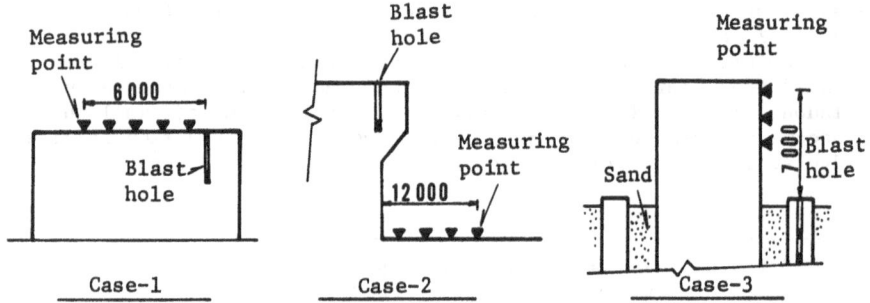

Fig. 2. Location of measuring points and blast holes.

Pick-ups were installed to buried type stainless steel bases at
locations within as close to blast source as 5m, and at farther locations, the iron plate installation bases were affixed with instantaneous adhesives to grounded concrete surface. The type of pick-ups
employed were the Piezo type acceleration transducers and strain gage
type acceleration transducers.

Since it was presumed that the frequencies of developing vibration
wave would be extremely high, around 10KHz, over the range of 1 ∿ 2m
from the blasting source, the measuring system was prepared so that it
was capable of measuring the frequencies up to 20KHz level.

2.3 Characteristics of concrete
The characteristics of respective concrete are shown in Table 2 below:

Table 2. Characteristics of concrete.

Job	Unconfined compressive strength (kgf/cm^2)	Primary wave propagation velocity (m/sec)	Static elastic modulus (kgf/cm^2)	Poisson's ratio
Case-1	323	4100	2.51×10^5	0.21
Case-2	320	4200	-	-
Case-3	240	-	-	-

In the Table 2, the unconfined compressive strength is not the
design strength but the actually measured value. The volume demolished by blasting in respective case was 71m³ in Case-1, 820m² in
Case-2 and 85m³ in case. The volume of reinforcement steel was
30kg/m³ with major reinforcement size having been 16mmϕ for Case-1
and about 100kg/m³ with major reinforcement size 22mmϕ for Case-2
and 93kg/m³ with major reinforcement size 16mmϕ for Case-3.

3. Result of measurement and its study

3.1 Features of acceleration waves
Result of measurement in Case-1, -2 and -3 are shown in Figs. 3, 4 and 5 respectively with their propagation distances shown in Table 3.

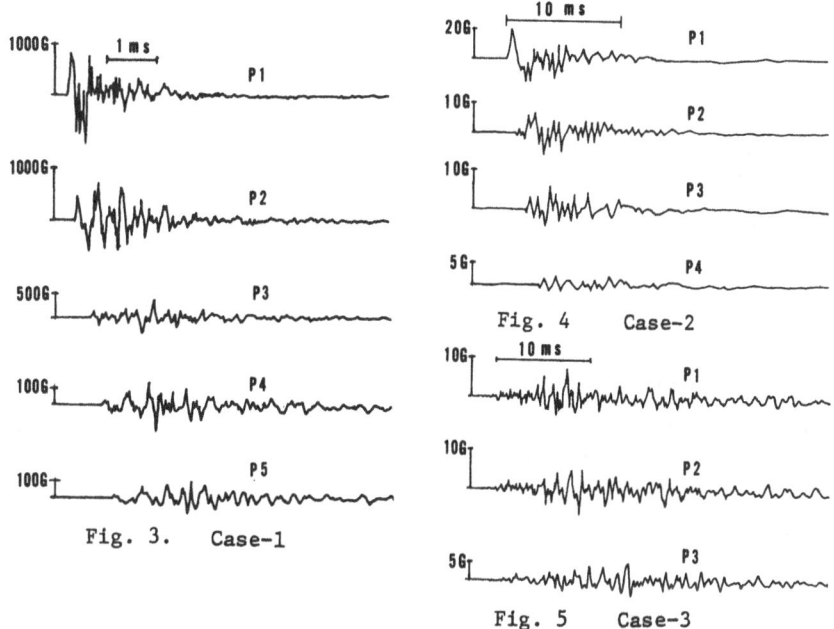

Fig. 3.   Case-1

Fig. 4   Case-2

Fig. 5   Case-3

Figs. 3, 4 and 5.   Result of measurement (Acceleration waveform)

Table 3.   Propagation distance

Measuring point Job	P1 (m)	P2 (m)	P3 (m)	P4 (m)	P5 (m)
Case-1	1.15	1.73	2.67	3.81	4.93
Case-2	3.4	6.6	10.2	15.4	-
Case-3	11.95	13.9	15.7	-	-

Graphs of Fig. 3, 4 and 5 respectively shows acceleration waveforms with time on axis of abscissa and acceleration (G) on axis of ordinate. The phenomenal waveforms for Case-1 show the waveforms at 5 measuring points, the propagation distances to which from source of blasting was P1 through P5 in such order. At the measuring point P1 (propagation distance 1.15m), a spearheaded peak wave appears at first and the phenomenon ends in about 3 milli-seconds. As the distance grows greater for P2, P3, ...P5, the value of acceleration becomes smaller and time of duration grows longer. And the spearheaded peak wave becomes gradually more gentle waveform and at the point P5 (pro-

63

pagation distance 4.93m), the maximum value is shown in the middle of waveform. While at P1 the waveform is "initially maximum value type," as the propagation distance grows greater at P2 and farther, a trend is seen for the maximum value to be delayed to appear. It can be regarded to be well indicative of the feature of elastic waves. Also it should be noted that in each of the waveforms, the phenomenon attenuates in extremely short period of time such as 3 to 8 milli-seconds.

The same applies to the phenomenal waveform of Case-2 as well. However, the duration of vibration at P1 (propagation distance 3.4m) is about 13 milli-seconds and at P4 (15.4m) about 20 milli-seconds which are slightly longer comparing with Case-1. It is considered to be attributable to the fact that the distance to measuring points being greater, the amplitude becomes smaller but the duration of low frequency is greater.

With the Case-3, different from Case-1 and -2, although the "initially maximum type wave form" does not appear, as the distance to measuring point grows greater, the waveform becomes more gentle and peak values becomes smaller. In the duration of vibration however, not much of change can be seen. It may be because of the propagation route that the waveform does not take the "initially maximum value type pattern. Although the vibration wave that traveled through rubber plate on the lower edge of cut-off wall did appear earlier than the wave traveling through filled bach earth, the vibration wave that went through such earth having been greater, it seems to represent the peak value. Also it is conceivable that the fact that respective distance to each of the measuring points P1, P2 and P3 had been so close, did affect as well.

3.2 Frequency characteristics
The acceleration waveforms in Fourier's spectrum are shown in Figs. 6 and 7.

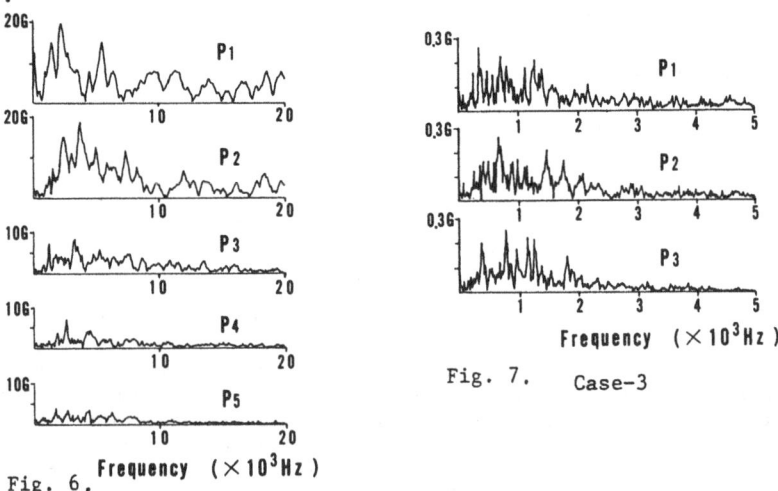

Fig. 6. Case-1

Fig. 7. Case-3

Figs. 6 and 7. Acceleration waveforms in Fourier's spectrum

64

The frequencies that develop at 1 ∿ 2m were anticipated to be predominant around the range between several KHz and 10KHz, and in the Case-1 and -2 shown in Figs. 6 and 7, such trend is observed around 3KHz. The fact that more than two predominant frequencies exist, indicates that the blasting vibration develops complicate phenomena. While it varies depending on the characteristics of object of blasting, procedure of blasting or propagation route, etc., the fact that more than two predominant frequencies exist can also be understood to be attributable to the fact that the relationship between the vibration velocity which is normally considered to be more or less directly related to the degree of damage caused by blasting and the acceleration, can only be represented by frequency.

Upon checking by the propagation distance to the measuring points, it is found that while in the places closer to the blasting source, the frequencies higher than 10KHz are frequently detected, such trend attenuates as the distance grows greater. This makes one assume that along with the propagation of vibration, the high frequency components are absorbed or eliminated in the concrete. In the Case-3, it attenuates as it is propagated through the earth between bridge pier and cut-off wall, causing the frequency around 1KHz to be predominant.

3.3 Distance attenuation of acceleration
The distance attenuation of acceleration is shown in Fig. 8 in reduced distance.

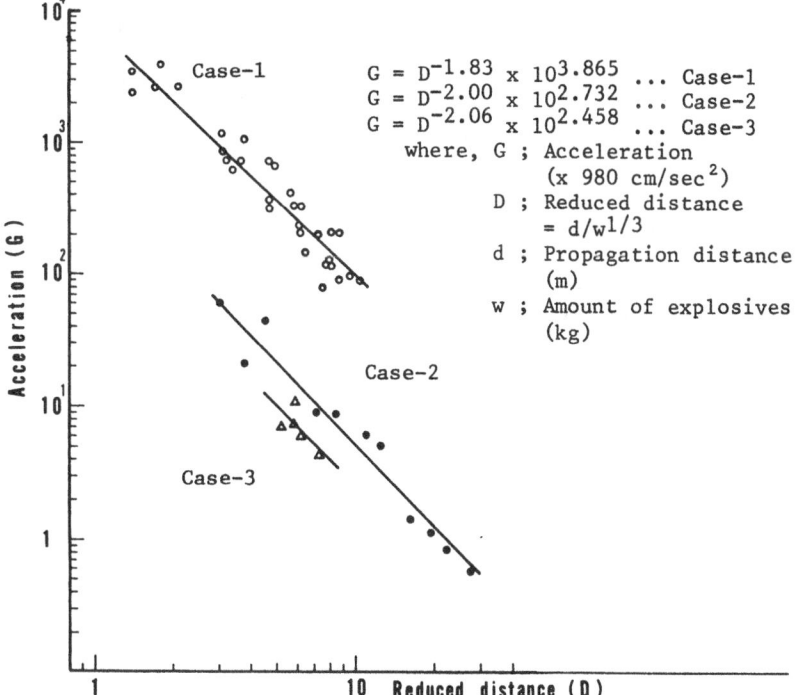

$$G = D^{-1.83} \times 10^{3.865} \quad \cdots \text{ Case-1}$$
$$G = D^{-2.00} \times 10^{2.732} \quad \cdots \text{ Case-2}$$
$$G = D^{-2.06} \times 10^{2.458} \quad \cdots \text{ Case-3}$$

where, G ; Acceleration
($\times$ 980 cm/sec^2)
D ; Reduced distance
$= d/w^{1/3}$
d ; Propagation distance (m)
w ; Amount of explosives (kg)

Fig. 8. Distance attentuation of acceleration.

As seen in Fig. 8, the inclinations in the graph of attenuation are not much different. Value of acceleration (G) grows smaller in the order or Case-1, Case-2 and Case-3. This is due to difference in the scale of blasting or route of propagation. The coefficient of correlation (r) was -0.953 in Case-1, -0.967 in Case-2 and -0.662 in Case-3 respectively.

3.4 Sketch of destruction
Condition of destruction in Case-1 is shown in Fig. 9. The horizontal distance from blast hole to the farther point where crack developed was 1.1m on concrete surface.

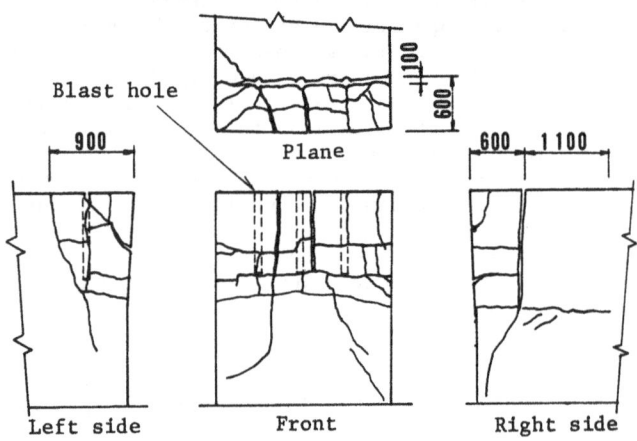

Fig. 9. Sketch of destruction

3.5 Distance attenuation of velocity waveform and assumption of destruction range
For an appraisal of degree of damage incurred with blasting vibration, generally the displacement velocity (V) is frequently used. Therefore in regard to the Case-1, we show the relationship between velocity and reduced distance in Fig. 10.

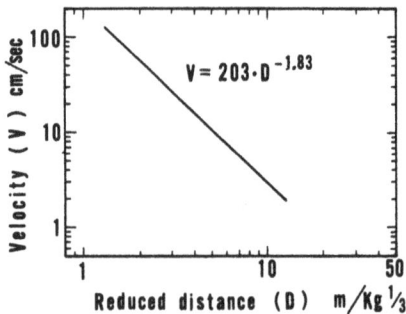

$$V = 203 \cdot D^{-1.83}$$

Fig. 10. Distance attenuation of velocity waveform

In order to obtain the distance attenuation of velocity waveform, we read the acceleration (G) and frequency (f) of peak value in the phenomenal waveform of acceleration and calculated the velocity (V) in accordance with an equation $G \times 980 = 2 \pi f V$. Further, while the relation between acceleration and reduced distance was given by equation $G = D^{-1.83} \times 10^{3.865}$, assuming that the same attenuation exponent (n) = -1.83 in this equation is also applicable to the velocity in the same manner, the Fig. 10 was attained. The equation appearing in the graph is given by $V = 203 \times D^{-1.83}$.

Futher, the reduced distance (D) being $D = d/w^{1/3}$, an approximate equation of propagation distance (d) should read, $d = 18.2 \times V^{1/-1.83} \times w^{1/3}$. In the Case-1, because the maximum length of crack was 1.1m as shown in Fig. 9 and amount of explosives at the time was 0.3kg, entering these values will give V = 46.9cm/sec. And assuming such an example as shown in Fig. 11 which is similar to the Case-1, gave the result of V = 48.1cm/sec. Although it is generally said that a damage to structure will develop when the vibration velocity (V) is 10cm/sec or greater, in our Case-1, still higher vibration velocity was observed.

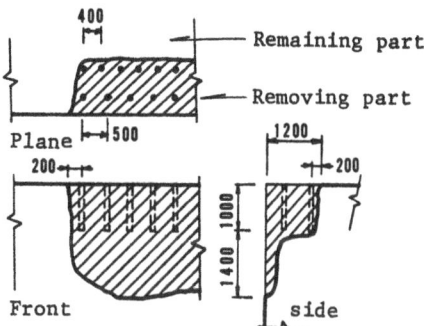

Fig. 11. Other example of destruction by blasting

4. Conclusion

With a continuous concrete, in immediate distance the waveform appears in an "initially maximum value type" form and attenuates with duration of 3 to 8 milli-seconds. As the distance grows greater, the duration becomes longer to 13   20 milli-seconds and shows the typical feature of elastic wave. If a medium of different quality such as earth exists in the propagation route, the "initially maximum value type" waveform does not appear. In respect of frequency, in immediate distance a few groups of higher component appear predominantly, and as the propagation distance grows greater, such high component progressively diminishes. In the case where different quality medium such as earth exists, the range of frequency appears to be around 1KHz.

Acceleration attenuates at the rate of approximately n = - second power against the reduced distance (D), and with the continuous concrete, it attenuates relatively irrespective of the traveling direction of the wave or the difference in shape of concrete that is blasted. With regard to velocity attenuation, if we consider the attenuation exponent (n) to be equal to that of acceleration, under such blasting conditions as we actually measured, we consider that it is possible to make an approximate assumption of the propagation distance (d) by means of following equation, that is:

Propagation distance (d) = $18.2 \times v^{-0.546} \times w^{1/3}$.

Although the assumption such as foregoing may not be quite generally applicable to broad range, we will be happy if it serves as an example for reference in the investigation of the effect to remaining part in immediate distance of blasting.

LOCALIZED CUTTING AND PARTIAL DEMOLITION IN CONCRETE WITH CAREFUL BLASTING

C. MOLIN
Building Technology, Swedish National Testing Institute

Abstract

It is important to increase and improve the range of methods available for localized cutting and partial demolition. Careful blasting with drilled-in charges has after research and practical experiments shown to be a feasible method for thicker structures, 250 mm and upwards. Strong slabs and walls, beams, columns, culverts, abutments quays are typical structures. The disintegration process, the effect of design parameters, the effect on the remaining structure and the surroundings are predictable and acceptable. Different kinds of detonating explosives can be used with a weight of 1-30 grammes and a spacing of 150-400 mm. It is possible to reduce the effect of the blasting by certain precautionary measures with regard to the remaining structure and surroundings. For example the effect of airborne shockwaves can be reduced 80-90 % by means of a simple light protective cover.

Key words: Concrete, Careful blasting, Partial demolition, Drilled-in charges.

1. Introduction

Localized cutting and partial demolition are frequently required in connection with modernization, conversion and repair of reinforced concrete structures. Alterations of this type are expensive and often create a troublesome working environment. Therefore it is important to increase and improve the range of methods available. The results presented in this paper are based on blasting operations carried out in an old concrete building and on slightly more than 100 blasting operations performed in full-scale trial slabs with different reinforcement, thicknesses and concrete quality. Different procedures for extra careful blasting have also been tested.

69

The research was funded by the Swedish Concrete
Research Foundation and the Swedish Council for Building
Research and carried out at the Swedish Cement and
Concrete Research Institute during the project period
1981-84.

A common drill-hole diameter and location of charge is
shown in Fig 1. Small charges, 10 g, with high detonation velocity were mostly used.

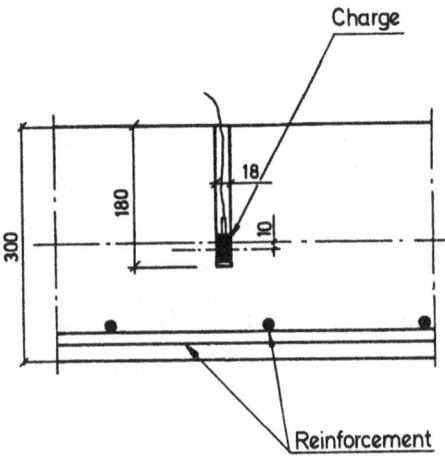

Fig 1   Location of charges at right angles to structure
         plane.

Interval initiation, 25 micro seconds delay, was
applied for the detonation. Location of charges in a
structure plane is shown in Fig 2. The first intervals
(the cut) and the last intervals (the free face) in the
opening have been studied in detail.

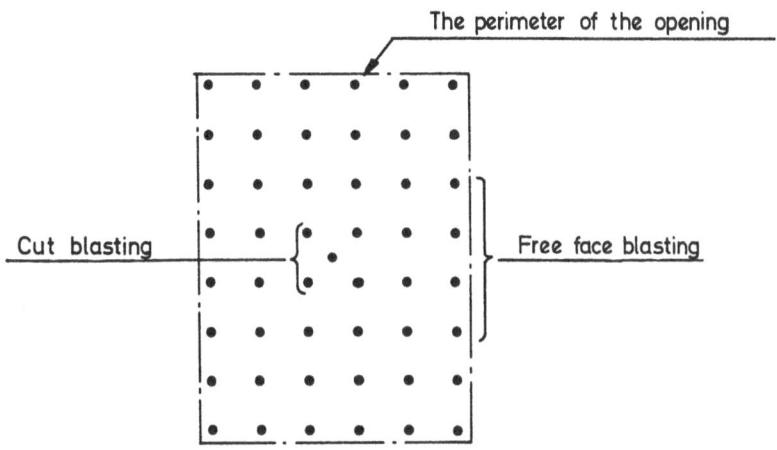

Fig 2   Location of charges in structure plane.

2. Results

The specific charge is defined as the charge weight
divided by the nominal blasted volume. The specific
charge as a function of the structural design parameters
so as to obtain precisely the breakout intended are presented in three following diagrams. Generally speaking,
a far lower specific charge was required for free face
blasting compared with cut blasting, the approximate
ratio being 1:5. The standard charge used, 10 g Primex,
proved to be too weak a charge for heavily reinforced
slabs and/or for slabs with considerable structure
thickness. Fig 3 indicates that an increased specific
charge is required for cut blasting when the degree of
reinforcement increases.

71

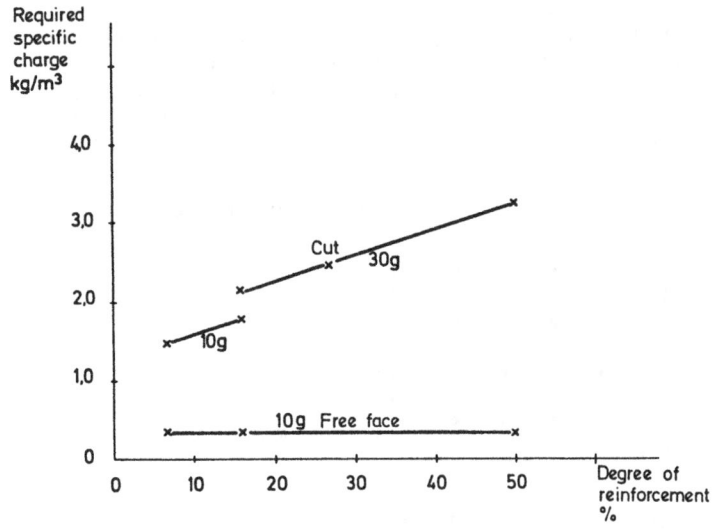

Fig 3   Required specific charge as a function of the
         degree of reinforcement.

    Fig 4 appears to show that there is minimum specific
charges for a certain weight, in this case 350 mm seems
to be an optimum thickness.

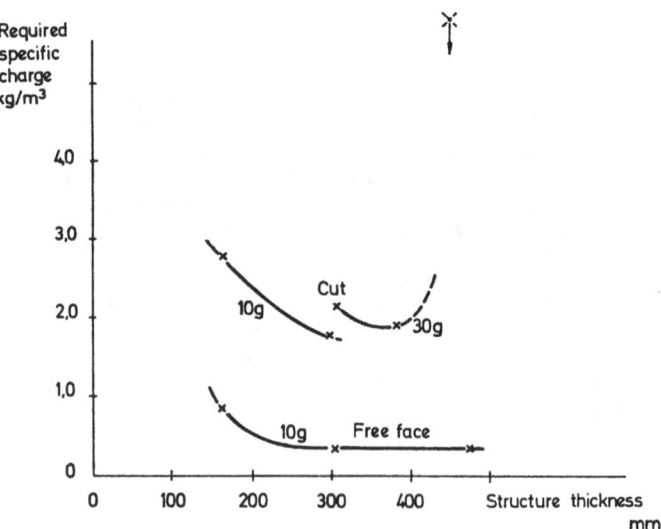

Fig 4   Required specific charge as a function of the
         thickness.

72

In Fig 5 it is shown that the required specific charge increases with higher concrete quality. However for strengths higher than about 50 MPa the blasting resistance decreases with increases in the strength, according to literature. In other words, there seems to be an optimum quality with regard to the capacity of the concrete to withstand disintegration.

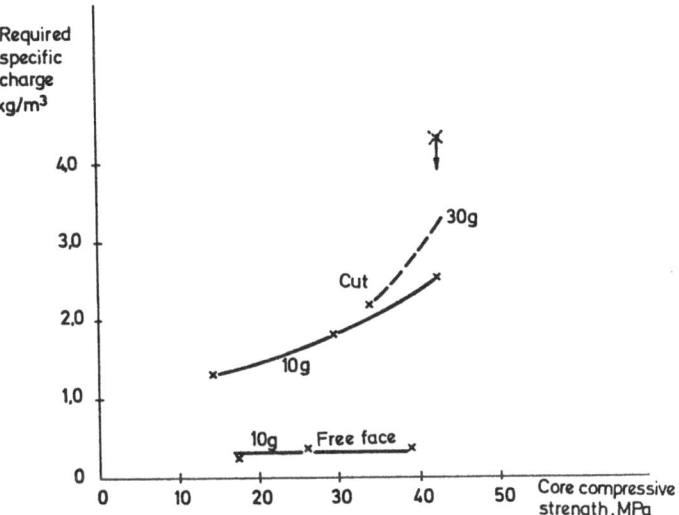

Fig 5   Required specific charge as a funtion of the compressive strength.

Some damage will occur on the remaining structure. The internal damage or crack propagation is measured by ultrasonic inspection. Coarse chamfering at the top and the bottom edge of the structure was also measured. In Fig 6 the distance to undamaged concrete for the various measures taken for extra careful blasting are presented. The mean damage distance varies between 150 and 250 mm. A large drill-hole (24 mm) and internal initiation gave the shortest average damage length. The degree of damage can be further reduced by combining various measures.

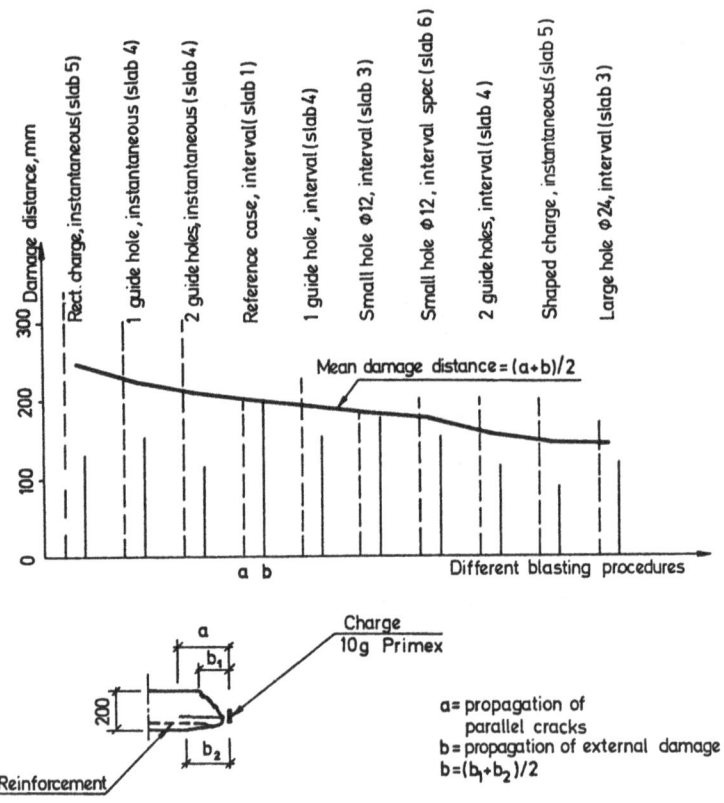

Fig 6   Crack propagation and external concrete damage for various blasting modes.

74

In Fig 7 the airborne shockwave's maximum pressure (unbroken line) and impulse (broken line) without damping are indicated in the barchart each with 100 % which corresponds to 6 kPa for the pressure and 3.2 Ns/m^2 for the impulse. It could be noted that the use of protection gave a marked reduction. This was not surprising for heavy covering materials, but the fact that even light materials such as mineral wool and celluar plastic had a marked effect was not expected. For example, 70 mm mineral wool plus one rubber mat gave a relative pressure and impulse level of only 10 %.

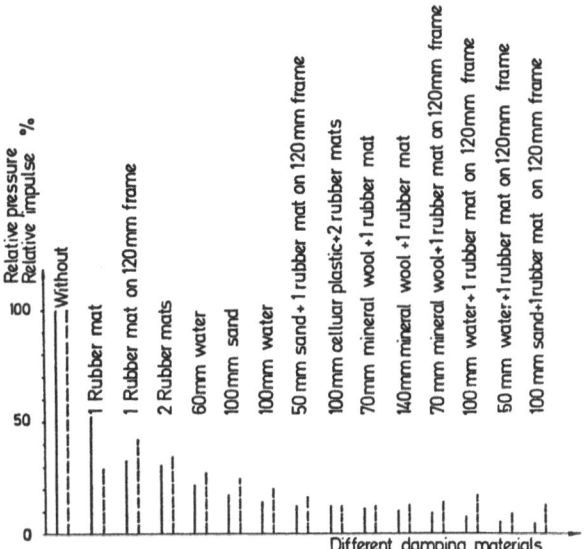

Fig 7   Relative pressure and impulse from a 10 g charge on a sand base covered with various damping materials.

3. Conclusions

Careful blasting with drilled-in charges has after research and practical experiments shown to be a feasible method for thicker structures, 250 mm and upwards. It is possible to reduce the effect of the blasting with regard to the remaining structure and surroundings to an acceptable level.

References

Bjarnholt, G., Holmberg, R. and Quchterlony F. (1981) Ett system för kontursprängning med styrd sprickinitiering. Bergsprängningskommittén. Protokoll från diskussionsmöte i Stockholm 29 jan 1981. Stockholm 1981. p. 313-336.

Bjarnholt, G., and Skalare, H. (1981) Instrumenterad bergsprängningsinledande försök i betongblock. Swedish Detonic Research Foundation. Report DS 1981:16. 40 p.

Chung, H.W. and Law, K.S. (1983) Diagnosing in situ concrete by ultrasonic pulse technique. Concrete international. 5 (1983) 10. p. 42-49.

First international symposium on rock fragmentation by blasting, Luleå, Sweden, August 22-26, 1983. Preprints. Luleå (University of technology) 1983. 2 vol. (814 p.)

Försiktig och skonsam sprängning under jord. Diskussionsdagar i Luleå 17-18 september 1980. (Careful and cautious blasting underground. Conference in Luleå 17-18 September 1980). Editor Agne Rustan. Luleå Univeristy. Technical Report 1980:80T.

Gustafsson, R. Bergsprängningsboken (Rock Blasting Manual). 7th ed. Stockholm (Swedish Building Centre) 1979. 322 p.

Granström, S.A. (1956) Loading characteristics of air blasts from detonating charges. Stockholm 1956. Kungl. tekniska högskolans handlingar 100.

Henrych, J. (1979) The dynamics of explosion and its use. Amsterdam (Elsevier) 1979. 558 p.

Johansson, C.H. and Persson, P.A. (1970) Detonics of high explosives. London (Academic press) 1970. 330 p.

Kiselov, O.I. and Rusakov, B.V The relation between concrete static and their explosion resistance. Beton i zhelezobeton. (1981) 5. p. 10-11.

Langefors, U. and Kihlström, B. (1978) The modern technique of rock blasting. 3rd edition. Stockholm (Almqvist & Wiksell) 1978. 438 p.

Molin, C. (1983) Localized cutting in concrete with careful blasting. Cement- och betonginstitutet. CBI forskning/research Fo 2.83. 252 p.

Molin, C. (1983) Försiktig sprängning av kantbalk. Fullskaleförsök på äldre betongbro. Cement- och betonginstitutet. Rapport 8368.

Molin, C. and Pettersson, E. (1984) Försiktig sprängning av hisschaktsöppning i befintligt trapphus. Förstudie i betongplattor. Cement- och betonginstitutet. Uppdragsfunktionen. Rapport 8413.

RILEM/CEB/IABSE/IASS-Interassociation symposium on
   concrete structures under impact and impulsive
   loading, Berlin (West), June 2-4, 1982. Berlin
   (Bundesanstalt fur Materialprufung) 1982-1983. 3 vol.
Rinehart, J.S. (1959) the role of stress waves in comminu-
   tion. Quarterly of the Colorado school of mines. 54
   (1959) 3. p. 61-76.
Robins, P.J. and Calderwood, R.W. (1978) Explosive
   testing of fibrereinforced concrete. Concrete 12
   (1978) 1. p. 26-28.
Rustan, A. (1978) Vibrationer och sprickbildning runt
   sprängborrhål. Litteraturstudie. Arbetarskyddsfonden,
   projekt 76/218:2, Försiktig sprängning i dåligt berg
   under jord, delrapport 3. Högskolan i Luleå. Teknisk
   rapport 1978:60T. 74 p.
Thum, W. (1978) Sprengteknik im Steinbruch- und Bau-
   betrieb. Wiesbaden (Bauverlag) 1978. 400 p.

BLASTING DEMOLITION OF MODEL REINFORCED CONCRETE PILLARS

T. SAITO   All Japan Association for Security of Explosives
N. KOBAYASHI   Department of Precision Machinary, Chuo University
T. YOSHIDA   Department of Reaction Chemistry, University of Tokyo

Abstract
Experimental blastings of model concrete pillars were carried out in order to obtain information on the blasting demolition of reinforced concrete pillars used by the Hausing Corporation apartment houses. The blastings were done in two steps : first, the concrete block was exploded by the blasting with internal loading of dynamite, and second, the remaining reinforcing rods were cut using cylindrical shaped charges of high detonation velocity. It was found that the blasting with plural bore holes are better for avoiding the blown out shot. Also, rubber sheets suspended near-by the concrete surfaces are effective for protecting from scattering of fragments. Lastly, using sand to cover the shaped charges can remarkably suppress the noise of blasting.
Key words : Blasting, Demolition, Reinforced concrete pillars, Dynamite, Shaped charge.

1. Introduction

The demolition of old buildings has been a recent topics in Japan. The All Japan Association for Security of Explosives organized a committee for the promotion of blasting demolition in urban areas. The committee carried out an experiment to blast model pillars made of reinforced concrete. An additional experiment was conducted for cutting the exposed reinforcing rods with shaped charges which were covered with sand. The results are described below.

2. Experimental method

2.1 *Model pillars*
A reinforced concrete structure model pillar typical of a Housing Corporation apartment house in Japan was used. The size was $800 \times 800 \times 2400$mm buriedd 400mm in the ground. The strength of the concrete(Fc) was 210kgf/cm², the reinforcing rods were of SR-24 with diameters being 22mmφ and 9mmφ for main and hoop rods, respectively.

2.2 *Explosives*
Explosives used in this study were No.3 Kiri dynamite(an ammonia gelatine dynamite) of 30mm diameter, and a cylindrical shaped charge of 22mm diameter composed of composition B(12g) and C4(20g). The liner of the shaped

charge was made of copper and the angle and thickness of the cone were 90° and 0.2mm, respectively. For initiating the shaped charge, a no.6 electric detonator and 1g Pentolite booster were used.

2.3 *Blasting the pillars*
Blasting patterns are shown in Figure 1. The diameter of the bore holes was 32mm. The length of the bore holes and the charge weight are also shown in Figure 1.

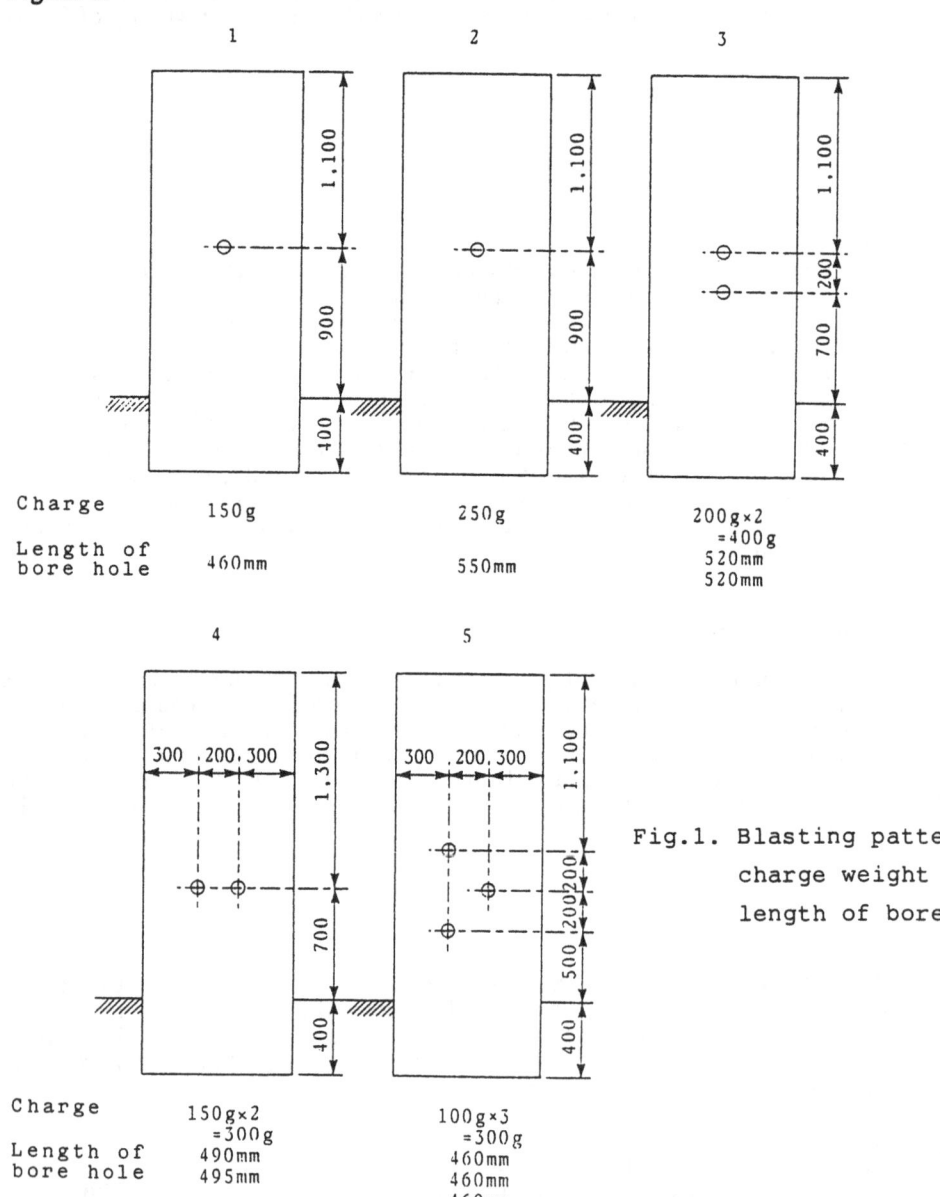

	1	2	3
Charge	150g	250g	200g×2 =400g
Length of bore hole	460mm	550mm	520mm 520mm

	4	5
Charge	150g×2 =300g	100g×3 =300g
Length of bore hole	490mm 495mm	460mm 460mm 460mm

Fig.1. Blasting patte
charge weight
length of bore

80

2.4 *Cutting the exposed reinforcing rods*

The set-up of the shaped charge on a rod is shown in Figure 2. All charges were covered with sand of 40cm thickness in order to reduce the blast noise.

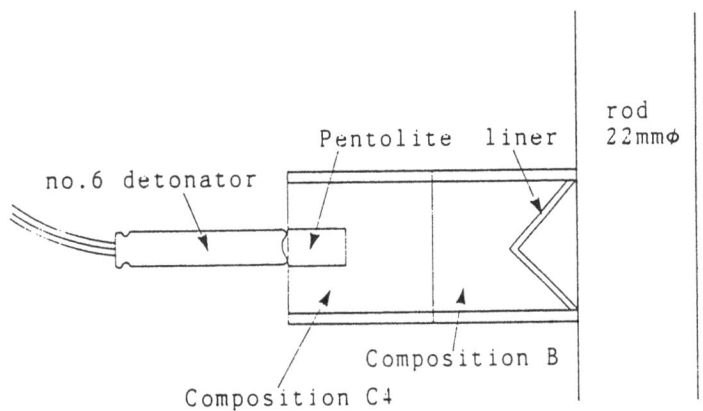

Fig.2. Set-up of the shaped charge on a rod.

2.5 *Measurements of crashed volume, noise and earth vibration*

Volumes of the crashed concrete were estimated from the geometry of the remaining structures after scraping the crashed concrete away. The blast noise levels and earth vibrations were measured by ordinary methods.

3. Results and discussion

3.1 *Blasting the model pillars*
3.1.1 *Results of the blasting*

The results of blasting the model pillars are shown and listed in Photo 1 and Table 1, respectively. In Photo 1(a),(b) and (c) shows the pattern of bore holes, and crashed pillars immediately after blasting and after scraping away the crashed concrete, respectively.

From these results, the single bore hole is not as effective for blasting concrete pillars with reinforcing rods when compared to double bore holes. The former is likely to achieve only partial crashing of the pillar. The best comparison is seen in run 2 where the rods were partly exposed and in run 4 where all rods were exposed, although the charges were similar.

Run

1

2

3

4

5

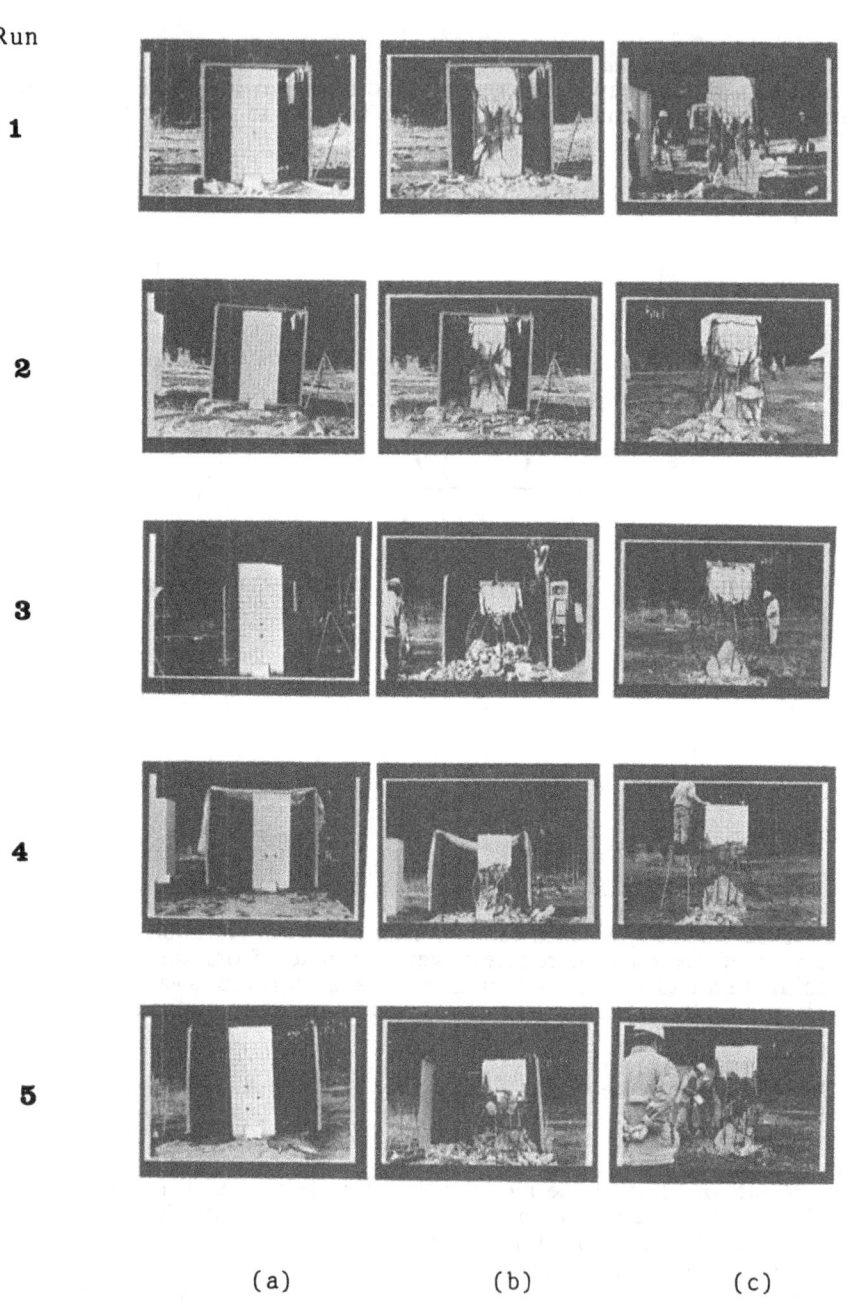

(a)   (b)   (c)

Photo 1. Photographs of blasting model pillars.
(a) before blasting, (b) immediately after blasting,
and (c) after scraping the crashed concrete away.

Table 1. Results of blasting model pillars.

Run	Blasting pattern	Weight of charge	Crashed volume	Note
1	⊡ (one hole)	150g	$0.19 m^3$	rods partly naked
2	⊡ (one hole)	250g	$0.35 m^3$	rods partly naked
3	⊡ (two holes)	400g	$0.64 m^3$	all rods naked
4	⊡ (two holes)	300g	$0.45 m^3$	all rods naked
5	⊡ (three holes)	300g	$0.58 m^3$	all rods naked

3.1.2 *Noise and vibration*

The observed noise and earth vibrations are listed in Table 2. Run 1 and 2 gave relatively high noise levels despite their rather small charge weight. This maybe attributable to the partial crashing of the concrete pillars. The gas energy of the explosions might not be used for blasting the concrete but escaping to the outside changing to noise. In Runs 3, 4 and 5, the blastings were successful with the blast noises rather small when compared to Runs 1 and 2. The noise levels in these successful blastings were approximately proportional to the charge weights. The vibration to y-direction was smaller than those in other directions. The earth vibrations at 2m distance from the blasting site contained higher frequency waves than those at the 20m distance. This could be due to the high frequency wave decreasing more quickly than that of the low frequency.

3.2 *Blasting the exposed reinforcing rods*
3.2.1 *Results of the blasting*

Photographs of the shaped charges attached to a rod, the charges covered with sand, the blasting and the result of blasting are shown in Photo 2.

A shaped charge with 28g high explosives could cut a steel rod of 22mm diameter, but not with 23g explosives. In this experiment for cutting 16 rods, we used 28g each of explosives for reliability; there was no optimization of the weight and shape of the shaped charge.

The noise level for this experiment is shown in Figure 3 along with other reference data for comparison. The blast noise was much less when the charge was covered with sand of 40cm thickness(Figure 3(1) and 3(2)). Internal charges also gave much less blast noise than external charges, but the sand coverage appered more effective in controlling noise level than the internal charge.

4. Conclusion

This study demonstrated that pillars of reinforced concrete with rods can be demolished by blasting with more than one bore hole to demolish the concrete. Also, following the initial blast, the exposed reinforcing rods can be cut using shaped charges. The noise level of blasting was significantly reduced by covering the charges with sand of 40cm thickness.

Acknowledgement : We thank the members of the Committee, peoples of Nippon Koki Co.,Ltd. and students of Chuo University who collaborated with the authors.

Table 2. Observed noise and vibration in the blasting of model pillars.

Run	Charge weight	Noise at 60m dB(A fast)	Distance from blasting	Earth vibration, velocity of vibration in kine				Direction of max. vibration
				x	y	z	$\sqrt{x^2 + y^2 + z^2}$	
1	150g	91.5	2m	2.10	1.09	2.80	3.36	z
			20m	0.13	0.07	0.29	0.30	z
2	250g	96.5	2m	2.56	1.25	2.60	3.43	z
			20m	0.17	0.07	0.27	0.28	z
3	400g	93.5	2m	—	—	—	—	—
			20m	0.33	0.15	0.53	0.54	z
4	300g	87.5	2m	5.74	0.69	5.30	7.76	x, z
			20m	0.23	0.14	0.46	0.46	z
5	300g	89.5	2m	5.74	1.56	5.52	6.92	x, z
			20m	0.25	0.11	0.40	0.46	z

(a)   (b)

(c)   (d)

Photo 2. Set-up of a shaped charge(a), covering with
sand(b), blasting(c) and result(d) of less
noisy blasting.

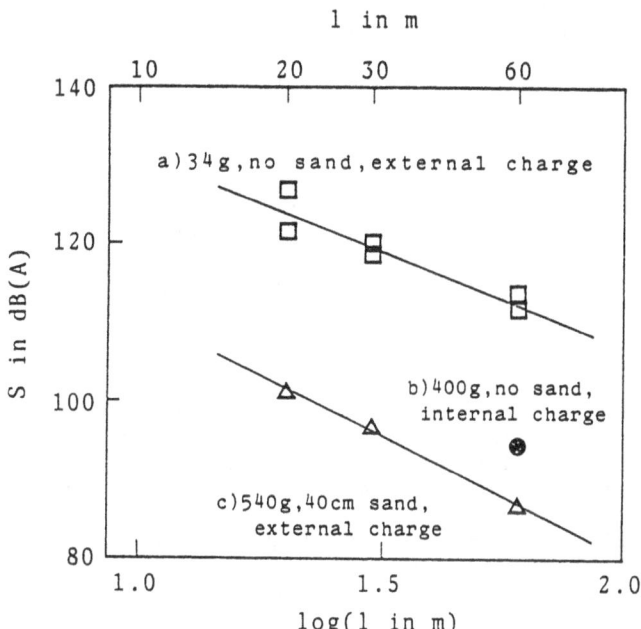

Fig.3. Plot of noise level(S) vs. log(distance(1) from blasting site).

FURTHER DEVELOPMENT OF "MINIBLASTING".

J.SCHNEIDER
Det Danske Spraengstofselskab A/S, Denmark.

Abstract.
Miniblasting technique requires the use of small quantity
of explosives inserted in small boreholes and the use of
stemming. Blasting of foundation piles requires furthermore covering of the pile when blasted. ROMEX^R a special
blasting mat has been developed and it can stand more than
1.000 blastings.

Micro cracks from the blasting are investigated and documentation is made. The Directorate of National Labour Inspection recommend blasting of pile tops as the best alternative to normal chopping by hand. Blasting of pile
tops is common in Denmark and has been introduced in Sweden and Germany. The principles of miniblasting are used
in many other types of blasting work today.

1. Introduction.

In 1985, at the EDA/RILEM conferences in Rotterdam "Miniblasting a Favourable Method of Careful Demolition of
Concrete" was presented by Mr. E.K.Lauritzen, ref 1.. The
principles of miniblasting has been used in blasting of
foundation pile tops in Denmark, Sweden and Germany, in
blasting of cantilever beams and in fragmentation of bad
concrete in a "pilot project subject renovation of Broend-
by Strand" a district in Copenhagen, and in several other
blasting works.

Blasting works are carried out in strong competition with
other methods of demolition. Miniblasting is economically
attractive, and the environment was never compromised. No
financial support was received from public funds.

The demolition work perfomed by miniblasting requires focus on the following subjects.
  1. Explosives, specifications and choice of the exp-
     losives.
  2. Drilling, borehole diameter, electrical or pneu-
     matic drilling machines.
  3. Stemming.
  4. Covering, development of strong and flexible
     lightweight blasting mats.
  5. Micro cracks in concrete, specification and docu-
     mentation for these.
  6. The public acceptance of the use of miniblasting
     in demolition.

In the following the miniblasting technique used on foundation pile tops will be discussed in details in relation to the above mentioned steps.

2. Different ways of removing the concrete from the reinforcement on pile tops.

Foundation pile tops are used for securing the foundation of buildings. Normally the piles are driven in, vibrated or drilled into the soil. The foundation piles are necessary in areas with bad carrying capacity in the soil, or if extra securing of the construction is necessary. To integrate the piles into the building, the concrete of the reinforcement is removed for at least 60 cm or more. The piles are weaved together with the reinforcement in the rest of the building foundation.

There are different types of foundation piles, some standard types are shown in figure 1.

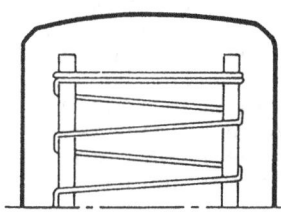
Dimension of piles:
20 x 20 cm, 25 x 25 cm, 30 x 30 cm
35 x 35 cm, 40 x 40 cm
Vertical reinforcement:
BSt 420/500, Ø 16 mm, Ø 18 mm
Ø 20 mm
Spiral reinforcement:
BSt 220/340, Ø 5 mm
Concrete:
$\sigma_{bk} > 40$ MN/m^2

Figure 1. Different types of foundation piles.

There are different demolition methods for removing the concrete in the reinforcement.

a. Chopping by hand.
   The advantages are that the method is well known and commonly used.
   The disadvantages are that the chopping by hand for a pile 25 x 25 cm takes more than 1 hour. The Directorate of National Labour Inspection recommends chopping by hand with a pneumatic hammer for only 0.5 hour per day. Therefore the method is recommended stopped and alternative methods should be found.
b. Cutting with a hydraulic scissors.
   The advantages are no vibrations into the pile and to the men handling the machine. 3 men can cut 4-6 piles per hour including the start up. The system is flexible.
   The disadvantages are that the equipment is heavy, see the weight for different types in figure 2. The hand-

ling of the machine demands 3 men. The machine can not tolerate dust, mud or moisture. The micro cracks into the piles have not been investigated. The price of the equipment is high.

c. Cutting with hydraulic scissors mounted on a machine. The advantages are that there is no manual work involved in removing the concrete. The method is fast, and the capacity is high.
   The disadvantages are that there must be a large amount of piles on the building area, and all the foundation piles must be cut fast, as the price of the machinery is very high. There must be some distance between the piles or the hydraulic scissors can not be used. The micro cracks in the piles have not been investigated, and there is a risk of breaking the pile if the hydraulic scissors are not handled with care.

d. Chopping by hydraulic hammer mounted on a machine. The advantages are that there are no manual work involved in removing the concrete. The method is fast, and the capacity is high. The machine can cut 6-8 piles per hour including moving around in the area.
   The disadvantages are that there can be difficulties with moving the machine around the building area. There must be many foundation piles. There must be some distance between the piles otherwise the hammer can not be used. The micro cracks have not been investigated, and there is a risk of breaking the pile if the hydraulic hammer is not handled with care. The hammer on the machine must be small or the risk of damaging the reinforcement is too high.

e. Cutting by blasting combined with chopping by hand. The advantages are that the method is fast, very flexible and there are no restrictions on the distances between the foundation piles. The chopping by hand with the pneumatic hammer is reduced to 2-3 min per pile effectively. Including the moving around on the building area the time is 8-10 min. The micro cracks resulting from the blasting have been investigated, see ref. 2,3,4 and 5.
   The disadvantages are that the safety distance to people in the building area is up to 10 m. There is a lot of paperwork and administration when using blasting. Blasting should always be made by a specialists.

Type	TBC-2	TBC-4	TBC-5
Max cutting capacity	25 cm	35 cm	46 cm
Weight	42 kg	80 kg	154 kg

Figure 2. Weight of different hydraulic scissors.

3. Procedure in blasting foundation piles.

The following procedures in blasting of foundation piles
have been developed through the last 4 years. In Denmark
more than 7.000 piles have been blasted since 1984, and
last year 1987 more than 4.000 piles. The percentage of
blasted piles out of the total number of piles is approximately 20 %.

3.1. Drilling
The piles are blasted with a small explosive charge inserted into a borehole. If there is a need, approximately
60 cm of reinforcement is removed, there is normally drilled one hole Ø 16-20 mm. The depth of the borehole is
shown in figure 3.

Type of foundation pile	Depth of borehole
20 x 20 cm	13 cm
25 x 25 cm	15 cm
30 x 30 cm	18 cm
35 x 35 cm	25 cm
40 x 40 cm	30 cm

Figure 3. Depth of bore holes in foundation piles.

The drilling is done with an electric or with a pneumatic
drilling machine. The drilling capacity is the same for
the mentioned machines, and the effective time for drilling one hole is 1 min. Including the time for moving a-
round the building area the time per hole is 4-5 min. The
holes are drilled not closer than 45 cm from where the final cut shall be.

3.2. Charging.
Normally the explosives used have a high detonating velocity, more than 7.000 m/s. The explosives shall be soft
and without nitroglycerin, nitrocellulose, ammoniumnitrate
or the like. In Denmark we use explosives with 85 % of
PETN and 15 % oil and wax. In foundation piles 40 x 40 cm
or bigger, a detonating cord 40 g/m is normally used as a
pipe charge. The different weight of charge depending on
the sizes of the piles are in figure 4.

Type of foundation pile	Bottom charge	Column charge	Total charge
20 x 20 cm	10 g	–	10 g
25 x 25 cm	13 g	–	13 g
30 x 30 cm	15 g	–	15 g
35 x 35 cm	18 g	–	18 g
40 x 40 cm	8+8 g	4 g	20 g

Figure 4. Explosive charges.

The charge is ignited by an electric blasting cap no 8.
The procedure for charging a pile is shown in figure 5.

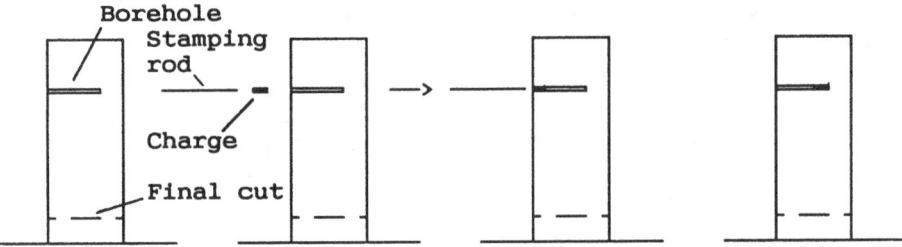

Figure 5. Charging piles.

3.3. Stemming.
The stemming is effectively made by moistured sand and gipsum.

3.4. Covering.
A special blasting mat, named ROMEX[R] has been developed.
The blasting mat is made of Kevlar fibres in a special
extra strong weaving and with a size of 1.4 x 2.4 m. The
weight of the mat is only 10 kg. The mat was developed by
ROBLON A/S and DEMEX A/S and has now been tested for the
last 3 years. The mat can today take more than 1.000 blastings, without compromising the safety of the blastings.
The ROMEX[R] is shown in figure 6.

Figure 6. Blasting mat, type ROMEX[R].

The result of blasting a foundation pile is shown in figure 7.

Figure 7. Blasted pile.

3.5.Micro cracks.

The micro cracks in the foundation piles have been investigated with special reference to unacceptable micro cracks in the concrete in the final cut of the piles. The micro cracks have been measured with a Pundit ultra velocity measuring equipment. The transmission velocity in the concrete has been measured before driving in the piles, after they are driven in (before blasting) and after blasting the piles. The measuring has been done as shown on figure 8.

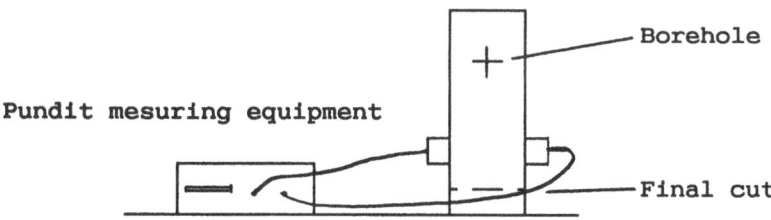

Figure 8. Principles in measuring transmission velocity in a pile.

The transmission velocity in a foundation pile 25 x 25 has been measured to 54 ms and with a standard deviation of +/- 2 ms. The criteria for accepting micro cracks from blasting has been set to the standard deviation on the measuring of the transmission velocity. Measuring the transmission velocity at a distance of 45 cm from the blasting hole, and setting the deviation into a diagram, a histogram shown below resulted, ref. 5.

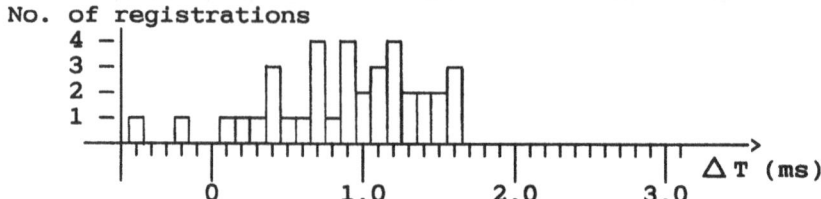

No. of registrations

Figure 9. Increase of transmission time, 45 cm from blasting hole.

The conclusion which can be drawn from ref. 2-5 is the same, and the recommended distance from borehole to the final cut is 45 cm.

3.6.The public acceptance for use of miniblasting of foundation piles.
The Directorate of National Labour Inspection has analysed the above described method, and has officially acceptet the method and when visiting new building areas, they recommend the method. Ref. 6.

4. Economical comparison of 3 different methods of cutting foundation piles.

The economical comparison of 3 different methods of cutting foundation piles can be made as follows. Based on experience from different companies in Denmark and experience from own work with cutting foundation piles.

Time of operation	Chopping by hand	Blasting and chopping by hand	Cutting by hydraulic scissors
Drilling	-	5 min	-
Charging, covering, blasting.	-	4 min	-
Chopping by hand.	60 min	9 min	-
Cutting by hydraulic scissors	-	-	36 min
Total time of operation	60 min	18 min	36 min
Material cost			
Drilling equipment	-	2,00	-
Explosives etc.	-	28,00	-
Compressor	62,50	10,40	-
Hydraulic scissors	-	-	27,50
Total costs	62,50	37,40	27,50

Figure 10. Time of operation, and cost of material in Danish kroner.

The condition for the mentioned time of operation, and the economical cost of the material are as follow.

- Drilling of a hole, including measuring of the height of the final cuts on the foundation piles, starting and cleaning of the machinery, the drill, the hammer etc.
- When charging, covering and blasting piles, 2 men are necessary. 2 men can normally charge, cover and blast 25 - 30 piles per hour, no matter how the access to the piles are.
- Chopping one pile by hand takes normally 1 hour. Chopping by hand after blasting the pile can vary depending

on the access to the piles. Normally one man can clean 6-15 piles per hour after they are blasted. Figure 10 calculate with 9 min. all included.
- Cutting by hydraulic scissors is normally done by 3 men. 2 men handle the scissors and one man handles the powerunit. 3 men can cut 4-6 piles per hour including starting and cleaning and access to the piles. Figure 10 calculates with 5 piles per hour.

The costs are calculated on the following conditions.

- As drilling equipment is used a HILTI TI 72, drilling diameter Ø 18 mm. Drilling of 98 holes per day. Cost of hire the machine per day 200,00 Danish kroner.
- Explosives, detonators and ROMEX[R] blasting mat costs as follows:

0,012 kg explosives á 120,00	=	1,40
1 pc detonator á 13,60	=	13,60
1 shot in ROMEX[R] 13.000,00:1.000	=	13,00
Total in Danish kroner		28,00

- Compressor and hammer for hire in one day 500,00 Danish kroner.
  (a) Normally chopping by hand, 1 hour per pile, 8 piles per day, cost per pile 62,50 Danish kroner.
  (b) Chopping by hand after blasting the pile, 6 piles per hour, 48 piles per day, cost per pile 11,00 Danish kroner.
- Cutting by hydraulic scissors. Machinery can be hired for 1.100,00 Danish kroner per day. Cutting of 5 piles per hour, or 40 piles per day, cost per pile 27,50 Danish kroner.

The above calculated times of operation and cost can easily be adjusted to conditions in different countries.

Finally the advantages and disadvantages of the different methods must be taken into account.

5. General development of Miniblasting.

In Denmark the principle of miniblasting is used in many different works. Blasting of foundation piles is now common and well accepted, but blasting of piles demands a knowledge of blasting technique and miniblasting technique, and access to the right types of explosives, safe covering like ROMEX[R] blasting mat and creative understanding of different types of demolition methods.

In Denmark we have succeed in making it a profitable way of cutting foundation piles by the use of miniblasting, and we are working with the method and principles from miniblasting searching for and working with new jobs in which we can use the experiences.

References:

1. Lauritzen, E.K. (1985), "Miniblasting, a Favourable Method of Careful Demolition of Concrete", in Proceedings "Demolition techniques", EDA/RILEM conference 1985.

2. Lauritzen, E.K. and Schneider, J. (1984), Blasting of foundation pile, DEMEX report 20-58, October 1984.

3. Christensen, K.K. and Kristensen, E.L., (1987), Sprængstoffers anvendelse til demolering af beton, AAlborg Universitet, January 1987.

4. Centrum Paele A/S, 1987.

5. Molin, C., (1984), Blasting of foundation pile with careful blasting, a study, Swedish Cement and Concrete Reseach Institute, Report No 8439, 1984-06-18

6. Scherling, P. (1987), Subject Blasting of Foundation piles, The Directorate of National Labour Inspection, 23th of January 1987.

DEVELOPMENT OF A DETONATOR WITH AN ELECTRONIC DELAY CIRCUIT

K. OCHI   Research & Development Section, Bibai Factory
                     Nippon Oil & Fats Co., Ltd.
M. HARADA   Harada Electronics Industry

Abstract
The common electric detonators used as their delay method delay powder, and therefore had a certain limitation in there task of improving
their delay accuracy. By building an electrical delay system, which
uses the latest electronic technics, into the electric detonator, we
have succeeded in improving its delay accuracy by a great amount.
Furthermore we have made thorough research also on its electric safety,
and therefore succeeded in developing an electric detonator which is
more improved on its electric safety than the commonly used ones. We
believe that our newly developed electric detonator is most suitable
for urban blasting.
Key words: Detonator, Delay, Electric, Electronic, Accuracy, Urban
blasting, Pyrotechnics.

1. Introduction

In general, a delay electric detonator is used for delay blasting.
However, the current delay electric detonator has only a limited time
precision because delay powder is used to get time delay. In addition, the type of a detonator changes the time delay depending on the
storage and ambient temperature. Thus it is improper to adopt the
current delay electric detonator to urban blasting and sequential
blasting which require high time precision for the sophisticated
blasting technics. If the type of detonator may be used practically,
the development of a delay electric detonator of high time precision
enables the blasting more effective than by the existing detonator.
    From the point of view, many attempts have been made to realize
delay electric detonator of high time precision by the use of the advanced electronic technics.
    The PDD (Precise Delay Detonator) is one of the delay electric
detonator of high time precision using the electrical delay system
described above.

2. Features of electrical delay system in PDD

Fig.1 shows the fundamental configuration of the electrical delay system in PDD. The delay system consists of energy accumulating

98

condenser for operating electronic circuits to fire a fusehead, a
timer circuit including a reference signal generating circuit and a
counter, a switching circuit detecting outputs from the counter to
discharge the charged energy to the fusehead, and a "reset circuit"
for operating the timer circuit stably. The inputs of the electrical
delay system are connected to blasting units through leading wires and
the outputs are connected to a fusehead or an instantaneous electric
detonator.

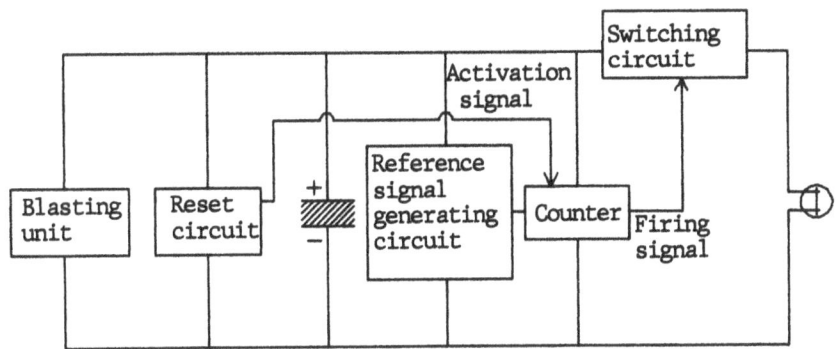

Fig.1   Block diagram of PDD electrical delay system

The time precision of the electrical delay system mainly depend on
the configuration of the timer circuit. The PDD adopts the digital
control method provided by a reference signal generating circuit and
a counter to the timer circuit and uses a crystal oscillator of high
frequency precision in the reference signal generating circuit. In
the past, the crystal oscillator could hardly be applied to the electrical delay system due to instability during activation in spite of
its high frequency precision. However, the PDD resolves the problem
by mounting a reset circuit and improves the time precision up to the
crystal oscillating precision (less than 0.1ms independent of the
length of time delay).

The reference signal generating circuit generates the signal with
the period of the minimum time difference in delay series. The delay
time can be determined by setting a value to the counter. The period
provided by the reference signal generating circuit is equivalent to
the step interval and the value set to the counter is equivalent to
the number of steps.

The exterior view of PDD and the blasting unit for PDD are shown
in Photo 1 and Photo 2.

3. Fundamental operation of PDD

The blasting unit for the PDD is powered by a 36VDC power supply.
After the power of the voltage is applied from the blasting unit to
the PDD, the energy accumulating condenser in the PDD electrical delay

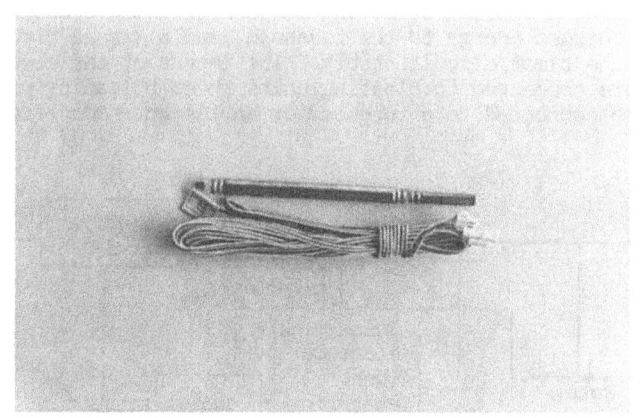

Photo 1   Exterior view of PDD

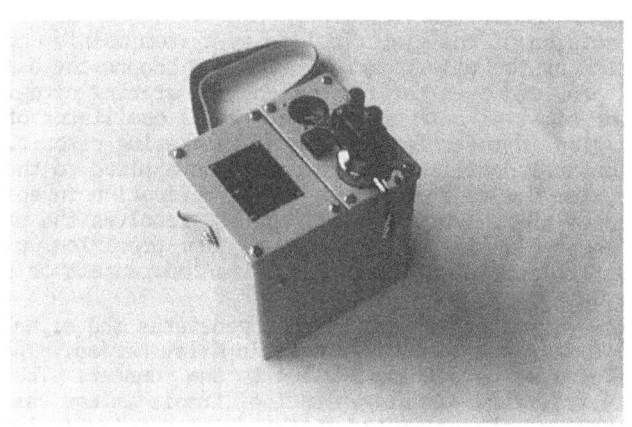

Photo 2   Blasting unit for PDD

100

system is charged for a few seconds and the PDD is activated. Then the reference signal generating circuit is energized by the energy charged in the condenser to start the oscillation. However, the oscillating pulses are not counted because the counter is reset by the reset signal. Next the break of the voltage from the blasting unit disables the reset signal for the counter to operate the timer. The PDD is triggered after the appropriate delay time.

4. Wiring of PDD

Electric detonators are generally used in series connection. However, if electric detonators including electrical delay system such as PDDs are connected in series, the following problems occur:

(a) The voltage level derived from multiplying the operating voltage of an electric detonator by the number of simultaneous blastings is required to result in the high voltage of the blasting unit. For example, the operating voltage of the PDD is 36VDC and, if 100 detonators are fired simultaneously in series connection, the blasting unit requires at least the voltage of 3600VDC.

(b) Because the operating voltage of each electric detonator cannot be set high, the energy must be maximized by enlarging the capacity of each energy accumulating condenser as much as possible, which is associated with the above problem. Thus the electrical delay system is made larger in size.

(c) The voltage applied to the blasting unit must be changed depending on the number of simultaneous blasting. It is because the electrical delay system may probably be destroyed due to excess voltage applied to each electric detonator if a blasting unit enabling a number of simultaneous blasting (by high voltage) performs a less number of simultaneous fires.

(d) The distortion of capacities in energy accumulating condensers varies the charging voltages of electric detonators. In general, a normal condenser has the distortion of 10 to 20%.

The above problems do not occur for electric detonator in parallel connection. However, a normal parallel connection is inferior to a series connection in operation efficiency. For the PDDs, the parallel connection shown in Fig.2 is used, which has nearly the same operation efficiency as the series connection.

A PDD has four leading wires, which is defferent from that of a normal electric detonator. As shown in Fig.2, two pairs of leading wires are connected in parallel to two input terminals of an electrical delay system. The ends of the wires are processed to the connectors by a pair.

Adjacent electric detonators are connected with the connectors. One of the two connectors remaining unconnected at the both ends is connected to a blasting unit to complete the parallel connection. In Fig. 2, both the connectors remaining unconnected at the both ends are connected to a blasting unit. This connection has the advantage that an invalid connection of connectors at a single position does not cause misfire. It is normally enough to connect either of the ends to

a blasting unit and the opposite side of connector can remain unconnected.

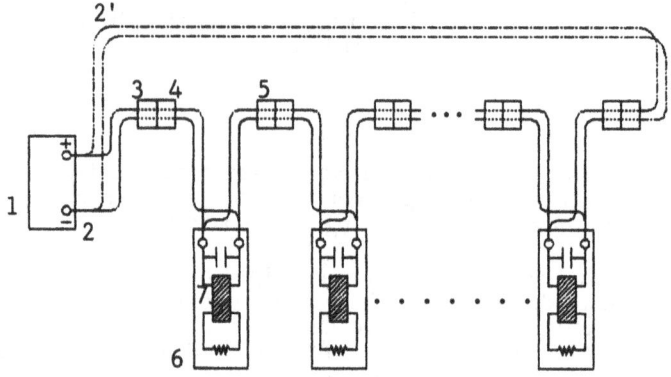

1: Blasting unit
2: Blasting wire
3: Connector of blasting wire
4: First connector
5: Second connector
6: PDD
7: Electrical delay circuit

Fig.2 Overview of PDD wiring

5. Time precision

Table 1 shows the comparison of the time performance of the PDD with
that of the previous delay electric detonator using delay charge.
They were measured in the precision of 0.1%. The data associated with
the existing delay electric detonator is the time from power-on to
initiating measured by a microphone. The measured data relating to
the PDD is the time from the break of power from the blasting unit to
the output of energy to fusehead. With the PDD, the actual delay time
must be obtained by adding a value in Table 1 to the initiating delay
time of the instantaneous electric detonator (mainly including the 1ms
firing delay time of the fusehead).

The time precision of the existing delay electric detonator is
about 3% in the coefficient of variation. Thus the dispersion becomes
larger as the time delay is longer. This means that failures such as
invalid sequence may probably occur as the number of steps increases
for constant step interval.

On the other hand, the unique electrical delay system adopted in
the PDD does not produce the dispersion of more than a single period
of the crystal oscillator. For example, if the most cost-effective
crystal oscillator for watch of 32.768kHz in frequency is used, the
dispersion is less than about 0.03ms, which is almost negligible.

The dispersion is always constant despite of the length of the time. Thus the delay time in the PDD is only dependent on the dispersion of the initiating delay time provided by the existing instantaneous electric detonator. In the other words, the PDD is a delay electric detonator with the instantaneous firing quality of the instantaneous detonator.

Table 1   Comparison between time precision of PDD and that of existing delay electric detonator

	PDD		Existing delay electric detonator	
Setup time (ms)	250	1000	250	1000
Measured value (ms)	250	1000	245	1013
	250	1000	264	977
	250	1000	258	1028
	250	1000	256	1056
	250	1000	242	963
	250	1000	259	1004
	250	1000	239	1034
	250	1000	244	1016
	250	1000	250	1014
	250	1000	263	1044
Average (ms)	250	1000	252.0	1014.9
Standard deviation (ms)	0	0	9.1	28.5
Coefficient of variation (%)	0	0	3.6	2.8

Fig.3 shows the results of measuring the time delay of the PDD and that of the existing delay electric detonator under various ambient temperature conditions.

The time delay of the existing delay electric detonator varies by about 3% depending on the change of the ambient temperature by 10°C. The longer time delay results in the larger variation.

The time delay of the PDD is largely affected by the frequency vs. temperature characteristic of the crystal oscillator. The maximum frequency change is extremely as small as 0.002%. Accordingly the time never varies by 1ms if the ambient temperature changes from -20°C to 60°C.

The PDD operates abnormally at the temperature of -30°C. This does not result in the destruction of the electrical delay system. The PDD starts operating normally after returning to the normal temperature.

It is known that the time delay of the existing delay electric detonator becomes longer by about 2% (or 1% to 2% in actual depending on the storage temperature) with the passage of time for 3 years. The change of the time delay of the PDD is now obtained only with the passage of time for a single year. However, there appears no remarkable change.

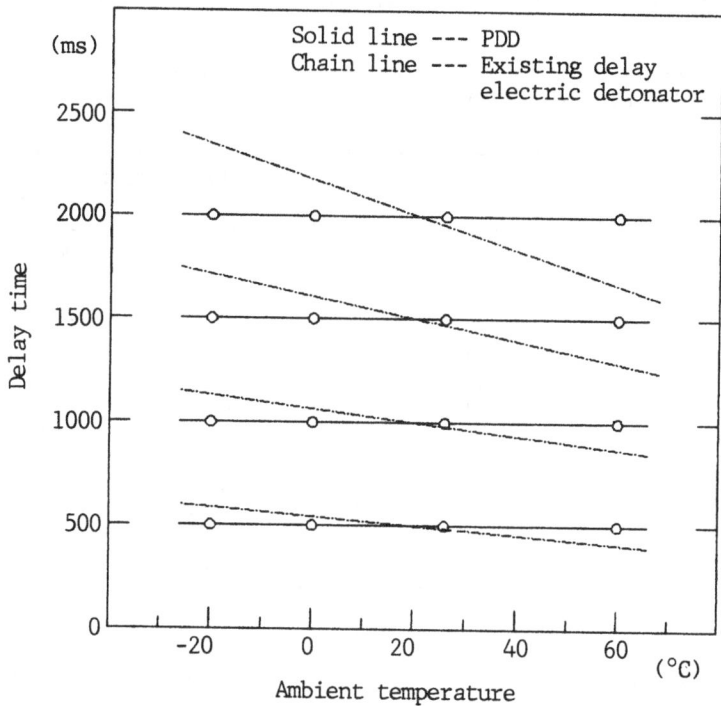

Fig.3 Variation of the time depending on ambient temperature

6. Electric safety of PDD

The electric detonators which adopt the electrical delay system for delay method are excellent in time precision, though there are some problems in their electric safety.

In the PDD, subsidiary circuits (not described in Fig.1) are mounted, and therefor succeeded in developing an electric detonator which is more improved on its electric safety than the commonly used ones.

The problems of electric safety and the methods for resolving the problems are given as follows:

(a) Premature firing caused by stray current may occur.

In the PDD, the voltage discriminating circuit is applied to the electrical delay system to prohibit the normal operation below 27VDC, and the supplied energy is consumed at the safety discharging circuit.

(b) The destruction of the electrical delay system caused by electrostatic discharge between wire-to-wire may occur.

In the PDD, a small gap structure is formed at the input terminal to obtain the prior electrostatic discharge. When the electrostatic energy is applied between wire-to-wire, the electrostatic discharge occurs at the small gap structure, and the electrical delay system is protected from electrostatic destruction.

(c) The breaking of a bridge wire cannot be detected.

In the PDD, the peculiar circuit for detecting the risistance of a bridge wire is applied. The resistance of a bridge wire can be detected by measuring the potential difference between a minus leading wire and a shell, on condition that the measuring voltage (not more than 27VDC) is impressed between wire-to-wire.

Table 2 Electric safety of PDD

Items	Properties of PDD
Against the stray current	No fire up to 27VDC
Against the electrostatic discharge between wire-to-wire	No fire up to 2000pF-15kV
Detecting the resistance of a bridge wire	Possible

7. The maximum number of the PDD in one round shot

As described above, a blasting unit drives PDD with the power supply of 36VDC. The battery mounted in the blasting unit can flow the current of more than 100A.

The PDD is configured so that the energy is charged in an energy accumulating condenser through a charge control resistor of 3k-ohm. The maximum current flow for a PDD is 12mA during start of charge. The energy accumulating condenser is fully charged for about 2sec.

As known from the above discussion and the connection of PDDs shown in Fig.2, the maximum number of the PDD in one round shot is determined by the maximum current that the leading wires of PDD located nearest to the blasting unit can accept. The current equal to the value obtained by multiplying 12mA by the number of the PDD in one round shot flows through the leading wires as well as through the blasting wire.

If a soft copper wire of 0.42mm diameter is used as the leading wires, up to 500 PDDs can be fired simultaneously under the condition described above. It is also assumed that the maximum current flowing through the leading wire is 6A during start of charging.

8. Conclusion

At last, the characteristics of the PDD can be summarized as follows:

(a) The PDD has extremely high time precision in comparision with other reported detonators utilizing electrical delay system as well as existing delay detonators using delay charges.

(b) With the PDD, the time delay scarecely change against the change of ambient temperature and passage of time.

(c) The PDD enables sequential blasting. Any existing delay electric detonator can only provide up to 20 sequences of blasting. The PDD can provide up to 127 sequences of blasting, assuming that the step interval is 15.625ms.

(d) A single blasting unit allows both fire of a single PDD and simultaneous fires up to 500 PDDs.

(e) The connection of bridge wire in the fusehead can be detected after product assembly.

(f) The electric safety between wire-to-wire (for electrostatic enrgy and stray current) is excellent.

NEW FIRING SYSTEM BY CORDLESS DETONATOR

K. KUROKAWA and Y. TASAKI and T. UEDA
Chemicals & Explosives Laboratory, Nippon Oil & Fats Co., Ltd.

Abstract
We developed MBS(electromagnetic induction blasting system) which
simplifies the handling and extensively increases the safety in electric blasting operations.

We have developed new firing system by use of cordless detonator.
Based on the principle of transformer, this system is similar to MBS
and operated as following; When high frequency alternating current
pulse (70-110 kHz) is applied to lead wire and connecting wire,
magnetic flux is generated in the first transcore and induced current
flows through the first loop-like wire. And then magnetic flux is
generated in the second transcore contained in the cordless detonator
and induced current flows through the second loop-like wire which
includes fusehead. Then the cordless detonators are ignited by the
induced current.

Because the first and second loop-like wires are looped and insulated, and also frequency band matching is limited, this system has
much more safety against stray current, leakage current, power transmission lines, radio frequency and static electricity et al.
Key word: Cordless detonator, MBS, Transcore, Stray current, Leakage
current, Static electricity, Electromagnetic induction.

1. Introduction

Recently a new firing system is developed in Europe and USA. We developed MBS which has been applied to several blasting sites such
as big metal mines and civil constructions.

Mew firing system by cordless detonator in this paper is based
on the technique used in MBS. In new firing system by cordless detonator, blasting operation is similar to MBS and a conventional electric blasting. But type of detonator and connecting method are different.

Owing to the structure of cordless detonator and characteristic
of the two type transcores, this system has much more safety in
electric blasting operations and simplifies the handling.

2. Principle of new firing system by cordless detonator

Fig. 1. shows the outline of new firing method by cordless detonator. Blasting machine generates high frequency alternating current pulse to lead wire to which is connected connecting wire. Connecting wire has some loop portions with which the first transcores are electro-magnetically connected. With the first transcores are also electro-magnetically connected the first loop-like wires.

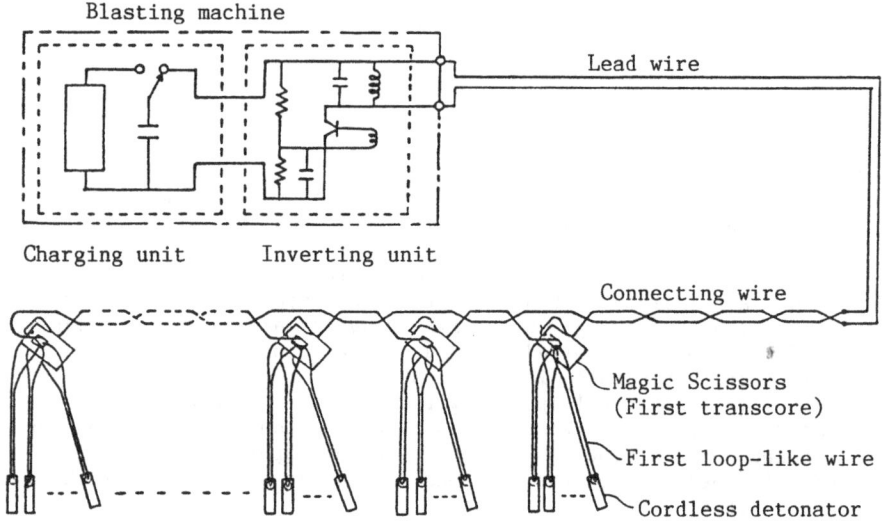

Fig. 1. Out line of new firing method by cordless detonator.

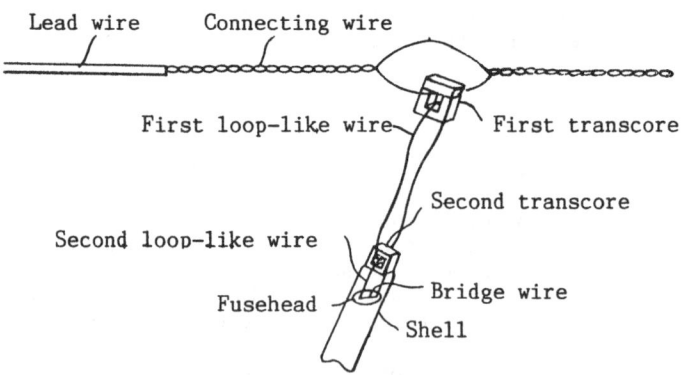

Fig. 2. Schematic view of this system.

Fig. 2. shows the schematic view illustrating an embodiment of the electric blasting method by this system. In order to facilitate the operation for connection the loop portion of connecting wire and first loop-like wire with the first transcore, the first transcore is formed into a square ring and one side block is movable with respect to the remaining block so as to form a space therebetween. After wires are passed through the space of the first transcore, the side block is moved to close said space. The first transcore is made of ferrite and is called Magic Scissors in MBS. Fig. 3. shows cross sectional views of Magic Scissors.

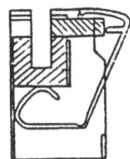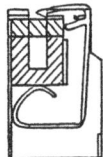

Before connection   After connection

Fig. 3. Cross sectional view of Magic Scissors(first transcore).

The first loop-like wire is electromagnetically connected with the second transcore which is provided integrally with the cordless detonator. With the second transcore is further electromagnetically connected with the second loop-like wire which is connected to a bridge wire around which a fusehead is provided.

When a blasting switch provided on blasting machine is actuated, high frequency alternating current pulse is supplied to lead wire and connecting wire and the current having the same frequency as that generated from blasting machine is induced in the first loop-like wire by the electromagnetic induction. Then, in the second loop-like wire high frequency alternating current of the same frequency via the second transcore is also induced. This current flows through bridge wire and fusehead of the cordless detonator is heated and fired.

3. Cordless detonator

Fig. 4. shows an embodiment of the cordless detonator. Fig. 4(a) is a front view, Fig. 4(b) is a tranversal cross section cut along a line A-A, Fig. 4(c) is a longitudinal cross section cut along a line B-B. A bridge wire made of a platinum wire, a fusehead applied around the bridge wire, primary explosive and base charge are inserted in a shell. The construction of the cordless detonator is the same as that of ordinary electric detonator. The second loop-like wire connected to the bridge wire is extended outside the shell through its opening, and then passed through a transcore serving as the above explained the second transcore. In this embodiment, the transcore is embedded in a plug. In the plug is formed a hole which is communicated with a central passage of the second transcore. Through the

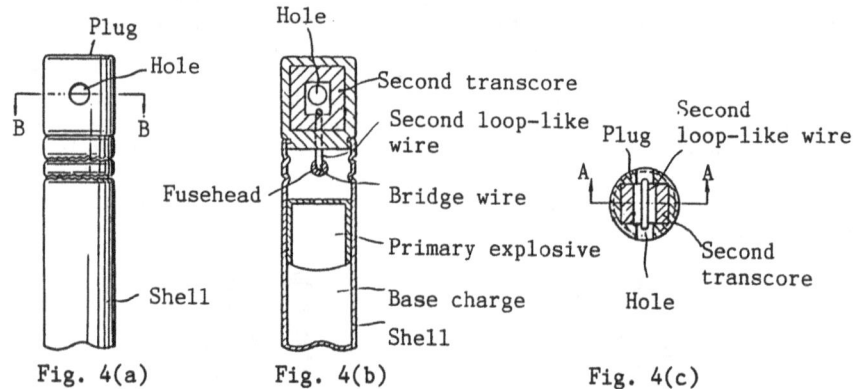

Fig. 4(a)   Fig. 4(b)   Fig. 4(c)

Fig. 4.   An embodiment of the cordless detonator.

hole of the plug the first loop-like wire can be passed through the transcore. As the transcore is embedded in the rubber plug, the transcore can be effectively protected against shock.

In the method according to this system, any undesired electric energy could never flow into the bridge wire of the cordless detonator, because the second loop-like wire connected to the bridge wire is not exposed outside, but is embedded within the detonator. Therefore, any unexpected or erroneous explosion of the detonator can be prevented positively. Further, the first loop-like wire can be easily connected with the detonator only by passing the wire through a central passage of the second transcore integrally provided in the detonator. After the wire is passed through the second transcore, both ends of the wire are connected with each other to form the loop.

Frequency characteristic of the first transcore contained in Magic Scissors and the second transcore contained in cordless detonator are shown in Fig. 5. Transmitting coefficient is the ratio of the current flowed through secondary circuit to that flowed through primary circuit. When frequency is less than 1 kHz, current flowed through primary circuit is hardly transfered to secondary circuit. Therefore direct current or commercial current is hardly transfered. And frequency band used for this system is 70-110 kHz.

4. Safety against environmental conditions

Using cordless detonator and Magic Scissors of MBS, this system offers various safeties over MBS and conventional electric blasting system. The safeties of this system are described in the following paragraph.

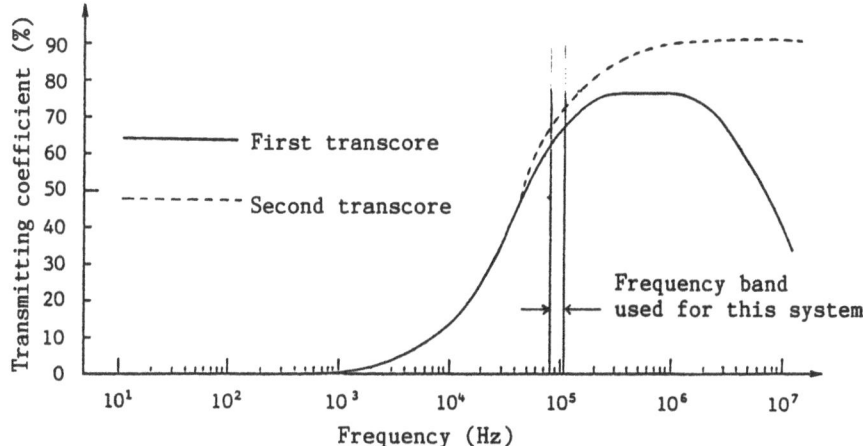

Fig. 5. Frequency characteristics of the transcores.

4.1 Safety against stray current

This system has safety against stray currents from direct current
and normal alternating current power sources because the detonator
looped leg wire is used from primary cartridge preparation to firing,
and there is no contact of naked conductor with the natural ground.
Moreover in this system, frequency of current flowed through primary
circuit is 70–110 kHz, and the transcore matches only that band.
Therefore, even if intense stray current of that frequency band flows
through primary circuit from connecting to firing, induced current
hardly flows through the first loop–like wire and bridge wire.

Table 1. shows results of safety test against stray alternating
current. Table 2. shows results of safety test against stray direct
current. The tests were conducted by connecting the cordless detonator with the blasting circuit by means of Magic Scissors, and applying the commercial alternating current (60Hz) or direct current to
primary circuit. In all experiment, one Magic Scissors was used, which
held one cordless detonator. The results of tests show that it is
impossible to fire the cordless detonator in this system with alternating current of up to 100 A or direct current of up to 120 A flowing
through primary circuit. This is well above the values of stray current which could occur through earth fault conditions on most machinery. It is also greater than the currents that could be induced in
primary circuit if blasting circuit is carried out close to overhead
power transmission lines.

In a conventional electric blasting method, the detonator was fired
if direct or alternating current of 0.5 A flows through the blasting
circuit for comparison.

Table 1. Safety against stray direct current

Stray direct current (A)	Induced current (A)	Result
30	0	Not fired
60	0	Not fired
120	0	Not fired

Table 2. Safety against stray alternating current

| Stray alternating current (A) | Induced current (mA) | | Result |
	First loop-like wire	Bridge wire	
10	60	0.09	Not fired
30	150	0.17	Not fired
50	230	0.26	Not fired
100	420	0.44	Not fired

4.2 Safety against leakage current

Most of misfire in a conventional electric blasting is caused by leakage current. Most arduous conditions in the use of electric detonators are underwater blasting and shaft sinking. In both types of operations, blasting circuit is often submerged completely or partially in a relatively high conducting medium, such as sea water. If standard electric detonators are connected in series, the resistance of the circuit is comparable with the resistance of the conducting medium. As a result, when the current generated by blasting machine flows through the blasting circuit, some current will pass through the conducting medium from any joints or breaks in the insulating to comparable joints elsewhere in the blasting circuit. Each joint in the blasting circuit, no matter how insulated, could act as one of these current leakage points. The leakage of current from the blasting circuit can lead to misfire due to insufficient current passing through individual detonators. Because of this problem of current leakage, a conventional electric blasting is seldom used in underwater blasting.

In this system, looped and insulated leg wire is used, and there is no naked joints in the blasting circuit. Moreover this system has its own low voltage power connection in the form of the transcores. Therefore, even if connecting wire and the first loop-like wire and the cordless detonators are submerged in the conducting

112

edium, there is no risk of current leakage. This system is suitable for firing in wet environment, for example, underwater blasting, shaft sinking and tunnelling especially where the rock is conductive and ground water is present.

Table 3 shows result of safety test against leakage current. Each test was conducted by firing 100 cordless detonators connected in accordance with normal way and lead wire, connecting wire, first loop-like wires and cordless detonators were submerged in various conducting mediums. And a conventional electric blasting method was also tested for comparison. All detonators could be fired reliably in this system, even totally submerged in various conducting mediums.

On the other hand, in a similar test using conventional detonators connected in series and totally submerging all connections in the same sea water, 89 % would fail.

Table 3. Protection against leakage current

Firing method	Medium	Number of detonators	Result
This system	Fresh water	100	No misfire
This system	Sea water	100	No misfire
This system	Dil. CH_3COOH	100	No misfire
This system	Na_2CO_3 Soln.	100	No misfire
Conventioal	Sea water	100(Series)	89 misfire

4.3 Safety against power transmission lines and radio frequency

Current induced by power transmission lines is influenced by voltage of current flowed through the lines, the distance between the lines and blasting circuit, and area of the loop formed by the blasting circuit et al.

In a conventional electric blasting, some hazards may be caused by induced current as the area of the loop formed by blasting circuit is large. Therefore some countermeasure to those hazards has to be considered. However in this system, not only lead wire but also connecting wire are used in parallel, the area of the loop formed by blasting circuit is extremely small. The frequency of current flowed through power transmission lines is 50 or 60 Hz. Therefore there is no hazards by the induced current.

This system has also much more safety than a conventional electric blasting against radio frequency hazards because each unit forms completly insulated circuit in itself.

4.4 Safety against static electricity

We have a conventional electric detonator which gives some degree of protection against static electricity hazards. But the protection (2000 pF x 8 kV) is assurance between end of leg wire and shell. Therefore when the end of leg wire is opened, a conventional electric detonator does not dispose sufficient protection against static electricity hazards.

In this system, the end of leg wire of the detonator is looped and insulated, so each unit forms completely insulated circuit in itself. Further, the cordless detonator has second transcore in the shell. Therefore this system virtually avoids accidental initiation from static discharges.

Table 4 shows results of safety test against static electricity. The test was conducted by connecting the cordless detonator in accordance with normal way to compare with a conventional electric blasting method. Static electricity energy (Capacity of condenser = 2000 pF) was applied to the primary circuit. The results of test show that it is impossible to fire the detonator in this system with static electricity energy up to 25 kV flowing through primary circuit. In a conventional electric blasting method, the detonator was fired if static electricity energy of 4.5 kV flows through blasting circuit.

Table 4. Safety against static electricity

Firing method	Discharging voltage (kV)	Result
This system	5.0	Not fired
This system	15.0	Not fired
This system	25.0	Not fired
Conventional	4.5	Fired

5. Conclusion

We have developed a new firing system by cordless detonator which simplifies the handling and extensively increases the safety compared with MBS and a conventional electric blasting.

We performed safety tests against many electric hazards and proved that this system has good safety against stray current, leakage current, current induced by power transmission lines and radio frequency and static electricity.

The cordless detonator has the second tanscore in a shell and frequency band matching is limited. Therefore, the cordless detonator has high safety against burglar because regular electric energy such as battery or commercial current can not ignite it.

Reference

K. Kurokawa, Y. Tasaki, M. Nakano, T. Ueda (1986) Development of
electromagnetic induction blasting system. Kogyo Kayaku.,
Vol.47, No.5
K. Kurokawa, T. Ueda (1986) Method of electrically blasting detonator
and cordless detonator for use in said method. Japan Patent Appl.
86-40227

NON-EXPLOSIVE DEMOLITION AGENT

K.SOEDA   Construction Materials Res. Lab., Onoda Cement Co.,Ltd.
S.YAMADA   Products Research Lab., Sumitomo Cement Co.,Ltd.
Y.NAKASHIMA Central Research Lab., Nihon Cement Co.,Ltd.
S.NAKAYA   Special Cement Additive Dept., Denki Kagaku Kougyo.K.K.
H.HANEDA   Tech. Development Dept., Yoshizawa Lime Ind. Co.,Ltd.
N.IZAWA   Chem. Products Sales Sect., Asahi Chemical Ind. Co.,Ltd.

Abstract
The accelerating redevelopment of urban and industrial areas in
recent years has led to a marked increase in the work of demolishing
old concrete structures. However, with increasing consideration for
the environment and safety, it has become essential to reduce the
levels of noise, vibration, and flying debris in demolition work.
Consequently, the use of heavy demolition equipments and explosives
has in most cases been restricted or prohibited. It was in these
circumstances that the world's first non-explosive demolition agent
(NEDA) was developed in Japan in 1979. The main component of NEDA is
CaO, and their demolition mechanism is expansion by CaO hydration,
which produces static pressure causing tensile fracturing. Demolition
work with this agent is therefore free from noise, vibration, dust
and flying debris. NEDA are available in three forms: bulk,
briquette, and capsule. General grades generate an expansive pressure
of about 30 MPa. at 24 hours after loading. High power, fast acting
grades have recently become available. NEDA are effective in
demolition work for all rock or concrete structures. This article
describes their expansive pressure generating mechanism, functional
characteristics and application.
Key Words: Non-explosive demolition agent(NEDA), Expansive pressure,
Calcium oxide, Tensile fracturing

1. Introduction

The accelerating redevelopment of urban and industrial areas in
recent years has led to a marked increase in the work of demolishing
and repairing old concrete structures. In the past, it had been used
explosives and heavy demolishing equipments in these demolishing
works. However, the use of explosives and heavy demolishing
equipments has in most cases been restricted or prohibited, because
these works create environmental problems such as noise, vibration,
dust and flying debris. In the circumstances with increasing
consideration for the environment and safety to solve these problems,
non-explosive demolition agent(NEDA) was developed in Japan in 1979.
The components of NEDA are calcium oxide and the other cement
minerals, and it generate cracks quietly in concrete and so on by
expansive pressure caused lime hydration. Demolition work with this

type of agent is therefore safe and free from the environmental problems. In this article, it is introduced the outline of NEDA used worldwidely in the demolition works.

2. Chemical composition

NEDA is a powder which consists mainly of specific inorganic lime compounds. Special organic and inorganic compounds are added for the purpose to control the reaction and to improve the workability. Although the chemical composition and mineralogical analysis are different according to manufacturers, the major component is calcium oxide (CaO). To take an instance of a commercial product A, it is a powdered clinker of $3CaO \cdot SiO2-CaO-CaSO4$ with about 50% CaO content. Crystalline particles of CaO are dispersed in crystals of $3CaO \cdot SiO2$ and interstitial materials of $CaSO4-3CaO \cdot Al2O3-4CaO \cdot Al2O3 \cdot Fe2O3$. These crystals and interstitial materials control the hydration reaction of CaO. Water reducing agent is added to improve the workability and to increase the demolition force.

Table – 1. Chemical composition (%)

	Ig. Loss	SiO$_2$	Al$_2$O$_3$	Fe$_2$O$_3$	CaO	MgO	SO$_3$	Total
A	1.2	8.5	2.3	1.0	82.0	0.7	4.0	99.7

When water is added to NEDA, the hydration reaction of the major component, calcium oxide (CaO), takes place gradually to form a hydrate. The hydrate, which expands the volume about three times if it is left free, generates an expansive pressure of about 30 MPa. in about 24 hours if it is constricted by a substance to be demolished.

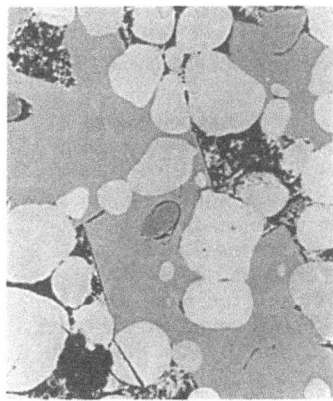

Fig-1 Before hydration

Fig-2 Hydration products

3. Demolition mechanism

The basic demolition mechanism of NEDA is illustrated in Fig-3. When the expansive pressure Pi of NEDA works on the inner wall of a thick cylinder (substance to be demolished), the circumferential tensile stress σ_θ and radial compressive stress σ_r of the following formula are generated at a distance r from the center. These formula are approximation on assumption that the substance be an elastic body.

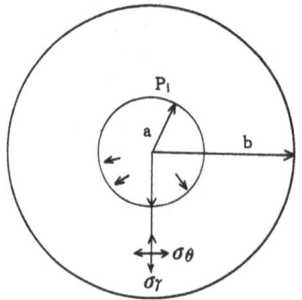

$$\sigma_\theta = \frac{a^2}{b^2 - a^2}\left[1 + \frac{b^2}{r^2}\right]P_i \quad \cdots\cdots (1)$$

$$\sigma_r = \frac{a^2}{b^2 - a^2}\left[1 - \frac{b^2}{r^2}\right]P_i \quad \cdots\cdots (2)$$

Fig-3　The basic demolition mechanism of NEDA (single hole)

The tensile strength of a brittle substance like concrete is as small as $1/10 \sim 1/20$ of its compressive strength, and is $2 \sim 5$ MPa.. Cracks which occur when σ_θ exceeds the tensile strength are the principal factor for demolition. Cracks may also be generated by the upheaval and shearing force as shown in Fig-4 and 5.

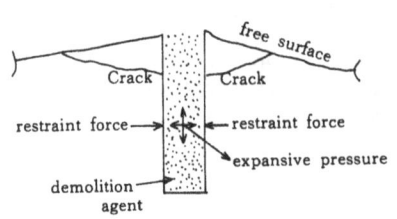

Fig-4　Upheaval

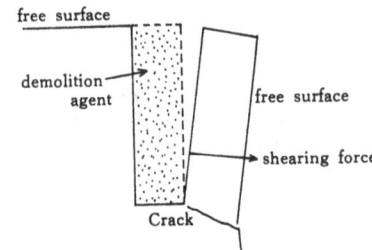

Fig-5　Shearing Force

When a single hole is used for demolition, a crack is generated at the weakest point of the substance near the hole. As the formula (1) indicates, the crack occurs at the hole wall, where σ_θ is largest, toward the external free surface near the hole. Since the expansive pressure of NEDA continues to develope after occurrence of the crack, more cracks are generated at the points where the stress becomes large. When plural holes are used for demolition, the tensile stress in a direction rectangular to the line between the adjacent holes grows larger due to the mutual influence of stresses at these holes, and a crack occurs to connect the holes. When two holes with radius r_1 shown in Fig-6 are filled with the NEDA, the tensile stress σ_θ at the middle point between the holes is expressed by the formula (3), where R is the distance between the

118

holes.

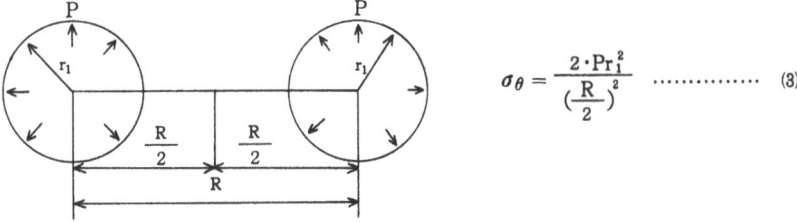

$$\sigma_\theta = \frac{2 \cdot P r_1^2}{(\frac{R}{2})^2} \quad \cdots\cdots\cdots\cdots \quad (3)$$

Fig-6 The basic demolition mechanism of NEDA(plural hole)

Accordingly, it is possible to develope a crack between two holes earlier by narrowing the distance between the holes.

4. Expansive pressure of NEDA

4.1 Measuring method of expansive pressure

In 1981, the Ministry of Construction has adopted "Development of demolition technique using expansive pressure of Non-explosive demolition agent" as one of annual themes which were selected in accordance with the estimating system about construction techniques based on the ninth article of the prescription with estimation about construction technique (Notification No.976 by the Ministry of Construction). It was general in Japan to use the method indicated in the above report for measuring expansive pressure of NEDA. Afterward, an improved method based on the above method was proposed by Association of NEDA after his additional experiences. At present, the improved method as shown in Fig-7 has been generally used. In this report, therefore, the improved method was explained as under.

(1) Each one of a paper strain gage was stuck on a central part of surface of steel pipe (JIS G 3454, Carbon steel pipe for high pressure, 40Axsch-80), having a steel plate welded on the bottom side, in opposite circumferential direction.
(2) The gages were waterproofed and then the steel pipe was vertically stood in a water batch kept a fixed temperature as shown in Fig-7.
(3) Before the test, a sample of NEDA and mixing water were kept at a fixed temperature.
(4) A bulk type of NEDA was mixed with water at a fixed ratio and became a slurry. In the case of a capsule type, it was immersed in water for a fixed time and was made to absorb enough water.
(5) The prepared slurry was poured into the pipe until the top. In the case of a capsule, the watered capsule was tightly loaded into the pipe by tamping each one with a rod up to the top.
(6) The amounts of strain of paper strain gages generated by expansive pressure with hydration of NEDA were recorded by strain meter after the setting.

119

(7) The values of expansive pressure were calculated from the amounts of strain by formula (4).

$$P = \frac{E}{2 - \nu} \times \frac{r_0^2 - r_i^2}{r_i^2} \times \varepsilon \quad \cdots\cdots\cdots\cdots \quad (4)$$

P : Expansive pressure(MPa)   E : Elastic modulus of steel (MPa) 2.1×10^5

ν : Poisson's ratio 0.3 r_0 : Average outer diameter of steel pipe (mm)

r_i : Average inner diameter of steel pipe (mm)

ε : Average expansive strain of pipe in circumferential direction

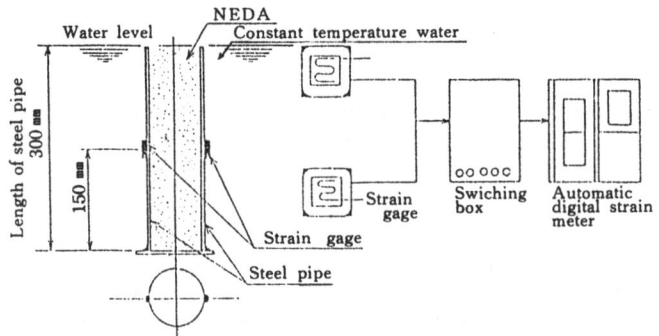

Fig-7   Measuring method of expansive pressure with NEDA by using steel pipe

4.2   Properties of expansive pressure

There are commercialized adequate NEDA according to ranges of working temperature, hole diameter and so on. There are some factors having influence on expansive pressure, for example, temperature of an object, water ratio, hole diameter and so on. So, general behaviors of expansive pressure were shown for instance as follows.

(1)   Influence of elapsed time and working temperature to expansive pressure

As shown in Fig-8, expansive pressure was gradually developed by hydration of lime (CaO) which was a main component of NEDA. The higher the working temperature was, the earlier the generating time of expansive pressure was. In the case of using the same kind of NEDA, therefore, the higher the working temperature is, the shorter the cracking time is.

(2)   Relationships between water ratio and expansive pressure

Relationships between water ratio and expansive pressure were plotted in Fig-9. The smaller the water ratio was, the larger the expansive pressure became. This relation means that expansive pressure is influenced by quantities of NEDA in a unit volume of slurry. Though expansive pressure is able to make larger by smaller water ratio, fluidability of slurry is lower and workability is worse. So, a recommended water ratio is usually from 28 to 30 % for mixing slurry.

(3) Relationships between hole diameter and expansive pressure
Relationships between hole diameter and expansive pressure was shown
in Fig-10. The larger the hole diameter was, the higher the
expansive pressure tended to be. This reason is that hydration of
NEDA is promoted with temperature rising of slurry because quantities
of NEDA is more in a larger diameter hole. However, it is important
to use the designated-size of hole diameter of each catalogue as a
dangerous blow-out phenomenon is afraid of occurring in the case
of using a oversize hole diameter.

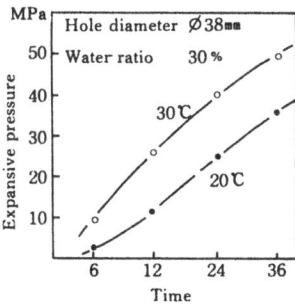

Fig-8   Change of expansive pressure
         of NEDA with elapsed time

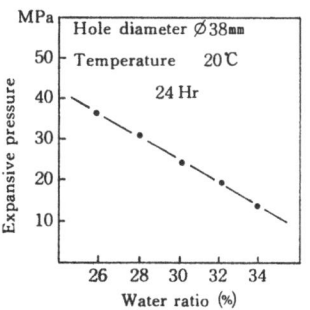

Fig-9   Water ratio vs expan-
         sive pressure

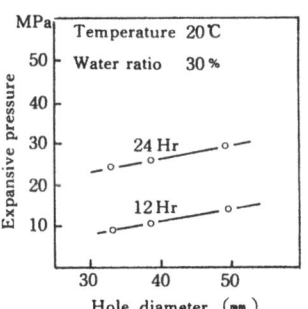

Fig-10   Hole diameter vs expan-
         sive pressure

5. Design

5.1 Procedure of demolition work

Some attention is necessary to faciliate the development of cracks.
Cracks generally develop more easily when free surface is increased
and resistance is reduced. The general procedure of demolition work
using NEDA is shown as Fig.-11.

121

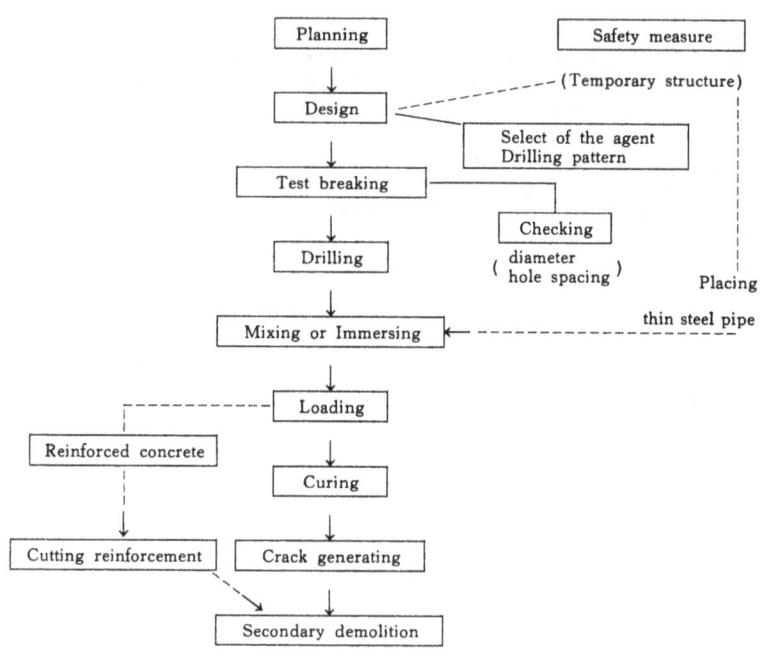

Fig-11   Procedure of demolition work

5.2 Drilling

Handhammer and crawler drill are generally used. In the case of a
temporary structure like foundation of crane, thin steel pipes are
placed as holes instead of drilling before placing concrete, so we
can eliminate labor and noise in drilling.

5.3 Hole diameter

In general, 38-42 mm bits are used for handhammer and around 65mm
bits are used for crawler drill.

5.4 Hole spacing

Shorter hole spacing produces higher expansive pressure, but too
short hole spacing reduces economy. Recommended hole spacing is
shown as Table-2.

Table-2   Example of hole spacing (∅ 40 mm)

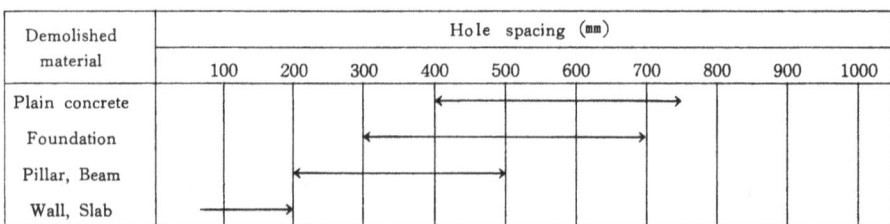

Demolished material	Hole spacing (mm)									
	100	200	300	400	500	600	700	800	900	1000
Plain concrete					←————————————→					
Foundation			←——————————→							
Pillar, Beam		←——————————→								
Wall, Slab	←——→									

5.5 Hole length
A greater effect is achieved with deeper hole, therefore, we should drill the hole as deeply as possible.

5.6 Examples of drilling pattern
Fig-12 shows examples of drilling pattern.

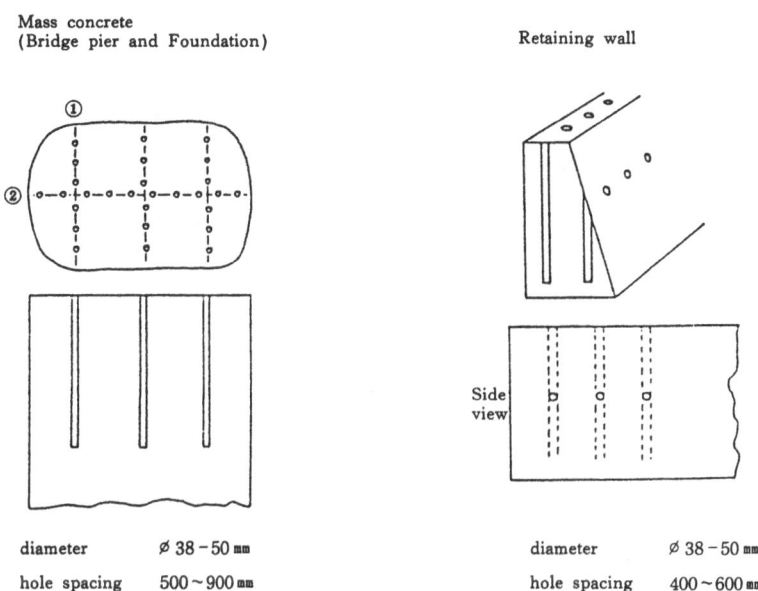

Mass concrete
(Bridge pier and Foundation)

Retaining wall

Side view

diameter	⌀ 38 – 50 mm	diameter	⌀ 38 – 50 mm
hole spacing	500 ~ 900 mm	hole spacing	400 ~ 600 mm

Fig – 12   Examples of drilling pattern

5.7 Mixing or immersing
Bulk type of NEDA should be mixed in a pail(10-20ℓ) with specified quantity of water (standard water ratio 30%) by electric mixer. After all NEDA is poured, it should be thoroughly mixed to obtain slurry. Capsule type of NEDA should be immersed in water for specified period of time.

5.8 Loading
As soon as the mixing completed the slurry should be filled to the top of hole. When the hole is vertical, the slurry is filled directly, however, otherwise the slurry is poured from a container such as an oiljack.

5.9 Curing
The holes after loading should soon be covered with sheets. These sheets should be weighed for prevention from blown away. After covered by the sheets, the area should be restricted from entering, and the holes should not be peeked for any causes by anybody.

123

5.10 Blow-out Phenomenon

Blow-out phenomenon of NEDA is caused by pressure of steam made by water contents of the slurry and heat from hydration reaction of CaO, the main component of the NEDA. For prevention of blow-out phenomenon right NEDA, right hole diameter and right circumstance conditions should be selected. The precautions in handling should be strictly observed.

6. Comparison with other method

Generally NEDA is used in the locations where explosives and heavy demolition equipments can not be utilized. NEDA method has the following merits in comparison with other methods. (Table-3)

(1) Except during drilling, noise and vibrations are almost none and influence to the surrounding circumstance is very small.
(2) Easy handling is obtained and limited and planned demolishing is applied by change of method.
(3) NEDA is not limited by laws such as handling prohibits for explosives and others, and any special license or application is not required.

Table-3 Comparison of NEDA Method and Other Demolishing Method

Items / Demolishing Method	Demolishing	Conditions during Demolishing					
		noise	vibration	dust	spreading stone	safety	* economy
NEDA	○	◎	◎	◎	◎	◎	○
Explosives	◎	×	×	×	×	×	◎
Concrete Demolisher	○	△	△	×	△	△	△
Jumbo Breaker	△	△	○	○	◎	○	△
Oil-pressure Breaker	△	◎	◎	◎	◎	◎	△

◎ Excellent (no pollution)
○ Good
△ Fair
× Poor
* Economy is subjected to the demolishing circumstances

Second phase demolishing by hand-breaker, oil-pressure breaker, lipper and others are generally necessary after cracks are made by NEDA. In this case the second phase demolishing is more expensive than the case of no NEDA used.

7. Applications of NEDA

Fig-13 and Fig-14 show the examples of the applications of NEDA in practical work sites.

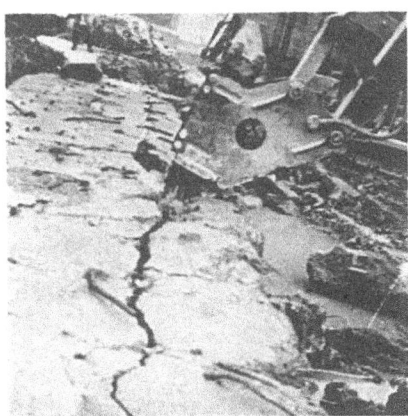

Fig-13 Reinforced concrete river
diversion bank at hydroelectric
construction site. Appearance
24hr after loading of NEDA.
Use of explosives precluded by
proximity of railroad and houses.
Drillholes: Approx. φ 40mm, 200mmL
Drillhole spacing: 200 to 750mm
Agent load: 11kg/m3

Fig-14 Plain concrete seawall
Cut faces resulting from cracks 8
to 10mm in width,48hr after loading
of NEDA. Use of explosives precluded
proximity of harbor construction site.
Drillholes: φ50mm,570 to 900mmL
Drillhole spacing: 400 to 500mm
Agent load: 4kg/m3

8. Conclusion

In this article it was summarized NEDA; safety, free from the
environmental problems and simple of handling etc,. With increasing
much consideration for the environment and safety in the future, the
demands for the use of NEDA have increased more and more. However,
there is only limitation on NEDA that it takes long time to generate
cracks and the breaking effect is very small compared to explosives.
Basic grade of NEDA generate cracks about 10 24 hours after loading.
It has therefore been expected to develope high power and fast acting
grade of NEDA which generate cracks in a short time and the
demolition technique using it. Recently some of these non-explosive
demolition agents have been developed and available.

Reference

The association of NEDA, Japan. (1984) Measuring method of expansive
     pressure of NEDA Explanation of it.
Shinobu bunryo, Japan. (1985) Outline of NEDA. J. Lime., 353, 5-18
The association of NEDA, Japan. (1987) Working method and pH of NEDA.

STATIC DEMOLITION OF CONCRETE BY NEW DEMOLITION AGENT

Y. YAMAZAKI
Central Research Laboratory, Nihon Cement CO., Ltd.

Abstract
It has been several years since static demolition agents were used
for breaking brittle materials such as concretes and rocks in Japan.
The demolition technique using them is free from the problems of
noise and vibration. But, it needs long time to break objects.
    Recently, new static demolition agents have been developed and
marketed. They are characterized by breaking more rapidly and generating higher power than the conventional agents.
    One of the new agents was examined and tested about the expansive
properties and the applications for breaking concrete.
Key words: Demolition agent, Expansive pressure, Breaking concrete,
Capping agent, Capsule

1. Introduction

The static demolition work using a chemical splitting agent has been
widely adopted for breaking concrete structures and rocks in urban
areas and near important plants. It is the reason that the static demolition agent has the virtue of non-explosion and non-vibration at
working. But, it has the demerits of slower acting and lower power
than those of an explosive. It has been still considered that it is
too difficult to accelerate more the reaction of it because a dangerous blowing-out phenomenon occurs by the sudden hydration of CaO
which is a main component of demolition agent.
    Recently, new static demolition agents of rapid and high expansive
pressure type have been developed and marketed in Japan.
    In this paper, one of the new agents was studied. The agent con-
sisted of a expansive compound capsule and a capping agent capsule.
Some properties of expansive pressure of it and applications for
breaking concrete were reported.

2. Experimental methods

2.1 Materials
New static demolition agent used in this paper was a capsule form.
It was produced and marketed by Nihon Cement, Nippon Oil & Fats and
Nichiyu Giken Kogyo Co., Ltd.. The commercial name of it was " High

126

Calmmite•30 " (abbreviated HC•30). HC•30 consisted of CaO clinker powder and additives in a permeable paper bag. Table 1 shows specifications of HC•30.

Table 1. Specifications of HC•30

Name (Type)	Size of capsule (mmϕ x mmL)	Weight of capsule(g)	Applicable temperature(°C)	Applicable hole diameter (mm)
W	30 x 250	300	0 to 20	38 to 42
S	30 x 250	300	20 to 35	38 to 42

A special capping agent " Cap Ace " was together used to prevent a blowing-out of HC•30 loaded into a hole. The form of Cap Ace was capsule (36mm x 100mmL, 100g/c) which consisted of a treated quick lime grain and a permeable paper bag.

2.2 Working procedure of HC•30 and Cap Ace
A flow sheet of working procedure of HC•30 and Cap Ace is drawn in Fig.1.

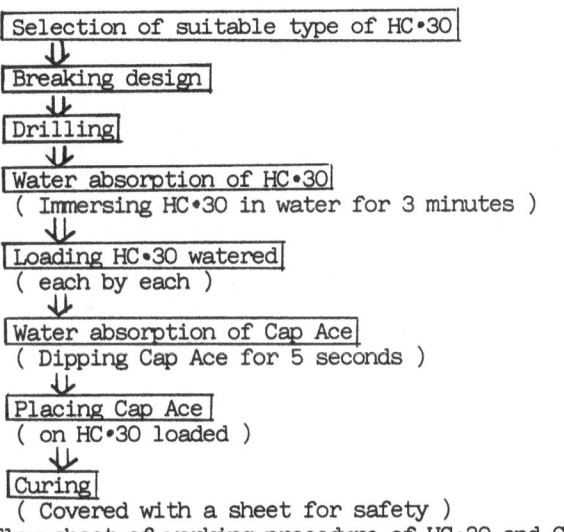

Selection of suitable type of HC•30
⇩
Breaking design
⇩
Drilling
⇩
Water absorption of HC•30
(Immersing HC•30 in water for 3 minutes)
⇩
Loading HC•30 watered
(each by each)
⇩
Water absorption of Cap Ace
(Dipping Cap Ace for 5 seconds)
⇩
Placing Cap Ace
(on HC•30 loaded)
⇩
Curing
(Covered with a sheet for safety)
Fig.1 Flow sheet of working procedure of HC•30 and Cap Ace

Each capsule of HC•30 watered was loaded by tamping it tightly 4 to 5 times with a tamping rod so that there was no space at the bottom or the side of the hole as well as between capsules. The hole was filled up to about 10 cm from entrance by HC•30.

After loading HC•30, Cap Ace was thoroughly dipped in water for absorbing water for 5 seconds. The Cap Ace dipped was taken out and inserted immmediately into the remaining upper part of hole and then contacted firmly with HC•30 by tamping it with rod. The expected works of Cap Ace were 1) releasing much heat which was liberated with the earlier hydration of it than that of HC•30 to remove free water as steam, 2) restraining and capping firmly the entrance of hole by volume expansion with hydration of it, 3) accelerating hydration of HC•30 by heat released with hydration of Cap Ace.

2.3 Measuring method of expansive pressure
Expansive pressure of HC•30 loaded was measured by means of an improved thick steel pipe method as shown in Fig.2 which was applicable up to high temperature of 300 °C.

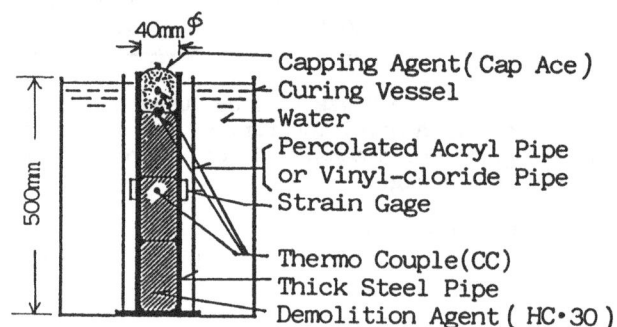

Fig.2 Device for measuring expansive pressure of HC•30

Thermo couples were set at three parts, central parts of Cap Ace and HC•30 ,and contact part of both agents. The most important improved point compared for a usual thick steel pipe method was that a specially percolated acryl pipe and a vinyl-cloride pipe were set around the steel pipe at curing temperature of 0°C to 10°C and 20°C to 35°C, respectively.The purpose of these plastic pipes was to insulate the steel pipe in which HC•30 and Cap Ace were able to hydrate at similar temperature rising to that of a actual working condition.
The magnitude of expansive pressure was caculated from the amount of strain of two wire strain gages stuck on the central part of steel pipe surface in opposite circumferential direction.

3. Results and discussion

3.1 Properties of expansive pressure
Fig.3 shows the typical relationship between the temperature rising and the developing of expansive pressure of agents loaded at curing temperature of 0°C. The hydration of Cap Ace scarcely progressed at early stage and occured suddenly about 45 minutes after loading. (curve 1) The hydration generated much heat by which temperature of a contact point of Cap Ace and HC•30 was rised. By the heating, upper parts of HC•30 began to hydrate rapidly. Consequently, two peaks were in the curve 2. The heat released by upper HC•30 conducted to lower part and promoted the hydration of it. The cycle of heat and hydration succeeded until bottom of hole.(curve 3) The expansive pressure with hydration of HC•30 was generated as soon as temperature rising. (curve 4) The maximum expansive pressure was 95 MPa about 60 minutes

128

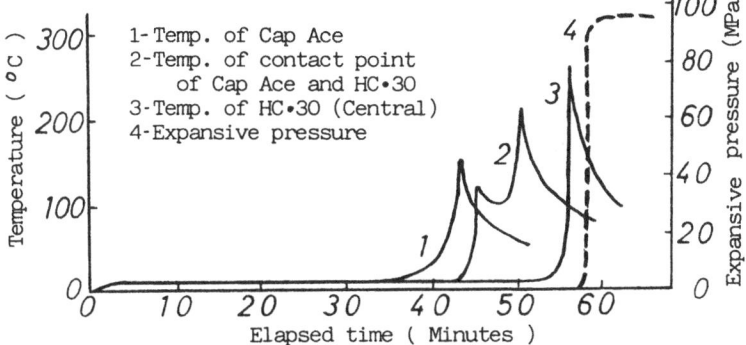

Fig.3 Temperature rising and developing of expansive pressure
of Cap Ace and HC•30 loaded at 0 °C

after loading.

A blowing-out did not occur at all in the whole examination even
though temperature of each part of hole became more than 100°C. It
is considered for the reason that the hydration of HC•30 progresses
succeedingly from entrance to bottom of hole and a firm sealing zone
is always formed at upper part before the hydration of lower part.

Fig.4 shows the time of maximum temperature of Cap Ace and the
time in which the maximum expansive pressure is obtained at each
curing temperature of 0 to 35 °C.

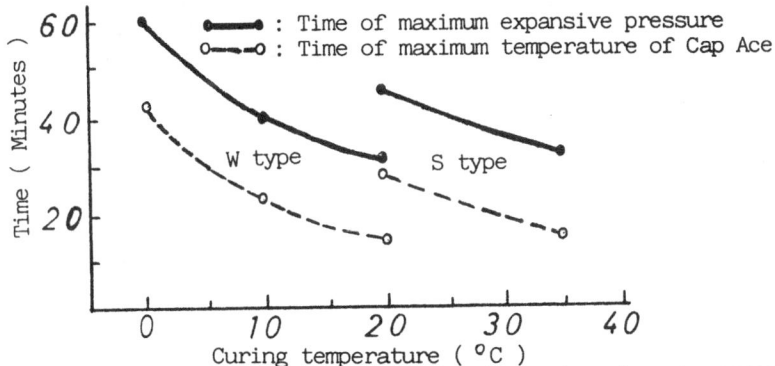

Fig.4 Times of maximum temperature of Cap Ace and maximum expansive
pressure of HC•30 at curing temperature of 0 to 35 °C

The higher the curing temperature, the shorter the time of temperature rising of Cap Ace became in both cases of W and S type. The
shortest time of temperature rising of Cap Ace was about 15 minutes
at the highest curing temperature. So, it is estimated that the working time is at least 15 minutes which are enough to load in safty the
both agents into hole.

From Fig.4, it is additionally mentioned that the maximum expansive pressure is obtained between 30 and 60 minutes after loading in
both cases of W and S type.

3.2 Applications

Fig.5 shows an example of breaking a cubic concrete block by HC·30 and Cap Ace. Photo-1 shows the cracks in 1 hour after loading them.

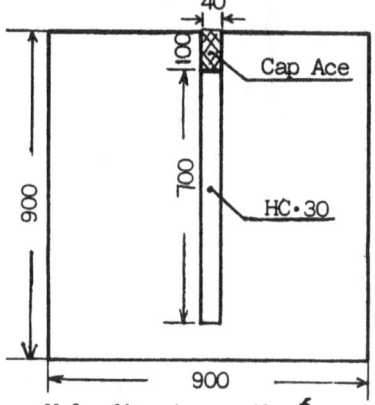

Hole diameter : 40mmϕ
Fig.5 Section of one hole in
   a cubic plain concrete block
   (900x900x900mmL)

Unite amount of agent: 2 kg/m³
Photo-1 Appearance of cracks on a
   cubic plain concrete block
   in 1 hour after loading

Fig.6 shows a rough sketch of unreinforced mass concrete foundation and a drilling pattern for cutting off a upper part of it on a horizontal plane by HC·30 and Cap Ace. Photo-2 shows the horizontal crack made on a expected place. It was very handy to load horizontally the both agents like this, because they were capsule form.

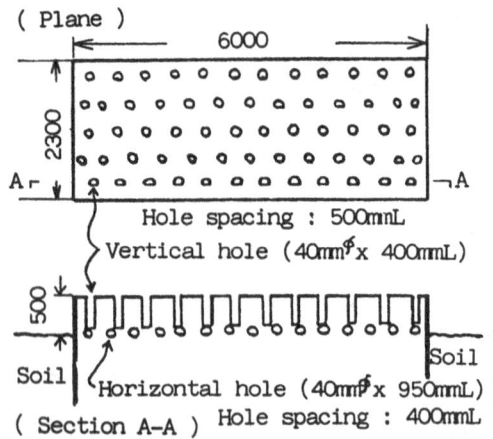

(Plane)

Hole spacing : 500mmL
Vertical hole (40mmϕ x 400mmL)

(Section A-A) Horizontal hole (40mmϕ x 950mmL)
   Hole spacing : 400mmL

Fig.6 Rough sketch of a drilling
   pattern used for limited demolition of unreinforced mass
   concrete foundation

Unite amount of agent: 19 kg/m³
Photo-2 Appearance of crack generated on horizontal direction in unreinforced mass
   concrete foundation

Photo-3 and 4 show an example of continious demolition of a pier. It was 30mLx6.5mWx1.5mH and was formed of five blocks which were all unreinforced concrete. Each block was bleaked one by one by using HC 30 together with a giant breaker. After cracks were generated in it by HC 30 and Cap Ace, secondary breaking was soon conducted by a giant breaker. The time in which secondary breaking work started was 1 to 3 hours after loading.

Hole diameter : 40mm⌀
Hole spacing : 500mmL
Photo-3 Appearence of cracks in a piar genarated by HC·30

Unite amount of agent : 4Kg/m³
Photo-4 Secondary breaking by a giant breaker

Photo-5 and 6 show a demolishing work about a old footing of a stalk by using HC·30 and Cap Ace. Secondary breaking was done by a hand breaker and a nibbling machine because the footing was closed to private hauses.

Hole diameter : 38mm⌀
Hole spacing : 400mmL
Photo-5 Secondary breaking of footing

Unite amount of agent : 6kg/m³
Photo-6 Breaked surface

4. Conclusion

New static demolition agent "High Calmmite·30" is efficient to break concretes rapidly and in safty. It has working time at least 15 minutes and generate high expansive pressure. The blowing-out of agents is prevented by means of using a special capping agent "Cap Ace" .

THE MECHANISM OF EXPANSIVE PRESSURE AND BLOW-OUT OF STATIC DEMOLITION AGENT

K.GOTO , K.KOJIMA and K.WATABE
Asahi Chemical Industry Co.,Ltd.

Abstract
For the constructor as well as the community, the basic
needs at the construction site are safe, fast, quiet,
pollution-free operation. This is particularly true for
breaking down and removing rock and concrete structures
without explosives. The static demolition agent (SDA)
meets these needs. This SDA allows a substantial shortening of the overall demolition operation, from initial
loading and breakage to final separation with pneumatic
hammers or breakers. The fast action of SDA also
facilitates supplementary or followup static demolition.
Key words : Static demolition agent, Expansive stress,
Blow-out, Pellet, Bed rock demolition, Reinforced conrete
demolition, Hydration, CaO.

1. Introduction
There are large number of reports on static demolition
agent. But we have a small number of reports on water-
action of static demolition agent. Aiming at this water
action, we studied caracteristics of the expansive
pressure and a mechanism of the blow-out.

2. Demolition agent

2-1 Mechanism of the demolition
The main reaction of this demolition agent is hydration of
lime with heat, which is expressed by:

$$CaO \ + \ H_2O \ \rightarrow \ Ca(OH)_2 \ + \ 15.2kcal/mol$$

Without confinement, the volume of the reaction product is
about twice that in pre-reaction stage. Under confinement,
this reaction rapidly generates an expansive pressure more
than enough to crack and break concrete and stone. As
this reaction rate varies with the ambient temperature, so
it is necessary to control the reaction in accordance with

132

the ambient temperature.

2-2 Reaction process
In the general, a few minutes after loading the hole with
demolition agent, vapor starts comming up from the brim of
the hole. Contained water in demolition agent near the
brim of the hole evaporates to dryness and demolition
agent begins to expand gradually from the brim to the
bottom of the hole. Pellets of demolition agent break to
powder without blow-out and filled out.

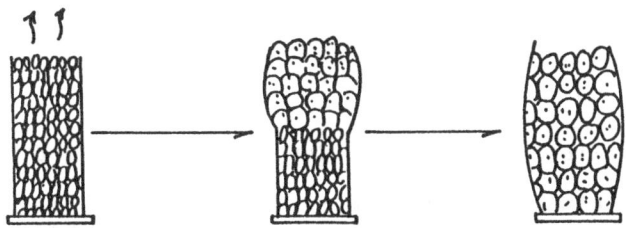

Fig.1 Reaction process in the general

2-3 Mechanism of the generation of expansive pressure
On the generation of expansive stress, we have Powder
Coagulation theory and Crystal Growth theory as shown in
Fig.2

○ Powder Coagulation ‥‥‥‥ · Wide vacant space between
   theory   microcrystallization
                                 (Powdering)
                                 · <u>Generation of expansive</u>
                                  <u>pressure</u>
○ Crystal Growth   ‥‥‥‥ · Narrow vacant space between
   theory   crystallization (Hardening)
                                 · Crystal growth
                                 · Wide vacant space between
                                   crystallization
                                 · <u>Generation of expansive</u>
                                   <u>pressure</u>

Fig.2 Mechanism of the generation of expansive pressure

2-4 Demolition mechanism by demolition agent
After loading the drillholes with demolition agent, its
expansive pressure goes up gradually and finally it
generates more than pressure of $300kg/cm^2$. Compressive
strength of the rock is generally 1,000 to $2,000kg/cm^2$
(300 to $400kg/cm^2$ for concrete), tensile strength of it
is 40 to $150kg/cm^2$(15 to $30kg/cm^2$ for concrete). It is
possible to break rocks and concrete by tensile stress of
the expansive pressure.
   Expansive pressure by the demolition agent works at

inside surface, compressive stress is generated radiately from the center to inside surface of the hole, tensile stress is generated perpendicularly to the direction of compressive stress, and cracks occur at the weakest parts of inside surface of the hole.(as shown in Fig.3)

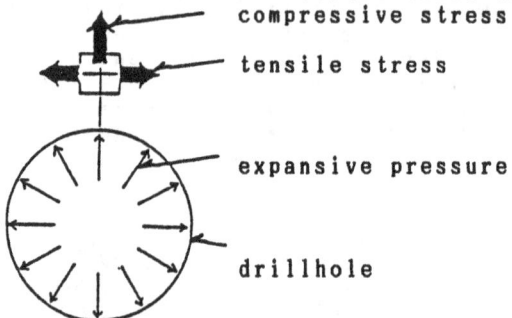

compressive stress

tensile stress

expansive pressure

drillhole

Fig.3 Demolition mechanism by demolition agent(1 drillhole)

In the case of two holes or more, cracks occur from hole to hole by tensile stress between holes, as shown in Fig.4.

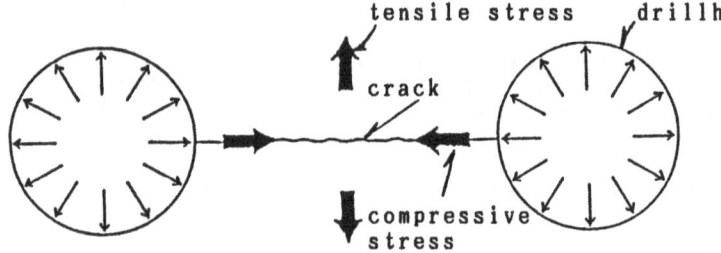

tensile stress   drillhole

crack

compressive stress

Fig.4 Demolition mechanism by demolition agent(2 drillholes)

As shown in Fig.5, in the demolition with explosives, Fragmentation zone, Compressive breaking zone and Tensile breaking zone are exist. As compared with this, in that with demolition agent, only Tensile breaking zone is in existence.

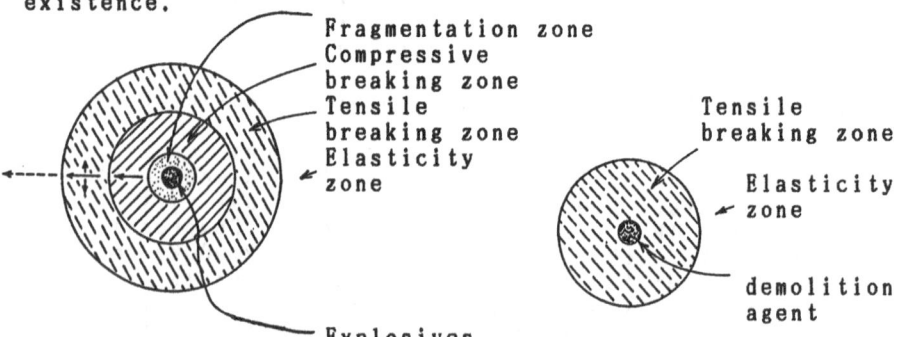

Fragmentation zone
Compressive
breaking zone
Tensile
breaking zone
Elasticity
zone

Tensile
breaking zone

Elasticity
zone

demolition
agent

Explosives

Fig.5 Differrence of explosives and demolition agent

3. Experimental methods

3-1 Experimental methods of hydrating condition and charactristics of expansive pressure

In order to make sure of the generation mechanism of expansive presure, we tested expansive presure under various testing conditions such as water content, temperature etc. Test method is using a pressure sensor for measuring expansive pressure. Pellets were broken by a crusher and were sieved. The size of pellets for this test was 1.2 ~ 14.0mm. Test result was very similar to one by strain gauge method, as shown in Fig.7. This result is reliable enough.

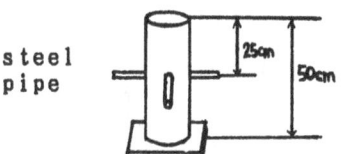

steel
pipe

· Steel pipe:inside diameter 40mm, thickness 10mm.
· Set 4 sensors on the steel pipe, as shown in fig.6.
· Point of each sensor reaches inside wall of the steel pipe.
· Pressure sensor:PGM-E 500kg/cm² by Kyowa Dengyo Co.,Ltd.

Fig.6 Situation of pressure sensor

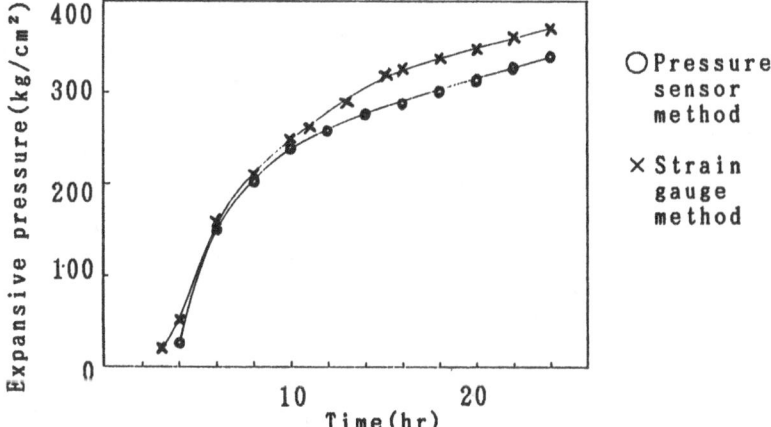

○Pressure sensor method

×Strain gauge method

Fig.7 Referrence of pressure sensor method and strain gauge method

3-2 Methods of observation of blow-out

We filled steel pipe, in which PVC pipe is sticked, with pellets of demolition agent. And we observed and studied its reaction process and blow-out.

4. Experimental result and discussion

4-1 Hydrating condition and characteristics of expansive pressure

As characteristics of expansive pressure vary with reaction

temperature of lime as main ingredient of demolition agent, it is particularly difficult to use only one grade product at every temperature. It is necesarry to control reaction of lime for temperatures. Fig.8 and 9 show expansive pressure of demolition agent appropriate for 10 to 20 ℃, expansive pressure become lower at inappropriate temperature. At lower temperature the reaction speed of lime is too slow, while on the other hand, at higher temperature it is too rapid to cause shortage of water. Under both conditions, expansive pressure goes down as compared with one under appropriate conditions.

There are two types of generation of expansive pressure. After reaching maximum pressure, one keeps its pressure level, the other decreases it. As shown in Fig.10, demolition agent with high water content increases its expansive pressure gradually, and keeps its maximum pressure level, the expansive pressure with low water content goes up rapidly and decrease after reaching its maximum level.

We think this is because of differrence of hydration. With high water content, the hydration is wet-hydration, and with low water content, the hydration is dry-hydration.

Wet-hydration is ionic reaction. Formulated ions binded each other to $Ca(OH)_2$, in this process super-saturated solution of $Ca(OH)_2$ is produced and cores of $Ca(OH)_2$ crystallizes. Expansive pressure by demolition agent with high water content is based on Crystal Coagulation Theory, but the reaction is slow because crystal coagulation is ionic reaction.

Dry-hydration does not needs so much water and its reaction is fast. The reaction is faster and faster, expansive pressure becomes higher and higher.

Generally, dry-hydration produces slacked lime aggregation more easily than wet-hydration because of the difference between reaction processes. In wet-hydration process, slacked lime crystallizes, crystals grow uniformly, and this process is influenced by the dispersion of water. In dry-hydration process, the reaction is radical. Crystals of slacked lime grow un-uniformly, the crystals are aggregated by surface energy generated secondarily and coarse grains of slacked lime.

Generation of expansive pressure by low water content demolition agent is based on Powder Coaguration Theory because coarse grains of slacked lime are produced. In dry-hydration process, the reaction is fast. As slacked lime is microcrystal aggregation, expansive pressure decreases by reconfigulation after reaching the maximum pressure.

136

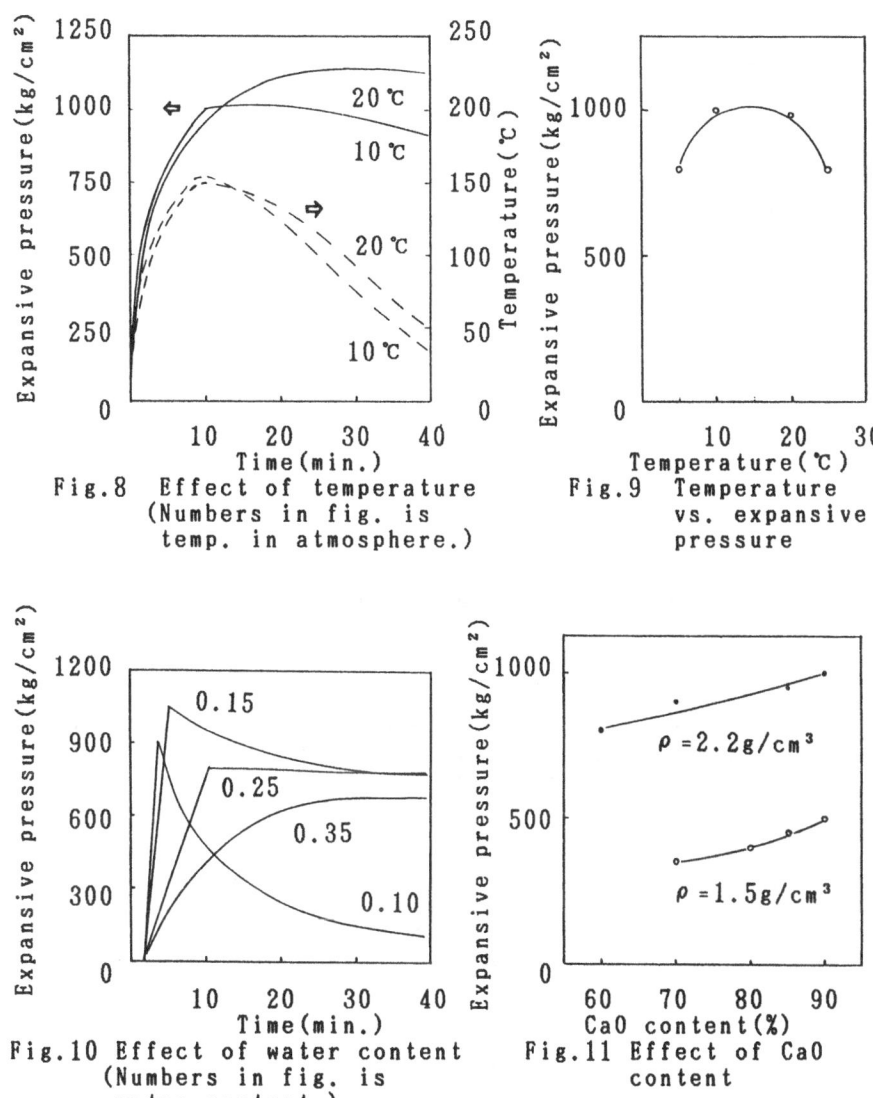

Fig.8　Effect of temperature
(Numbers in fig. is
temp. in atmosphere.)

Fig.9　Temperature
vs. expansive
pressure

Fig.10 Effect of water content
(Numbers in fig. is
water content.)

Fig.11 Effect of CaO
content

As shown in Fig.11, CaO content is more and more,
expansive pressure becomes higher and higher. It is easy
to understand this result because reacting quantity of CaO
becomes more and more.

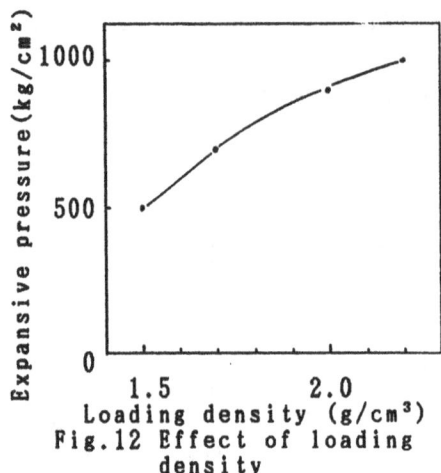

Fig.12 Effect of loading
         density

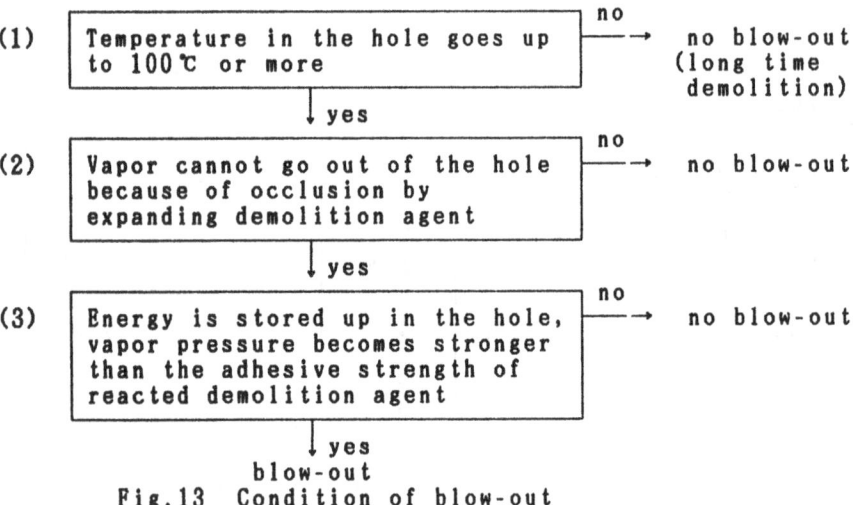

(1) │ Temperature in the hole goes up │ ──no──→   no blow-out
    │ to 100 ℃ or more   │   (long time
                                                   demolition)
                    ↓ yes

(2) │ Vapor cannot go out of the hole │ ──no──→   no blow-out
    │ because of occlusion by   │
    │ expanding demolition agent   │
                    ↓ yes

(3) │ Energy is stored up in the hole,│ ──no──→   no blow-out
    │ vapor pressure becomes stronger │
    │ than the adhesive strength of   │
    │ reacted demolition agent   │
                    ↓ yes
                 blow-out
        Fig.13 Condition of blow-out

Loading density is higher and higher, expansive
pressure becomes higher and higher, as shown in Fig.12.
Under high loading density condition void ratio is very
low from the beginning of the reaction and immediately
demolition agent starts to expand and generate high
expansive pressure.

4-2 Mechanism of blow-out
Fig.13 shows conditions in which blow-out may occur.
When the temperature in the hole is 100 ℃ or less, little
vapor generates. Under this condition no blow-out occurs,
but it takes long time to demolish because demolition
agent reacts slowly. In case that vapor comes out of the
hole, little vapor pressure works inside the hole and no
blow-out occurs. In case that it is difficult for vapor
to go out of the hole and vapor pressure is stronger than
the adhesive strength of reacting demolition agent,
blow-out may occur.
    In addition, in the case of high density loading it has
a tendency not to blow out because adhesive strength of
demolition agent is strong and vapor pressure is weak with
a little water and air in the hole.
    Fig.14 shows the relation between temperature and
expansive pressure. At 100 ℃ and 200kg/cm², 80kg/cm², no
blow-out occurs. At 100 ℃ and 20kg/cm², blow-out occurs.
It may be easy to blow-out when temperature rise more
rapidly as compared with expansive pressure rise.

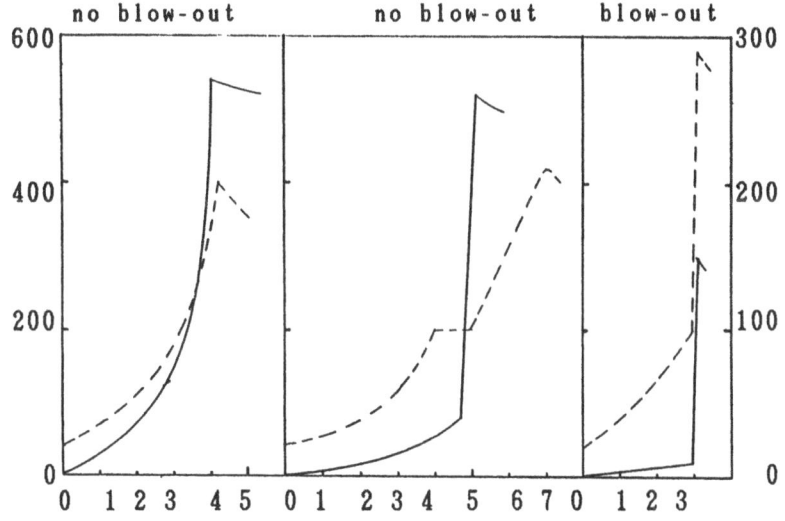

Time(min.)
Fig.14 Relation of expansive pressure and temperature
       (——— Expansive pressure)
       (········· Temperature)

5. Conclusion

We discussed the mechanism of expansive pressure and blow-out of demolition agent. As the result of this discussion, we conclude that loading density and water content has large influence on expansive pressure and blow-out. In order to use demolition agent safely and advantageously, the most important point is how to control loading density and water content.

References

Y.Kasai;(1970) Demolition work of concrete structures (in Japanese), Nikkan Kougyo Shinbunsya
Issued by Technical Comittee of Japanese Lime Association; (1979) Hand book of manufacturing lime (in Japanese), Japanese Lime Association

SPLITTING MECHANISM OF ROCK AND CONCRETE UNDER EXPANSIVE PRESSURE

Jin Zhongzhe, Liao Hong and Zhu Wen
Research Institute of Building Materials, Beijing, China

Abstract
Splitting mechanism of rock and concrete under the slowly increasing expansive pressure by expansive splitting agent is studied in this paper. The splitting strength and expansive pressure in holes of different brittle solid materials and concrete are obtained through experiments. In this paper we proposed that expansive pressure be related to elastic modulus of material. Finally, we deduced two equations. One is about free surface effect, another is for calculating the optimal distance between two holes. After doing experiments in concrete we have obtained results which are in consistent with those of the theoretic calculation.
Key words: expensive pressure, compressine strength, tensile stress, compresive stress, crack, split, failure, strain, non—linear, propagation, FEM, medium.

1. Introduction

Since the emergence of static cracking technique, the damage mechanisms of concrete and rock under expansive pressure were proposed[1]. All these explanations are based on the linear elastic theory or elastic plastic theory. They cannot explain the variation of expansive pressure in different materials, i.e. when expansive agent is in different materials, the expansive pressure increases as its resistance increases. Also, they cannot explain the effect of free surface and adjacent hole on the damage of concrete and rock.

After measuring characteristics of expansive pressure and doing experiments on concrete models, damage mechanism and splitting mechanism of concrete are discussed in this paper. Formulas about calculating expansive pressure, distance to free surface and distance between two holes are propossed.

2. Experiment

2.1 Measurement of expansive pressure
Expansive agent SCA — 2 (soundlessly cracking agent) is studied and produced by Research Institute of Building Materials. In our experiments, the water／cement ratio of concrete is 0.28. Four kinds of tube are

141

used. They are steel tube (elastic modulus E=200 GPa), copper tube
(E=100 GPa), ceramic tube (E=450 GPa) and plastic tube (E=0.5 GPa).
The inner diameter of these tubes is 20 mm and the outer diameter
is 40 mm. Length of the tubes is 500 mm. These tubes are filled
with expansive agent, and YJ-5 strian meter is used for measuring
strain of tubes. Strains in every ten hours and their corresponding
expansive pressure are showed in Fig.1. Ceramic tube is broken at
the time of three hours. Its corresponding expansive pressure is
36 MPa. Volume increase of expansive agent in plastic tube is more
than two times, and its expansive pressure is smaller than 1 MPa.

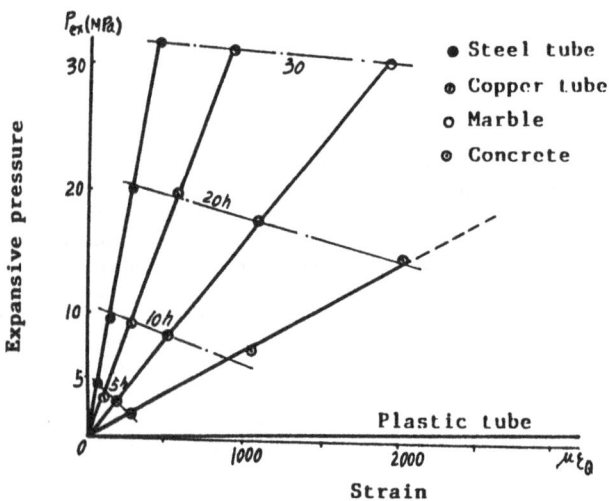

Fig. 1   Relation among expansive pressure,
strain of medium and time

2.2 Splitting experiment on concrete models

In order to measure and analyse cracking process and optimal distance
between holes as well as free surface effect, large concrete models
are used. The size of the models is 450x750x50 mm. Their compressive
strength is 40 Mpa. The diameter of the hole in them is 30 mm. The
depth of the hole is 400 mm. There are two concrete models with
one round hole, two with two round holes. The distance between the
center of two holes is 300 mm. And another two models with
one rhombus hole. There are 20 to 30 strain gauges around the hole.
Interval between them is 50 to 150 mm.

Type DTU-102 straining meter is used in measuring longitudinal
strain and tangential strain, the time interval of measuring is
1 hour. Fig.2 shows the stress distribution and experimental results
near the round hole.

3. Discussion

Under expansive pressure, the damage process of concrete and rock
can be divided into three periods.

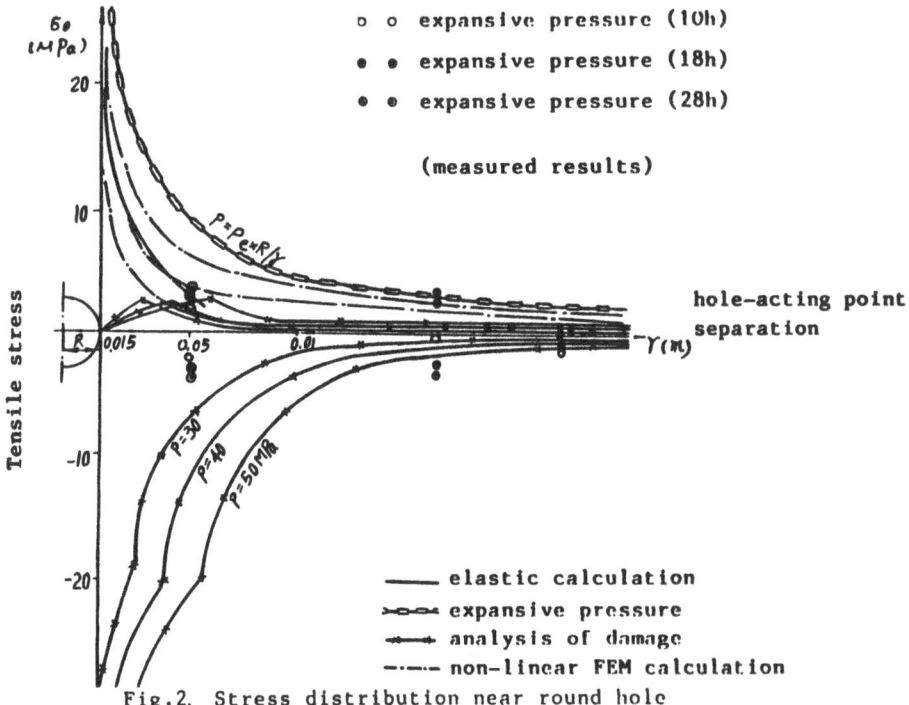

Fig.2. Stress distribution near round hole

The first period:

Fill the expansive agent in the drilled hole, when the expansive pressure increases from 0 to p in concrete or rock, radial compressive stress σ_r and tangential tensile stress σ_θ are produced near hole. When the stresses reach limited value, cracking or damage occurs. Generally, microcracking zone (σ_r microplastic deformation zone) is caused by tensile stress σ_θ, it is called damaged zone or failure occurence zone. And this period is called microcracking period. At the begining of this period, stresses around hole is linear, but as the expansive pressure increases, the stresses around hole become non—linear.

The second period:

As time increases, the expansive value increases, i.e. the expansive pressure is getting higher and higher and it propagates outwards through the damaged zone. This period is called expansive pressure propagation period.

The third period:

When expansive pressure propagates to free surface, cracks near free surface propagate to it first and make concrete split. This period is called splitting period.

The characteristics of every period in concrete and rock splitting under the effect of expansive pressure are discussed as follows:

Microcracking period:

1. Stress distribution near drilled hole in linear elastic period.

Microcracking period is divided into linear elastic period and non-elastic period. At the time, when SCA—2 effects on the concrete

143

from 0−10 hours, the materials around the hole is linear elastic. So stresses can be calculated from thick wall tube stress equation in elastic mechanics.

Radial stress and tangential stress is

$$\begin{Bmatrix} \sigma_r \\ \sigma_\theta \end{Bmatrix} = P_{ex} \frac{R^2}{(R+d)^2 - R^2} \left[1 \mp \frac{(R+d)^2}{r^2} \right] \qquad (1)$$

where P_{ex} is expansive pressure, R is the inner raidus of tube, d is the thickness of wall.

After some deduction, we can obtain when r=R+d

$$P_{ex} = E (2R+d)d/2R^2 \qquad (2)$$

When d is very small

$$P_{ex} = E \cdot \xi_\theta \, d/R \qquad (3)$$

When E, d, R are known, after measuring ξ_θ , we can obtain P_{ex} from (3).

When $(d+R) \to \infty$,

Stress distribution in equation (1) is showed in Fig.2.

2. Microcracking zone (damaged zone)

When expansive pressure increases beyond a certain value, stress-strain relation of concrete and rock becomes non−linear. Under tensile stress and compressive stress, stress strain relation is unsymmetric[3]. This kind of problems can not be solved analytically, Zhu Wen and her colleagues solved the problem by using non−linear FEM calculation[3]. The results are shown in Fig.2.

The above linear elastic and non−linear elastic stress solution are accurate enough for calculating the stresses in expanded materials for a certain period. But when expansive pressure is very high, materials begin to be cracked and developed into damaged zone. The stress distribution at the beginning of forming damaged zone is deduced as follows:

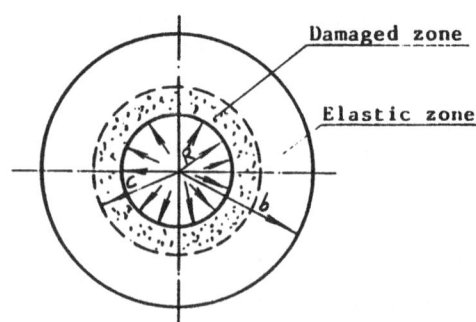

Fig.3 Damaged zone

Suppose when $R \leqslant r \leqslant c$, r is in damaged zone.
When $c \leqslant r \leqslant b$, r is in elastic zone. (See Fig.3).
The balance equation is

144

$$\frac{d\sigma_r}{dr} + \frac{\sigma_r - \sigma_\theta}{r} = 0 \qquad (4)$$

According to the failure condition

$$\frac{\sigma_\theta}{\sigma_t} - \frac{\sigma_r}{\sigma_c} = 1$$

and boundary condition

$$\sigma_r | \ r = R = -P_{ex}$$

we obtain radial and tangential stresses

$$\sigma_r = (\frac{r}{R})^{-r}(\sigma_t/K - P_{ex}) - (\sigma_t/K)$$

$$\sigma_\theta = (\sigma_t/\sigma_c)(\sigma_c + \sigma_r) \qquad (5)$$

where $K = 1 - \dfrac{\sigma_t}{\sigma_c}$.

In the linear elastic zone $c \leqslant r \leqslant b$, when $r=b$ and the stress at r just reaches failure critical condition, we obtain stress distribution as follows:

$$\sigma_r/\sigma_c = (2 - b^2/c^2)c^2/2b^2$$

$$\sigma_\theta/\sigma_c = (1 + b^2/c^2)c^2/2b^2 \qquad (6)$$

stress distributions expressed by equation (5), (6) are showed in Fig.2.

4. Expansive pressure propagation period

4.1 Propagation of expansive pressure
After the forming of damaged zone, there is stress relaxation in damaged zone, but as time increases, expansion value increases, so the expansive pressure increases and it propagates outwards through medium. When the expansive agent is filled in a hole with radius R and it expands slowly, its static pressure is P_{ex}, after t hours the expansive pressure propagates to a certain distance r and its value is P. If the loss of pressure is neglected, then

$$P = P_{ex}R/r \qquad (7)$$

r varies from R to failure occurence zone boundary c and then to elastic zone b. i.e. the expansive pressure propagates near the free surface. There are some kind of damage, crack propagation and splitting in this process. Equation (7) can be proved in our experiment. When time are 10, 18, 28 hours, the expansive pressure is measured respectively and the results are also showed in Fig.2.

4.2 Crack propagation
Many microcracks produced after occurring of damaged zone. The maximum microcrack propagates and results in failure.

145

In fracture mechanics, fracture toughness of the round hole with two edge cracks under inner pressure is

$$K_I = FP\sqrt{\pi a} \tag{8}$$

when $a/R = 1 \sim 1.4$, $F = 0.1 \sim 0.34$

where a is crack length i.e. the distance between the center of hole to the crack tip. R is radius of the hole, crack propagation condition is

$$FP\sqrt{\pi a} \geqslant K_{IC} \tag{9}$$

or expansive pressure needed for crack propagation is

$$P_{ex} \geqslant K_{IC}/F\sqrt{\pi a} \tag{10}$$

for concrete K_{IC} is 0.5 to 1.3 MPa $m^{\frac{1}{2}}$.
for rock K_{IC} is about 0.9 to 2 MPa $m^{\frac{1}{2}}$.

5. Splitting period

5.1 Free surface effect

Expansive pressure propagates to the material near free surface is the last period of damage or splitting. Here we discuss the free surface effect.

In elastic, non-elastic and elastic-plastic stress analysis, when the distance f between the hole center and free surface is greater than 4R (R is the hole radius) or the distance 2L between two holes is greater than 4R, we can take the medium as an unlimited solid material, i.e. free surface and adjacent holes have little effect on the stress distribution of the drilled hole[3]. But in explosion or expansive breaking engineering, free surface and adjacent hole have great effect on the stress distribution.

Here we deduce the distance f between free surface and drilled hole by using the stress concentration condition caused by expansive pressure propagates near the free surface.

Suppose that the effect radius of expansive pressure P be r, when f = 1.5 r, stress distribution along the free surface is showed in Fig.4.

When $f/r = \sqrt{3}$, tangential stresses at D and A' are the maxmium, we can get $\sigma_{\theta D}/P = \sigma_{\theta A'} = 2$, and $\sigma_{rD} = 0$ [2].

consider failure condition $\quad \sigma_\theta/\sigma_t - \sigma_r/\sigma_c = 1$

we obtain $\quad f = 2\sqrt{3}RP_{ex}/\sigma_t$

In our experiment, when $P_{ex} = 30$ Mpa, f = 0.31 M.

5.2 Equation about distance between two holes

The distance between two holes (2L) is calculated according to the failure condition, when the expansive pressure propagates distance L in time t, the expansive pressure at 2L's median point C can be obtained by adding stresses of the two hole together. See Fig.5.

146

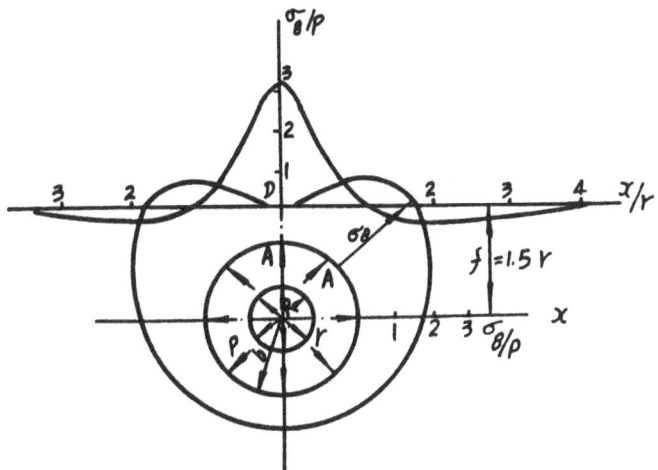

Fig.4 Stress distribution on free surface

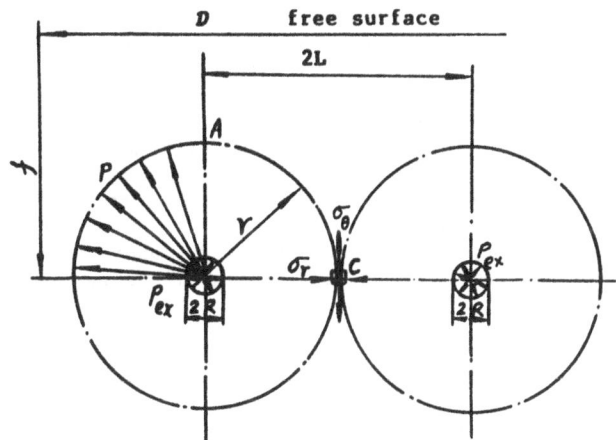

Fig.5 Stress interaction between two holes

Suppose that radius is r, then $P=P_{ex} 2R/L$, stresses at point C are

$$\left\{ \begin{array}{c} \sigma_\theta \\ \sigma_r \end{array} \right\} = \pm P \tag{11}$$

when failure condition $\sigma_\theta /\sigma_t - \sigma_r /\sigma_c =1$ the optimal distance between two holes is

$$2L = 4R \cdot P_{ex}(\sigma_c +\sigma_t)/\sigma_t \cdot \sigma_c \tag{12}$$

In our experiment $P_{ex}=30$ Mpa, $2L=0.3M$.

6. Discussion on expansive pressure

The general work done by expansive agent is related to the quantity of the agent, time of the expansion, temperature, and expansive pressure as well as properties of medium (mainly the elastic module).

When time and temperature are certain, if the work done by expansive agent in different medium is a constant, we can get $P_{ex}=f(d,t,E)$.

where   d is radius of hole;   t is expansion time;
        E is elastic module of medium.

According to the experiment results and fig.1, we can obtain the relation approximately

$$P_{ex} = W d t^4 E \qquad\qquad (13)$$

where w is a constant related to expansive agent and medium properties. For SCA-2, W=2 3.

7. Conclusion

In this paper, splitting mechanics of concrete and rock under the effect of expansive agent is described. The expansive pressure is related to properties of damaged material. After forming damaged zone near drilled hole, when expansive pressure propagates near free surface, stress concentration on the free surface lead to the failure of material. In this process, concrete and rock are splitted by crack occurence and propagation. In this paper, the equations for calculating free surface distance and distance between two holes are valuable both in practical application and theory.

In splitting engineering of concrete and rock, crack control and decreasing cracking time are the main problems to be solved. The new method of splitting and breaking concrete and rock with expansive agent is widely used in mining industry and dismantling concrete structure.

References

Wang Yan Sheng et al. "The properties and application of soundless cracking agent" (1981).
G.B. Jeffery, Phil. Trans. Roy. soc. London, A Vol.221, P.265 (1921).
W.Zhu. Z.Z. Jin et al., "Non-Linear FEM Stress Analysis of Holes in Rock and Concrete (to be Published).

APPLICATION OF SOUNDLESS CRACKING AGENT IN CHINA

WANG YANSHENG, YOU BAOKUN and ZHANG GUIQING
Research Institute of Building Materials, Beijing, China

Abstract
This paper deals with the characteristics of high efficient Soundless
Cracking Agent, the design of drilling hole, the equations for
calculation of spacing of holes and the minimum resistant distance,
the arrangement of drilling hole, the relation between parameters
of spacing holes and cracking time as well as various kinds of
engineering examples. Practice have proved that the operation
of soundless cracking technique is simple, safe and non—pollutant.
It is most suitable for the cracking and demolition engineering
of reconstruction and extension in cities and other construction
under complicated conditions. It is particularly suitable for
a construction site where space is narrow, work can't be interrupted
or using explosive is not permitted. Soundless cracking technique
is an important supplement and development to explosive technology.
It is even more effective when used jointly with controlled explosion
and mechanical demolition.
Key words: Soundless cracking agent, Demolition, Cutting, Drilling
hole, Hole spacing, Non—pollution, Minimum resistance line.

1. Introduction

Silient non—explosive demolition agent which has developed in
recent years is a supplement and development to explosive technology.
It realizes to a great extent the aspiration of safe cracking
for which people yearn day and night.
   Study on Soundless Cracking Agent (SCA) in China begun in 1981.
It was developed successfully in Research Institute of Building
Materials (RIBM) and put into commercial production in 1983. Afterward
we made further research on applied technology and cracking mechanism.
In recent years, RIBM has developed a kind of highly efficient
soundless cracking agent (HSCA). Its expansive pressure is raised
from 30—60 to 60—90 MPa and the demolishing efficiency has greatly
increased. Up to now, China has produced SCA of about 10,000 tons
and demolished various kinds of concrete and quarried marble and
granite of about 650,000 m^3. SCA has now been extended for application
throughout the country and begun to sell to Malaysia, Singapore,
U.S.A., Canada, Austrilia and HongKong.

149

2. Characteristic of HSCA

HSCA is an expansive material. Its main expansive origin is from CaO. When mixed with water and stirred into slurry, HSCA reacts with water and originate expansive crystals which generate expansive pressure in the holes drilled in rocks and concrete, and exerts compressive stress σ_r and tensile stress σ_θ on the walls of drilled hole (Fig. 1). When the tensile stress of SCA exceeds the tensile strength of rocks or concrete, cracks initiate and gradually become greater as the expansive pressure increases continously. Finally cracking and spillting of rocks or concrete are performed under without any vibration, noise, flyingrock and harmful gases.

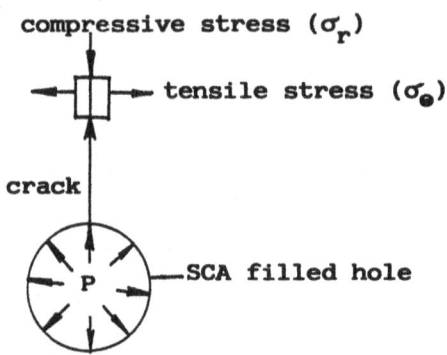

Fig. 1   Digram of the cracking principle of SCA

There are 3 types of HSCA and their properties are shown in Table 1 and Fig. 2.

It is shown in Fig. 2 that the expansive pressure of HSCA-I is $\geq$40 MPa in 8 hours and $\geq$90 MPa in 24 hours. Therefore in different seasons defferent types of HSCA should be used. The experimental results show that for the same type of HSCA the higher the temperature, the higher the expensive pressure; the expensive pressure increases with the decrease of W/HSCA ratio and the increase of hole diameter.

Table 1   Types of HSCA

Types	Season for working	Working Temperature (°C)	Diameter of dirlled hole (mm)
HSCA — I	Summer	25 — 40	30 — 50
HSCA — II	Spring, Autumn	10 — 25	30 — 50
HSCA — III	Winter	− 5 — 10	30 — 50

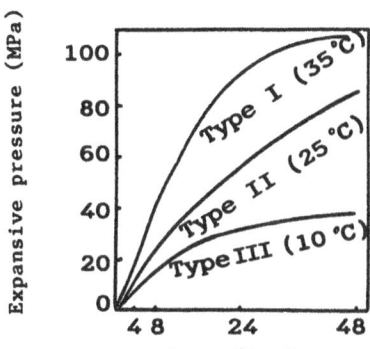

Fig. 2   Expansive pressure of HSCA

3. Deisgn and Construction of Soundless Cracking

3.1 Selection of drilled hole parameter

There are two main factors to obtain an ideal cracking effect, one is higher expansive pressure and another is reasonable parameters of drilled hole. Jin Zongzhe et al derived equations for calculation the hole spacing (A) and the minimum ressistance distance (W) and we have made some correction as follows:

$$W = (2\sqrt{3}\ RP/\sigma_t)B \dotfill (1)$$

$$A = 4R(\frac{\sigma_c - \sigma_t}{\sigma_c \sigma_t})PB \dotfill (2)$$

where, R — radius of drilled hole (m)

σ_c — compressive strength of material to be cracked (MPa)

σ_t — tensile strength of material to be cracked (MPa)

P — expansive pressure of HSCA (MPa)

B — empriical corrected coefficient.

B=1 (rocks), B=0.5 (plain concrete) and B=0.3 (reinforced concrete).

In the calculation, provided that R=0.02 m, P=40 MPa (8 hrs) and the properties of the substance to be cracked as shown in Table 2.

Table 2   The properties of the substance to be cracked

Name	Compressive strength (MPa)	Tensile strength (MPa)
Cranite	148.4	10.3
Siliceous marble	77.77	6.19
Marble containing calcite	69.7	4.5
Concrete	35.0	2.9

151

According to the equation (1),
Granite: W=2√3x0.02x40/10.3=0.27 (m)
Plain concrete: W=(2√3x0.02x40/2.9)x0.5=0.48 (m)
Reinforced concrete: W=(2√3x0.02x40/2.9)x0.3=0.29 (m)
According to the equation (2),

Granite : $A = 4 \times 0.02 \times (\frac{148.4-10.3}{148.4 \times 10.3}) \times 40 = 0.29$ (m)

Plain concrete : $A = 4 \times 0.02 \times (\frac{35-2.9}{35 \times 2.9}) \times 40 \times 0.5 = 0.51$ (m)

Reinforced concrete : $A = 4 \times 0.02 \times (\frac{35-2.9}{35 \times 2.9}) \times 40 \times 0.03 = 0.30$ (m)

The results calculated according to the equation (1) and (2) are shown in Table 3.

Table 3. W and A of the substance to be cracked

Name	W (m)	A (m)
Granite	0.27	0.29
Siliceous marble	0.45	0.48
Mable containing calcite	0.62	0.66
Plain concrete	0.48	0.51
Reinforced concrete	0.29	0.30

We recommend the parameters of drilled hole as shown in Table 4 according to the experience of practical working and the theoritical calculation.

Table 4.   Parameters for drilling holes

Name	Hole diameter (mm)	Minimum resistance distance (cm)	Spacing of holes (cm)
Soft rock cracking	30 − 50	40 − 60	40 − 70
Medium hard rock cracking	30 − 50	15 − 25	20 − 30
Rock cutting	30 − 40	15 − 25	20 − 30
Plain concrete cracking	30 − 50	25 − 40	35 − 50
Reinforced concrete cracking	30 − 50	20 − 30	25 − 35

The hole depth (L): for cracking or cutting virgin rocks, L=(1+0.05)H, for isolated stones or concrete foundation, L>0.7H and for reinforced concrete, the holes should be drilled as deep as possible, but not penetrated thourgh.

3.2 Deisgn of drilling holes and cracking time
They are close relations between cracking time of the substance

to be cracked (T) and hole diameter (d), spacing of holes (A), minimum resistance distance (W) and hole depth (L). In order to study the influences of the above factors on cracking time, let 3 of the factors remain unchanged and vary one factor to study its effect on cracking time. The results are shown in Fig.3–6.

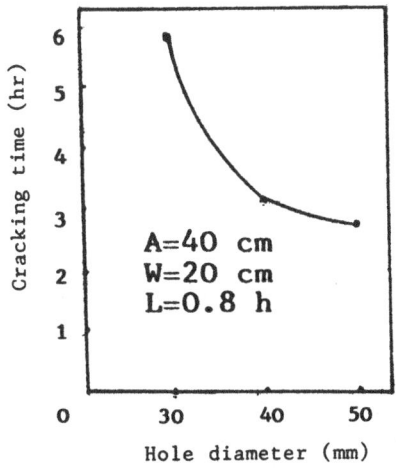

Fig.3 The relation between hole diameter (d) and cracking time (T)

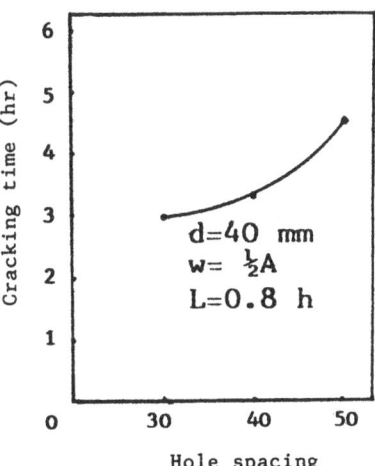

Fig.4 The relation between spacing of holes (A) and crack‐ing time (T)

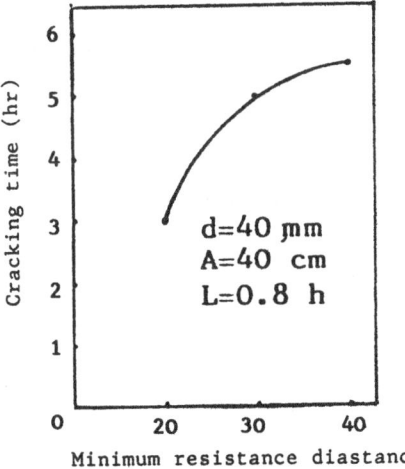

Fig.5 The relation between minimum resistance distance (W) and cracking time (T)

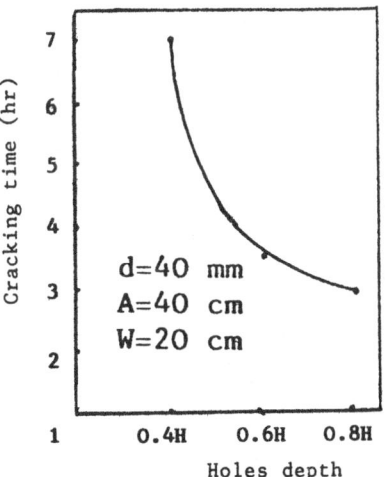

Fig.6 The relation between hole depth (L) and cracking time (T)

153

It can be seen from Fig.3 – 6, the larger the hole diameter, the samller the hole spacing, the smaller the minimum resistance distance and the deeper the hole depth, the shorter the cracking time will be.

3.3 Arrangement of drilling holes

The importance thing of cracking design is to decrease the resistance of substance to be cracked as much as posible and set up free surface as larger as posible.

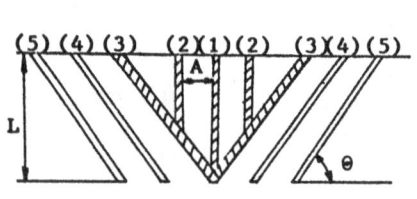

Fig.7 'V' shaped drilling
A=20—35 cm, 2R=30—50 cm,
$\theta=40° - 60°$
Fill in holes (1)(2)(3) and
then holes (4) and (5) after
6 hours

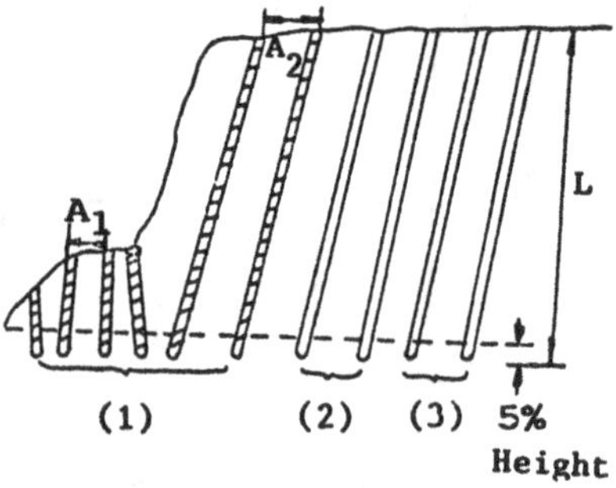

Fig.8 Cracking of virgin crack
Hole diameter=30—50 mm.
A_1=25—30 cm, A_2=40—60 cm,
L=adding 5% of height.
Fill in holes (1) and than holes
(2) and (3) after 6 hours

If the substance to be cracked has only one free surface, 'V' shaped drilling is required. It should be dug into trenches in order to make 2 free surfaces (see Fig.7). If more free surfaces are available, then larger hole spacing can be used, and vice versa.

If the rock to be cracked has two free surfaces, it is necessary to remove the base part of the rock first. Two rows of holes near the free surface are filled with SCA and then the others after 6 hours (see Fig.8).

In the case of reinforced concrete, closely drilling hole and filling with SCA should be near the inner layer of steel bars.

3.4 Working of HSCA

HSCA is mixed with water in W/HSCA=0.28 and stirred into a slurry by using mechanical mixing or mixing by hand with gloves on and then filled in drilled holes. Don't look directly at any hole for at least 2 hours after pouring HSCA into a hole to prevent spurt in case.

Practice show that cracking efficiency can be raised by combining soundless cracking technique with controlled explosion, hydraulic gun and mechanical demolition. Also, the efficiency of quarrying stones can be greatly increased by combining soundless cutting technique with steel cable quarrying machine.

4. Engineering Examples

4.1 Zhuo country construction brigade of soundless cracking using HSCA-I. Demolished the foundation of air compressor in Shijiazhuang Fertilizer Factory. Every foundation measured about 3.4 (long) by 1.8 (wide) by 2.0 (high) m. Both sides of the foundation closely near the water well and cable well, and they should not be damaged during working. The vertical steel bars of the foundation were double row (Ø12 mm) and the spacing was 20 cm. The horizontal ones were double raw too (Ø10 mm) and the spacing was 30 cm. The compressive strength of the concrete was 25 MPa. A Ø4 cm pureumatic drill was used drilling holes and the spacing of holes was 30 cm. The water/HSCA ratio was 0.28. Cracking occurred in 3 hours after pouring the HSCA slurry into the holes. The length of cracks was found to be 1—1.5 cm. It was easy to clean away and there was no damage at all to the water well and the cable well.

4.2 A situe concrete basement of a highrise library in a certain Shanghai university was found to have some problem in part of the engineering in 3 months after construction. Therefore, a part of the separating wall had to be demolished but steel bars in the wall should not be damaged in order to repair concrete. Concrete of 120 m³ were reqired to demolish. The project was very urgent and working conditions very difficult. Finally using HSCA to demolish the seperating wall was determined.

Vertically and horizontally accumulative length of the separating wall was 100 m, thickness 60 cm and height 240 cm. The lower part of the wall should remain 160 cm. The demolition part was double faced and the steel bars were arranged in two ways. The diameter of the vertical steel bars was 16 mm and that of the horizontal ones was 12 mm. The spacing of holes was 20—25 cm, hole depth 0.9H and W/HSCA = 0.28. Cracks occured in 6 hours and the width of the cracks was 3 cm in 24 hours after filling the holes with HSCA slurry. The maximum crack width was 5—6 cm. It was easy to demolish the separating wall with the assitance of pneumatic picke.

4.3 The organosilicon workshop of a certain factory in Tianjin was a five storey building of reinforced concrete frame and wanted demolishing. The height of every storey was 4 m so the total height was 20 m. There was 37 cm long brick wall between the pillars. A overhead main steampipe was 2 m near the building. A paint workshop was in the side of the building and the distance between them was 20 m. The factory is a major fire prevent unit and the combustible and explosive must be strict controlled. Precracking treatment using soundless cracking technique was carried out first for ensuring safety. According to the calculation, when the building collapses and its former part touches the ground, the incidence will be 40° and the end velocity 0.8 redian/sec. These were still not enough to disintergrate the whole building, so it was necessary to weaken the rigidity of the location to touch the ground by using SCA in order to collapse the building completely. SCA was used to fill the holes and their number all together was 200.

Pouring SCA into holes was on the day before the explosion which enables the building to become surfficient pre—cracking but it was not instable before integral explosion. Then the explosion of the lower pillar was performed. The total number of the explosion holes was 70 and 35 kg explosive was used. The explosion untirely reached the anticipation of the deisgn and surrounding buildings and pipelines were safe. The extent of the disinterfral pieces was below 3 t.

4.4   Cutting grainte: Qingdao Stone Factory already has 40 year history of producing granite. The useful product rate is only 20—30% using the old method in the past, it has increaseed to 95% and a very good economic benefit has been obtained since SCA was used in 1983.
    "Downstream" cutting (cutting along stone cleavage veins) a qurry stone, 8 m long, 5 m wide and 5 m high has a volume of 200 m^3 was required along single face into two pieces. 16 hours were drilled. The hole diameter was 42 mm, the hole spacing 45 cm, the hole depth 4 m, L/H=0.9 and working temperature 22° C. 130 kg of type I ordinary SCA was used and W/SCA=0.32. The consumption of SCA was 3.25 kg/m^2 of cutting area. A crack occurred in 12 hours and the width of the crack was 3 cm in 20 hours after filling with SCA. The section was smooth.

4.5   Cutting marble: We went to New Caledonia and performed the marble cutting in July, 1987. A quarry stone of 3.4 m long, 2.84 m wide and 1.5 m high was required to cleave in two. 11 hours were drilled. The hole diamter was 25—35 cm and the hole depth 80—110 cm. The consumption of SCA was 6.6 kg/m^2. A crack occurred in 7 hours and the marble automatically fractured into two pieces in the 16 hours after filling with SCA. Another quarry stone measured 3.7 m (long) 2.4 m (wide) and 1.45 m (high). It was required be cut into 3 parts in the direction perpendicular to the cleave. Holes was drilled in two rows. The hole diameter was 3.5 cm, the hole spacing about 20 cm and the hole depth 0.8—1.32 m. The holes were filled with ordinary SCA—I and the consumption of SCA was 5.7 kg/m^2. Cracks occurred in 7 hours and the crack width was 6—8 cm in 16 hours after filling with SCA. The sections were smooth.

5. Conclusion

It is simple, safe and non—pollutant in operation to crack rocks or concrete constructions by with HSCA. It is suitable for the cracking and demolition engineering of reconstruction and extension in cities, the technical transform of old enterprises and other constructions under complicated conditions. It is particularly suitable for a construction site where space is narrow, work can't be interrupted or using explosive is not perimitted. It is easy to demolish into big pieces when cutting rocks and concrete with HSCA and the big pieces are convenient to reuse. HSCA is also suitable for cutting granite and marble. Experiences show that soundless cracking technique is an important supplement and development to explosive technology. It can play a geater role when used jointly with cotrolled explosion and machine tools.

Reference

Jin Zongzhe et al. "The cleavage mechanism of rocks and concrete under expansive pressure", To be published.

Pang weitai and You Baokun, Communication of Soundless Cracking Engineering (in Chinese), No.10 (1986).

Wang Yansheng, You Baokun, zhang Guiqing and Jiang Yun'an "The application of soundless cracking agent", Construction Technique (in Chinese), No.6 (1984).

NUMERICAL SIMULATION OF ROCK FRACTURING PRODUCED BY STATIC
PRESSURE

MASAAKI YAMAMOTO
Industrial Explosives Section,Explosives Laboratory,
Asahi Chemical Industry Co.,Ltd.

Abstract
Breakage caused by explosive gas pressure and astatic
demolition agent is considered to be caused by static
pressure. In this papers, we try to predict the progress
of a fracture zone by means of the stress transfer method—
one of the techniques of finite element method—in order to
get basic data about drill pattern and the expansive pressure
of the static demolition agent required.
Key words: No tension solution,One free face,Static
demolition agent.

1. Method of solution

No tension solution is one of the techniques of the stress
transfer mothod advocated by O.C.Zienkiewicz (1968). The
principle of this no tension solution is as follows:

   (a) Analyse the problem as an elastic one and compute
the principal stress in each element, adding any initial
stresses which may be present before the loading operation
began.

   (b) At the end of stage (a) it will be found that
certain tensile principal stresses have developed. As the
material is assumed incapable of sustaining tensions, they
are eliminated with out, houever, permitting any point in
the structure to displace. In order to maintain
equilibrium 'restraining' forces have to be temporarity
applied to the structure at this stage.

   (c) As the restraunung forces do not in fact exist
their effect has to be removed from the structure by
superposition of equal but opposite nodal forces. The
structure is now reanalysed for the effect of such forces
and the stress computed are added to those pertaining at
the end of stage (b) (when tension have eliminated).
During the application of the 'de-restraining' forces the
structural behaviour is again assumed elastic and when the

158

principal stresses are computed it will be found that tensions may still develop. These tensions will, however, be much reduced compared with the previous stage.

(d) If at the end of stage (c) principal tensions are still in existence step (b) and (c) are repeated until all stresses are reduced to a negligiable figure.

And the conditions on the stress field after fracture are proposed as follows:

(a) The element which has been fractured or yielded, cannot sustain tensile stress.

(b) The stresses which should be redistributed after compressive fracture (under compressive principal stresses) are determined depending on the states of stress beyond the fracture surface.

(c) For combined fracture (under both tensile and compressive principal stresses), firstly tensile stress is redistributed and then the compressive stresses over the fracture surface are redistributed according to condition (b).

The flowchart of the program coded under this principle is shown in Fig. 1.

Next we introduced a continuous boundary condition in order to observe the structure to be analysed as part of one free face structure [see Fig. 2]. According to the theory of two dimensional elasticity when a pressure "P" is loaded into a bore hole with a radius of "a" which is located in an infinite body, the radial displacement "Ur" which is generated at a place with a radius of "r" is described as Formula (1).

$$Ur= \frac{1 + \nu}{E} \frac{a^2}{r} P \qquad ---(1)$$

where, "E" is Young's modulus, "ν" is poissons ratio.

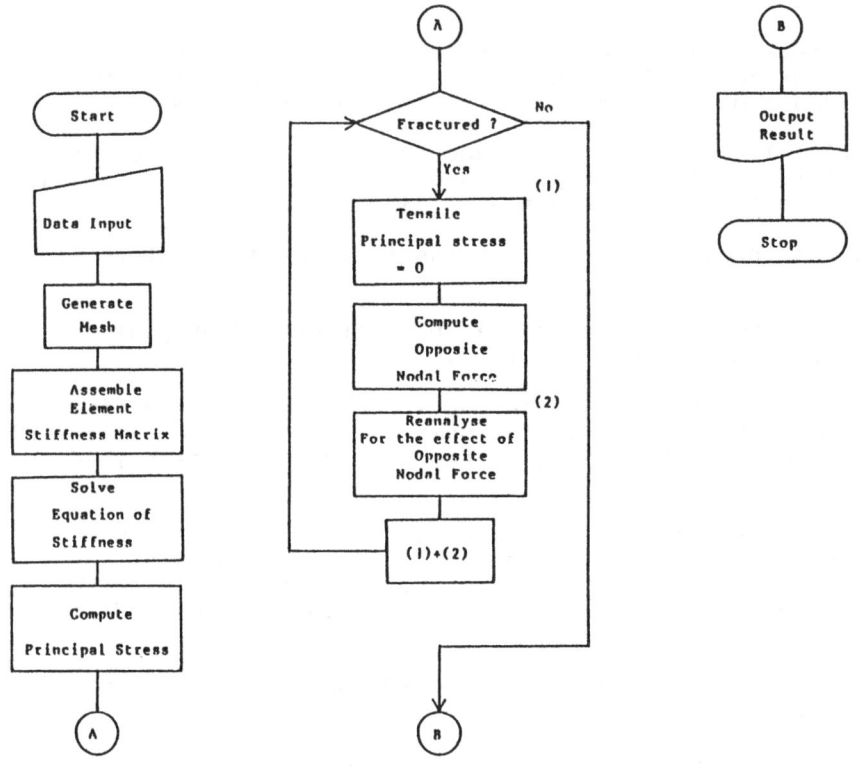

Fig. 1. Flowchart of the program.

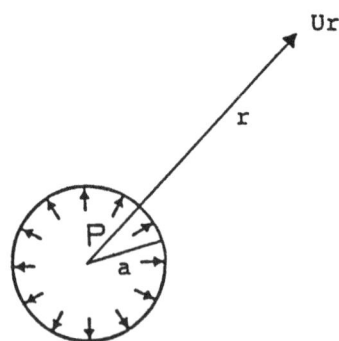

Fig. 2. The infinite body which has a bore hole with a
radius of "a" which is loaded expansive uniform
pressure "P".

2. Model for analysis

The model for analysis is presented in Fig. 3. This is a
cylindrical structure with a radius of 75 which has a bore
hole with a radius of 1 in the center of this structure.
The radial displacements "Ur" (given by Formula (1)) are
given to outer boundary in order to obserbe the structure
as   part of a one free face structure.
The elastic constants used for this analysis are as
follows:

    Young's modulus = 20.6 GPa
    Poissons ratio = 0.25
    Static uniaxial compressive strength = 29.4 MPa
    Static uniaxial tensile strength   = 2.94 MPa

where these constants are general constants of Concrete.

The configuration and mesh used in solution are shown in
Fig. 4.   The number of elements and nodes are each 224 and
128. The Mohr's envelope which was considered as a parabora
was used as failure criterion.

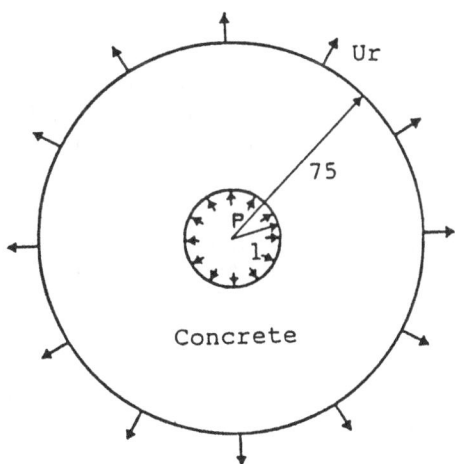

Fig. 3.   Model for analysis.

161

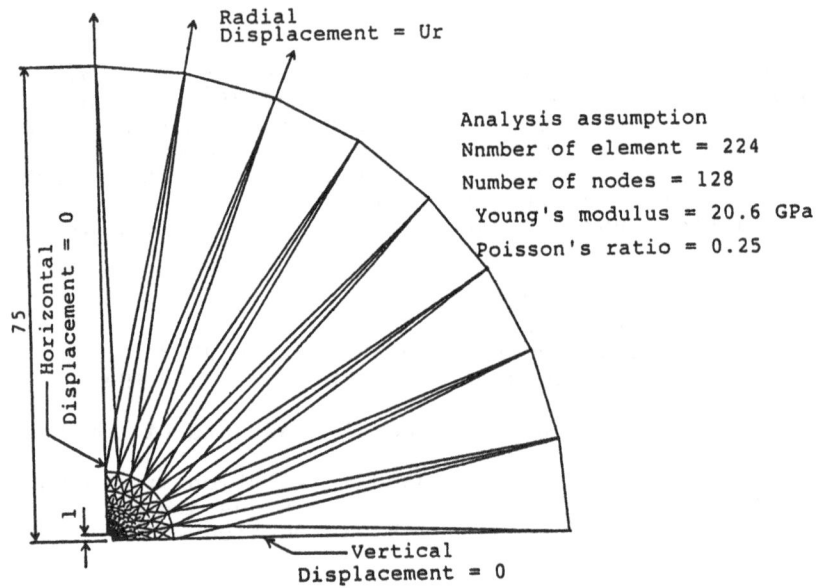

Analysis assumption
Nnmber of element = 224
Number of nodes = 128
Young's modulus = 20.6 GPa
Poisson's ratio = 0.25

Fig. 4.   Configuration of mesh.

3. Results

Fig. 5 to Fig. 11 respectively show the final 'no tension'
solutions with all regions, load step by load step.   The
dark shadings on the figures indicate fissured regions
produced by pressure loaded into bore holes. The
pressures loaded into bore holes were calculated by
multiplying the static uniaxial tensile strength by the
number of load step.   The fissured region that appeared on
the longitudinal axis at load step 1 was given as initial
crack in order to control the direction of crack
propagation.
On load step 1 expansive pressure as large as the static
uniaxial tensile strength "St" was loaded into the bore
hole, only the initial crack on the longitudinal axis was
present [see Fig. 5].
On load step 2 expansive pressure as large as "2St" was
loaded into the bore hole, a new fissured region was
produced near the transversal axis.   This crack was produced
by the stress concentration nearby the transversal axis
which was caused by the stress releasa around the initial
crack [see Fig. 6].
On load step 3 expansive pressure "3St" was loaded and the
next crack appeared halfway between the longitudinal and
transversal axes.   This crack was produced by the stress

162

concentration caused by the release of stress around the longitudinal and transversal axis [see Fig. 7].
On load step 4 expansive pressure "4St" was loaded, each of the three cracks expanded and the longest one reached 2.75 [see Fig. 8].
On load step 5 expansive pressure "5St" was loaded but the cracks did not expand.
On load step 6 expansive pressure "6St" was loaded and the crack on the longitudinal and transversal axis reached 4. At this load step the crack between the two axes stopped growing, i.e., stress became concentrated at the crack tips on the axes [see Fig. 9].
On load step 7 expansive pressure "7St" was loaded. Only the crack on the transversal axis expanded and reached 5.5 [see Fig. 10].
On load step 8 expansive pressure "8St" was loaded and this time, only the crack on the longitudinal axis expanded and became as long as the one on the transversal axis. After this load step these two cracks continued to expand dependant of one another [see Fig. 11].
On load step 9 expansive pressure " 9St" was loaded but the cracks did not expand.
On load step 10 expansive pressure "10St" was loaded and the cracks did not expand, too.

Fig. 12 shows the length of cracks as a result of the load conditions. In this figure, ○ shows the result calculated from the theory of two dimensional elasticity. □ shows the result calculated from the cylindrical model with a radius of 15 which has a bore hole with a radius of 1. △ shows the result calculated this time. The data from one free face calculation compared with that from the theory of two dimensional elasticity, almost corresponds to under load condition = 5. But where the load condition is over 5, the data value from one free face calculation increase slightly more than the data obtained by the theoretical calculation. Next we compared the data from the multi free face calculation with those from the theoretical calculation. There are noticeable differences between the former and the latter. This indicates that it is difficult to predict the length of cracks using the theory of two dimensional elasticity with respect to multi free face.

References

O.C. Zienkiwicz, et al., "Stress analysis of rock as a 'no tension' materials", Geotechnique,18(1968),pp.56-66.
T. Kawamoto and T. Saito, "Stress and stability analysis of underground openings taking post-failure behaviour of rock into consideration", Numerical Methods in Geomechanics, ASCE, 1976, pp. 791-801.

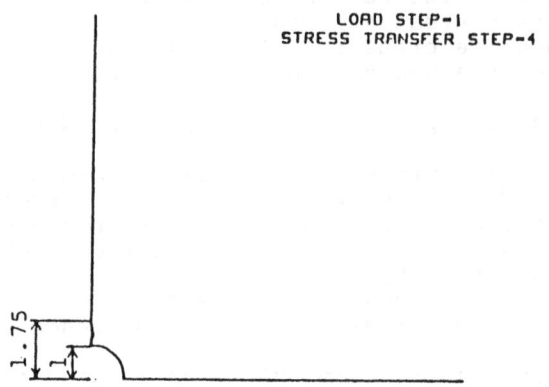

Fig. 5. Final 'no tension' solution at "P" = "St".

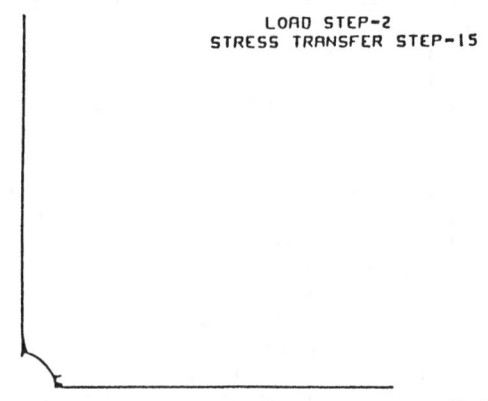

Fig. 6. Final 'no tension' solution at "P" ="2St".

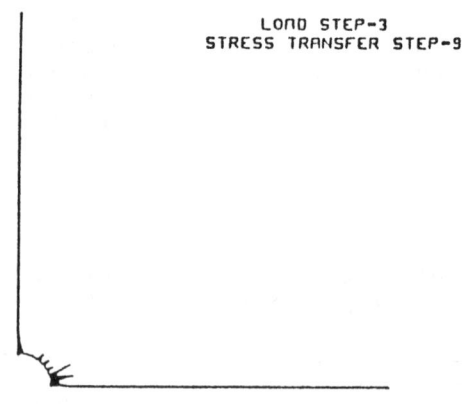

Fig. 7. Final 'no tension' solution at "P" = "3St".

164

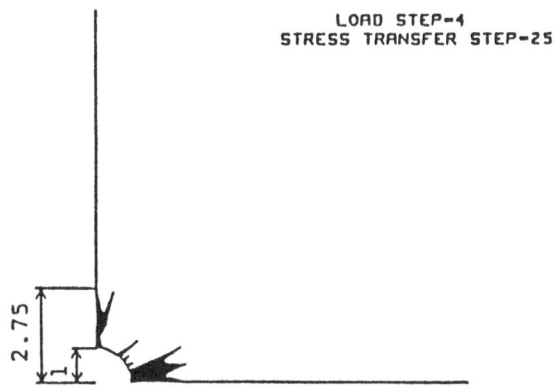

Fig. 8. Final 'no tension' solution at "P" = "4St".

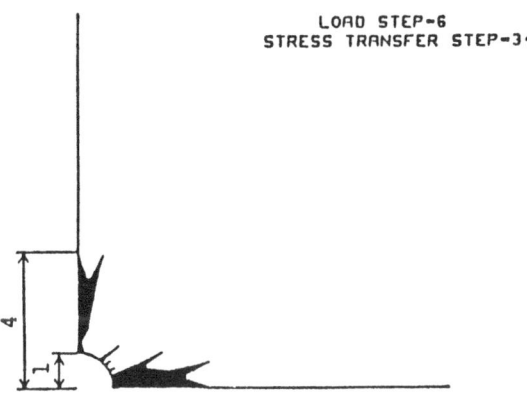

Fig. 9. Final 'no tension' solution at "P" = "6St".

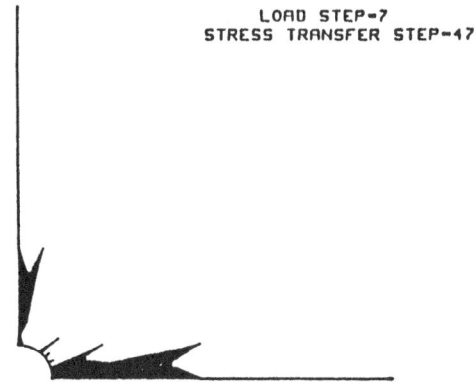

Fig. 10. Final 'no tension' solution at "P" = "7St".

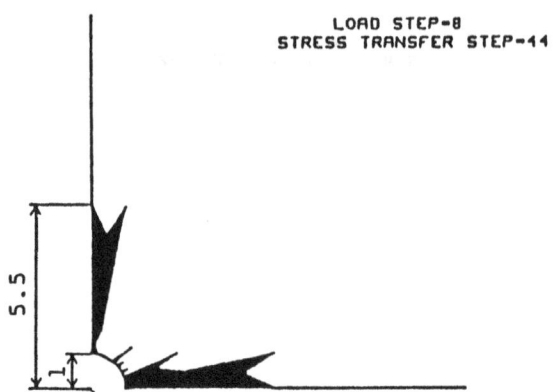

Fig. 11. Final 'no tension' solution at "P" = "8St".

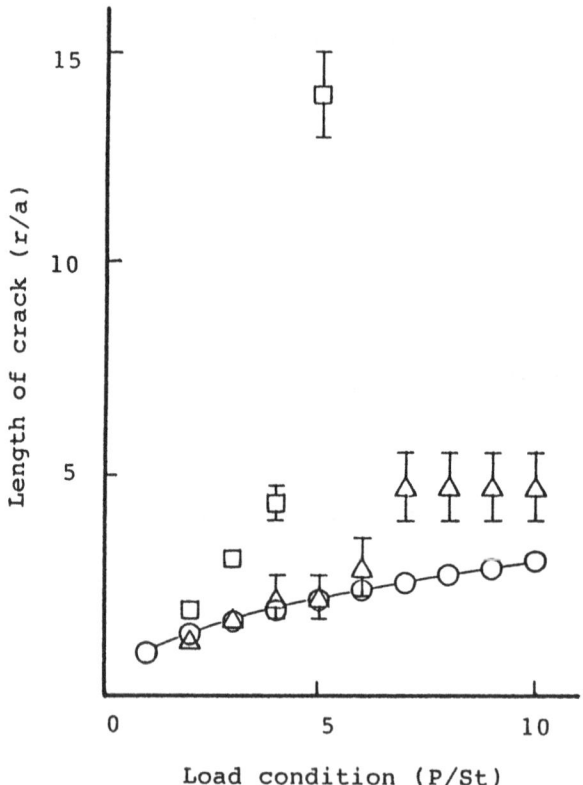

Fig. 12. Relations between the length of crack and
the load condition.

APPLICATION OF NON-EXPLOSIVE DEMOLITION AGENT FOR REINFORCED CONCRETE STRUCTURE

SHIRO ISHII and KOICHI SOEDA
Construction Materials Research Laboratory, Onoda Cement Co., Ltd.

Abstract

Recently, the demand for pollution-free demolition work of old reinforced concrete structure has rapidly increased as the redevelopment of urban area has been accelerated. Many demolition works of rocks and concretes using non-explosive demolition agent have been reported since this agent was developed in 1979. However, in terms of efficiency, only a few applications for demolishing reinforced concrete structure have been reported using this agent.

The object of this report is to establish the effective demolition method for reinforced concrete structure using non-explosive demolition agent. The efficiency of demolition work of reinforced concrete is expected to depends on several factors such as density of reinforcement, its arrangement and shape of the structure. Therefore, we carried out demolition experiments for several parts of reinforced concrete structure at first and then we applied these results for practical jobs.

Two types of non-explosive demolition agent were used in this experiment, i.e., powder type and granular type. Powder type agent mixed with water can generate cracks in 10-16 hours after poured into the hole, while granular type agent can generate cracks in only 30 minutes without blown-out shot after loaded into the hole.

Key words: Non-explosive demolition agent, Demolition work, Crack, Reinforced concrete, Expansive stress

1. Introduction

Non-explosive demolition agent is a substance which is used for cracking and demolishing brittle materials like concrete structure safely without environmental pollution. Up to the present, the heavy demolishing equipments such as iron balls and hydraulic breakers had been used in demolishing concrete structure. The use of them, however, has in most cases been restricted because these works create environmental problems such as noise, vibration, dust, and flyrock etc.. Therefore the development of safe and pollution-free technology for demolishing concrete structure has been required. In these circumstances, non-explosive demolition agent was developed with our own technology in 1979 and paid much attention to in many fields, and the applications of the agent have been spread worldwidely now.

The agent mixed with an appropriate quantity of water is filled into holes drilled or prepared by thin steel pipes before-hand. Then, hydrating gradually, it hardens and expands with the passage of time and creates cracks connecting between holes at about 10 hours after filling. These cracks make the secondary demolition and removing works easier. Therefore, the demolition method using this agent has superior points compared to other methods in terms of safety and environmental problems.

The fracture mechanism is as follows. After filling the agent into holes, the expansive stress is exerted onto the holes with the passage of time and compressive stress is produced. Simultaneously, the tensile stress in the horizontal direction to the compressive stress occurs. When this tensile stress exceeds the tensile fracture stress of the substance, cracks initiate. The tensile fracture stress of the brittle materials like concrete is from one tenth to one fifteenth the compressive fracture stress. So, if the hole design is proper with consideration of the restrict condition of the materials to be demolished in actual works, we can demolish most of concrete structures using the agent which produces more than 300 Kgf/cm2 of expansive stress. However, in demolishing reinforced concrete structures, the larger density of reinforcements and the larger restriction due to its arrangement make it difficult for cracks to propagate further.

The purpose of these experiments is to establish effective demolition methods for reinforced concrete structures using non-explosive demolition agent. In these experiments, we examined the situation of cracks and the efficiency of the secondary demolition, changing the degree of restriction by reinforcements.

2. Demolishing plain concrete

2.1 Experiment method for demolishing plain concrete
First of all, we examined the situation of cracks using plain concretes as specimens to obtain basic data.

We used "Bristar 100"(B-100) and "Super Bristar 1000"(SB-1000) produced by Onoda Cement Co., Ltd. as non-explosive demolition agent. Fig-1 shows the relationship between the expansive stress of the agents and elapsed time. The mix-proportion, the strength, and the size of concrete specimens used in this experiment are shown respectively in Table-1 and Table-2. Holes were drilled into the center of the concrete specimens. Hole diameter was ϕ40mm and ϕ60mm, and hole length was 80% of specimens' height.

We measured cracking time after filling the slurry of the agent.

Table-1 The mix proportion and strength of concrete specimens

max. size (mm)	sl (cm)	air (%)	w/c (%)	s/a (%)	weight of unit (kg/m³)					28 day strength (kgf/cm)	
					W	C	S	G	A·D	compressive	tensile
20	12	4.0	45	44.8	184	409	763	984	0.82	451	37.5

Table−2　The distance index to hole diameter

size dia.	40	50	60	70	80	90	100	110	120	130
∅ 40	5.0	6.25	7.5	8.75	10.0	11.25	12.5	13.75	15.0	16.25
∅ 60	3.33	4.17	5.0	5.83	6.67	7.5	8.3	9.16	10.0	10.83

distance index = the distance from the center of the hole to free surface/hole diameter

2.2 The results

The results of this experiment were shown as Fig−2. In the case of using B−100, cracks occurred at 10−20 hours when K value (the distance index to hole diameter) was 10−12, and when K value was over 12, cracking time was delayed. In the case of using SB−1000 cracks occurred in several minutes when K value was 7−8, however, when K value was over 8, cracking time was extremely delayed. The reason of this is that SB−1000 produces high expansive stress in very early stage, but the progress of it is smaller compared to B−100.

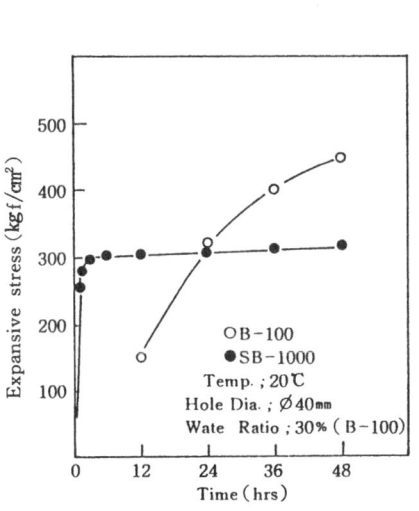

Fig 1　Change of the expansive stress with elapsed time

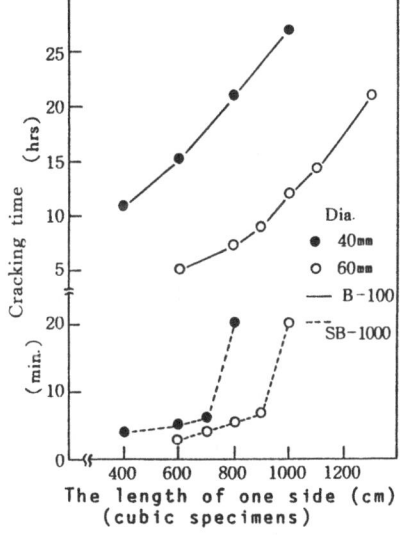

Fig 2　Cracking time (penetrating crack)

3. Demolishing reinforced concrete

3.1 Experiment method for demolishing reinforced concrete

Table−3 and Fig−3 show steel ratio and outline figures of concrete specimens used in this experiment. In demolition method for reinforced concrete, we decided to cut the specimens horizontally on assumption that we cut out the parts of structure such as pillar and beam into blocks. Therefore, we experimented with hole design shown

169

steel pipes in concrete, and then holes were obtained by pulling them
out after concrete hardened. Hole length was 80cm. The mix-proportion
of concrete was same as Table-1. and we carried out the experiment at
28 days after placing it using B-100.

The measuring items were observation of crack situation (cracking
time and propagation of it with elapsed time), and the strain change
of reinforcement due to expansive stress produced by B-100. We stuck
the strain gages on the main reinforcements and hoop reinforcements
before experiment.

The second demolition was carried out using hand breaker at 1 week
after cracks occurred. and we measured the efficiency of the second
demolition both on cracked parts by B-100 and the parts without
crack.

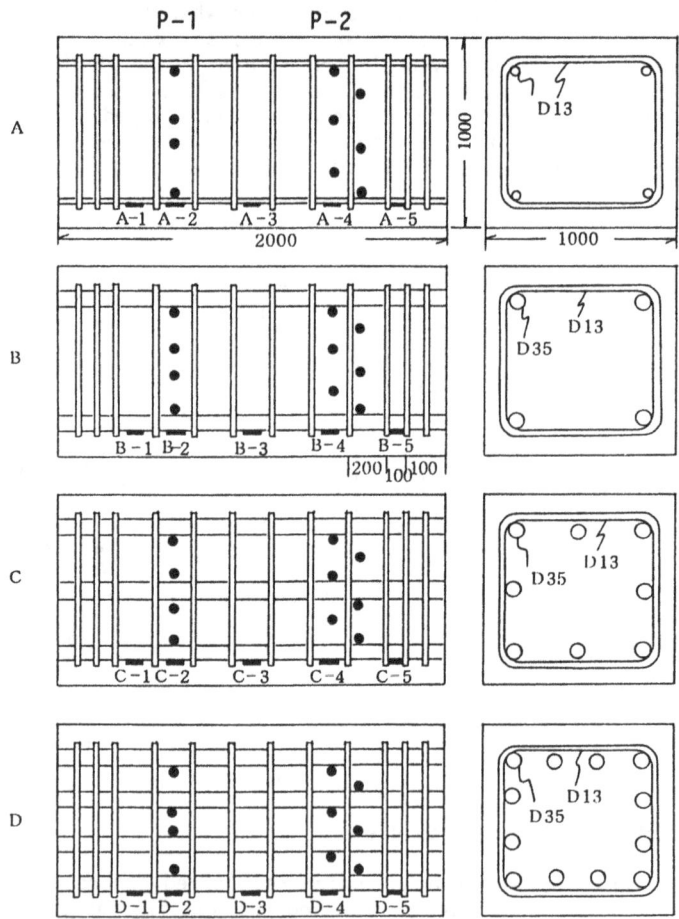

Fig 3   The outline figures of specimens
(situation of reinforcement, drilling pattern, position of strain gages)

170

Table−3 Steel ratio and density of reinforcement of concrete specimens

specimen	steel ratio (%) *	density of reinforcement (kg/m³)	comment
A	0.05	23.0	size ; 1000×1000×2000 mm
B	0.38	49.1	volume of concrete: 2.0 m³
C	0.76	79.2	making hole ; placing thin steel pipes
D	1.15	109.2	hole diameter ; 40mm , hole length ; 800mm

$$*\text{steel ratio} = \frac{\text{area of the main reinforcements}}{\text{gross area of concrete specimen}}$$

3.2 The results

Fig−4 shows the situation of cracks at 1 day after filling B−100. When the hole design was same, the situation of cracks was not changed, even though density of reinforcement was varied. The cracks generated as following way, and they were independent on steel ratio or hole design.

(1) The initial crack occurred diagonally from the closest hole to free surface toward the free surface, pushing out the main reinforcement.

(2) After that, cracks propagated to connect between holes.

(3) The width of the cracks generated around the outside holes tended to be slightly larger than that of cracks inside.

(4) crack of side surface initiated vertically and then horizontal cracks occurred radially.

The cracking time is shown as Table−4. Changes of the strain of main reinforcements and hoop reinforcements are shown as Fig−5 and 6. According to them, the larger steel ratio of specimens was, the later the cracking time became and the smaller both the width of crack and the strain of reinforcement became. The strain of reinforcement tended to reduce rapidly as the distance from the center of the hole became larger. Therefore, it is estimated that the extent of demolishing using B−100 would be 20−30cm from outside hole when density of reinforcement exceeds 30Kg/m3. This estimation coincided with the results of crack observation.

In specimens C, removing covering concrete upheaved by B−100, we exposed reinforcements and cut them. Then, we found out that the inner concrete can be completely demolished due to expansive stress by B−100.

Especially in the case of demolishing concrete with large density of reinforcement, since the width of cracks doesn't propagate further though the cracks occurred, the following procedure as shown Fig−7 is recommended to improve the efficiency of demolition work.

(1) The non−explosive demolition agent is filled into the holes just inside the reinforcement.

(2) The covering concrete is removed.

(3) The restriction due to reinforcement is released by cutting the exposed reinforcement

(4) The inside concrete is demolished using non−explosive demolition agent again.

171

In the case of the holes drilled in parallel to axial direction of the main reinforcement, the restricting reinforcement is not the main reinforcement but the hoop reinforcement located vertically to the holes, however, we can also crack because the hoop reinforcements have relatively weak restriction force.

Table − 4　Cracking time

(hours)

specimens	A		B		C		D	
hole pattern	P − 1	P − 2	P − 1	P − 2	P − 1	P − 2	P − 1	P − 2
initial crack	0.5	0.5	1.0	1.0	2.0	1.5	2.0	1.5
penetrating crack	2.5	2.0	2.5	2.0	4.0	3.0	5.0	4.5

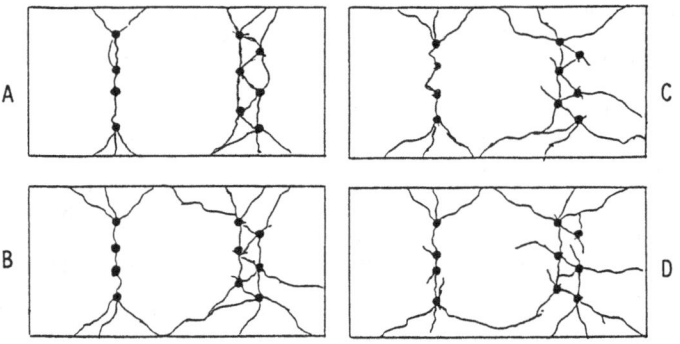

Fig 4　The situation of crack (after 24 hours)

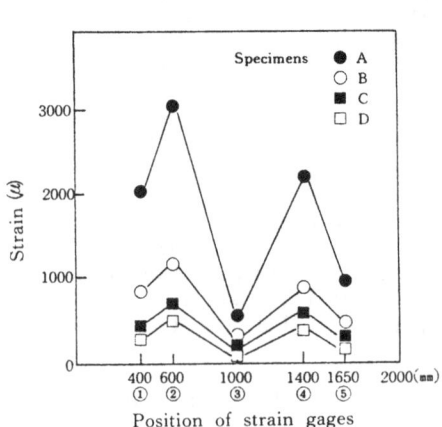

Fig 5　Strain distribution of main reinforcement
(after 12 hours)

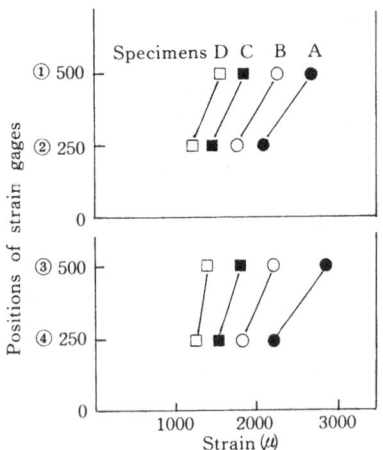

Fig 6　Strain distribution of hoop reinforcement
(after 12 hours)

4. Examples of the practical demolition work using "Bristar"

Generally, cracks develop more easily when free surface is increased and resistance is reduced. We need to design the demolition work to facilitate the development of cracks with consideration of free surface and resistance. Especially, it is necessary to examine density of reinforcement, its arrangement and the shape of structure, and in addition to them, we should consider about the purpose of demolition, required effect, methods of secondary demolition, and demolition period, etc..

"Bristar" has been used for demolishing reinforced concrete structures in various fields. We introduce several examples of demolition works using "Bristar".

4.1 Removing work of temporary foundation of steel tower

In removing work of temporary foundation of steel tower located in urban area, B-100 was used because it was necessary to reduce noise.
(1) Shape : Length(L) X Width(W) X Height(H) = 16m X 16m X 2m
(2) Reinforcement : D32 (JIS G 3112)   @ 300mm
                    thickness of covering concrete 100mm
(3) Density of reinforcement : 50Kg/m3
(4) Drilling, Hole diameter : placing thin steel pipes, ϕ 50mm
(5) Hole spacing : 600mm
But, as shown in Fig-7, two extra holes (hole spacing 2m) were drilled just inside of the outside reinforcement in order to remove covering concrete and cut the reinforcements.
(6) Amount of usage of B-100 : 8.9 Kg/m3
(7) Result
Cracks with 2-5mm width occurred at 15 hours after filling B-100, therefore we removed the covering concrete and cut the reinforcement.(at 2m interval) Consequently, the width of cracks propagated to 10-20 mm after 3 hours, and we carried out the secondary demolition using large hydraulic breaker. We could demolish 512m3 of reinforced concrete in 2 days.

4.2 Demolition work of rigid cutoff wall

In demolition work of rigid cutoff wall accompanied with new building construction work, B-100 was used in order to affect no serious damages on underground structure and obtain enough demolition efficiency. .
(1) Shape : L X W X H = 37,000mm X 800mm X 2,000mm
(2) Reinforcement : main reinforcement D35, hoop reinforcement D13
(3) Density of reinforcement : 58Kg/m3
(4) Drilling, Hole diameter : placing thin steel pipes, ϕ 40mm
(5) Hole spacing : @ 400mm
(6) Amount of usage of B-100 : 14Kg/m3
(7) Result
1-2mm width crack occurred connecting between holes at 12 hours after filling B-100 and propagated to 3-4mm at 24 hours. When we carried out secondary demolition by 20Kg-class hand breaker, we could obtain 2m3/hr of demolition efficiency, which was 4-5 times that of the method without B-100.

4.3 Demolition work of reinforced concrete bridge

The repairing work of damaged bridge pier had been carried out by demolishing damaged parts and replacing with new pier, suspending floor slab of road by large hydraulic jack. At first, small hydraulic splitter was used because large demolition equipments and explosives could not be used, but only small demolition efficiency could be obtained. So we decided to use B-100 to improve efficiency of demolition work.

    (1) Density of reinforcement : 113 - 165 Kg/m3
    (2) Volume of concrete to be demolished : approx. 30m3
    (3) Drilling, Hole diameter : handhammer (horizontal hole), ϕ 40mm
    (4) Amount of usage of B-100 : 14 - 16.5 Kg/m3
    (5) Secondary demolition : hand breaker
    (6) Result

1-2mm width cracks occurred at about 10 hours after filling B-100 and immediately secondary demolition was carried out using hand breaker. The results are shown as Fig-8. From these result, we could find out that the demolition method using hydraulic splitter required 91 persons, while the method using "Bristar" required only 32-51 persons to demolish 26m3 concrete. The demolition efficiency improved to the extent of 1/2-1/3 that of non-Bristar method.

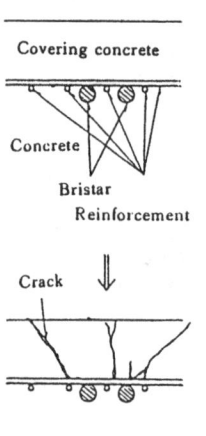

Fig 7 The demolition method for reinforced concrete structure using Bristar

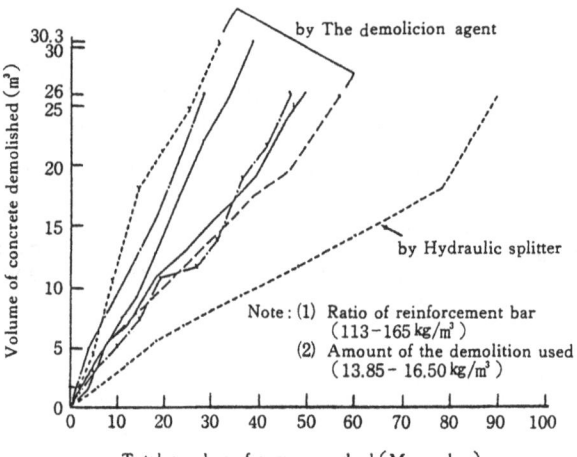

Fig 8 Breaking efficiency of reinforced concrete

4.4 Demolition of temporary reinforced concrete structure

B-100 was used for the demolition work of temporary reinforced concrete structure in an exhibition pavilion in order to improve demolition efficiency and reduce noise and vibration.

    (1) Shape : shape of bottom plane 8,000mm X 6,000mm ⌐
              height   5,200mm   ] trapezoid
              shape of surface plane 3,000mm X 3,000mm ⌐
    (2) Density of reinforcement : 100 Kg/m3

(3) Drilling, Hole diameter : thin steel pipes, ϕ 50mm
(4) Hole spacing : @ 500mm
(5) Amount of usage of B-100 : 13kg/m3
(6) Result

The crack situations at 16 hours and 24 hours after filling B-100 are shown as Fig-9. From this result, cracks occurred connecting between holes in 16 hours, and the width of it were 2-3mm. The cracks slightly propagated to 4-6mm in 24 hours. And then when we carried out secondary demolition by large hydraulic breaker, we could demolish concrete in only 4 hours.

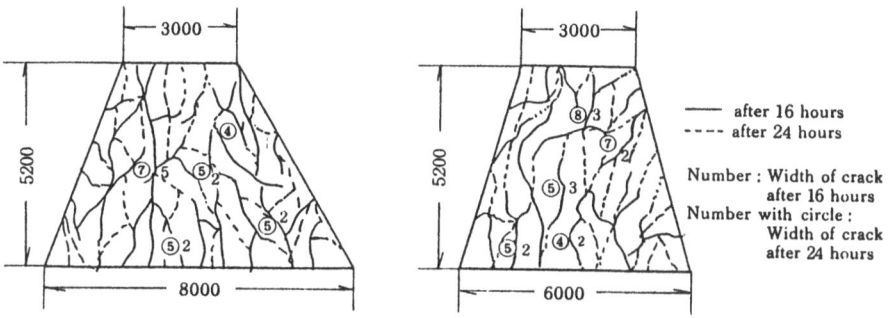

Fig 9 Demolition of temporary reinforced concrete structure

4.5 Demolition of floor slab and wall

In planning partial demolition work of slabs and walls to improve comfortableness in architecture, preserving the function of other parts of structure, B-100 was used to reduce noise and vibration.

a. Planning demolition of floor slab
(1) Density of reinforcement : 95Kg/m3
(2) Drilling, Hole diameter : core drill, ϕ 28mm
(3) Hole spacing : @ 100-150mm
(4) Drilling pattern

Since the thickness of slab was only 15cm, we drilled holes diagonally to obtain larger effective length and drilled partially to demolish as we planned.

(5) Result

Slab concrete was demolished upheavally at 10 hours after filling B-100 Immediately we removed concrete by hand breaker and exposed reinforcements and cut them.

b. Planning demolition of wall concrete
(1) Density of reinforcement : 60Kg/m3
(2) Drilling, Hole diameter : core drill, ϕ 40mm
(3) Hole spacing : @ 300-400mm
(4) Drilling pattern

As shown in Fig-11, we drilled continuous holes at the center of wall at first, and then drilled the several holes around the central hole's. We took a method that we drilled just above reinforcements using reinforcement surveying machine and cut them.

(5) Result

2-4mm width cracks occurred connecting between holes at 15 hours

175

after filling B-100. Breaking the concrete by hammer, we could easily penetrate the wall as we planned.

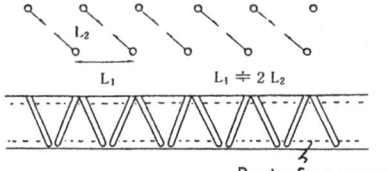

Fig 10   The drilling pattern in slab

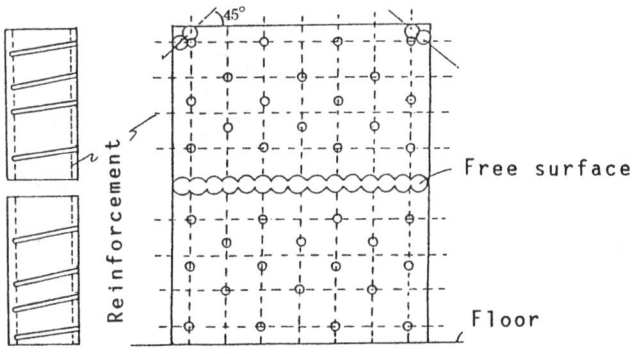

Fig 11   Drilling pattern in wall

5. Examples of demolishing other reinforced concrete structures

In addition to above-mentioned works, "Bristar" has been used for various kinds of demolition works of reinforced concrete structures, such as foundation of machines, retaining wall, abutment, head of concrete pile, pillar, beam, etc..

6. Conclusion

The demolition works of reinforced concrete using non-explosive demolition agent have been carried out often, but this method has also inferior points compared to other methods in aspect of economy and cracking time.

However, new type of non-explosive demolition agent and the demolition method, which can enormously shorten demolition period, have recently been developed, and this method has received good reputation steadily as more realistic demolition method.

In order to develop this safe and pollution-tree demolition method further more, we hope the people concerned with this method cooperate mutually.

Reference

Toshio Kawano, Japan.(1982) Gypsum & Lime, No.176, 41-48.

ABRASIVE WATERJET METHOD FOR CUTTING REINFORCED CONCRETE STRUCTURES

K. KONNO
Taisei Corporation

Abstract
This system is a method to cut hard material (reinforced concrete,
etc.) efficiently with an abrasive jet stream by mixing abrasive in an
ultra-high speed waterjet stream. Macroscopically, it is a kind of
tool with multi-blades, but there are several complex elements in
controlling cutting ability. The following three items are important
elements: 1) Special characteristics of the ultra-high speed jet
stream, 2) abrasives, and 3) cutting conditions. We clarified these
elements through the results at job sites and experiments at our
laboratory. This paper outlines the method, equipment, elucidation
of the elements and future subjects.
Key words: Abrasive waterjet method, Ultra-high pressure jet stream,
Abrasive, Cutting ability, Low level vibration, Low level noise.

1. Introduction

In recent years, there have been numerous instances where a part of a
building (wall, floor, etc.) has been dismantled to renovate an
existing building, update plant facilities, etc.

Where such dismantling is performed using the conventional method,
the effects on residents in the neighborhood of the consequent
vibration, noise, dust, etc., have been a problem.

Particularly in the case of medical institutions, hotels, etc.,
which operate 24 hours a day, the conventional method is difficult to
use for partial dismantling.

Therefore, in these facilities demand is increasing for dismantling
work using the abrasive waterjet method, which features a low level of
vibration, noise, and dust.

Our institute has been promoting the reduction of dismantling
costs, upgrading of dismantling technology, the establishment of a
dismantling organization system, and the application for practical use
of the abrasive waterjet method, hereinafter referred to as the A.W.J.
(Abrasive WaterJet) method.

This report describes the technical aspects of the practical use of
the AWJ method.

2. AWJ method

In this dismantling method, an abrasive is mixed in an ultra-high pressure water jet stream. The water streams at twice the speed of sound for cutting. The method makes use of the high cutting ability of the abrasive and has the following advantages:

1) With a sound arrester, it is possible to cut at a low level of noise.

2) A free alignment at an arbitrary position (horizontal, vertical).

3) It can be used in water or in air without touching the cutting object.

4) As the nozzle (type of the cutter) is light and the reaction of the jet stream is small, it can be used at an elevated location and in a relatively narrow space.

On the other hand, however, the method has the following disadvantages:

1) Dismantling cost becomes more expensive and the durability of the device is slightly poor.

2) Because of the killing and wounding power of the method, the greatest possible care must be exercised to secure safety during dismantling.

3) Slag (white wash, abrasive, etc.) must be disposed of.

3. AWJ cutting device

As shown in Fig. 1, the AWJ device consists of the ultra-high pressure jet stream generation device, the abrasive supply device, the control system, the travelling device, and the scattered material recovering device. Table 1 shows the specifications of the ultra-high pressure jet stream generation system.

Table 1. Specifications of ultra-high pressure jet stream generation device.

Specification / Type	40 EQ	40 ED	55 ET
Water pressure generated	2,800 kgf/cm^2	2,800 kgf/cm^2	3,800 kgf/cm^2
Discharge volume	21 liters/min	11 liters/min	11 liters/min
Motor output	150 kW (440V)	56 kW (440V)	94 kW (440V)
Weight	2,700 kg	1,700 kg	2,600 kg
Dimensions (length, width, height)	2,380 x 1,220 x 1,320 mm	2,400 x 920 x 1,100 mm	2,380 x 1,220 x 1,300 mm

Know-how is particularly important when dealing with the material, the processing accuracy and of the orifice, nozzle body, etc., of the nozzle section that is to be installed on the travelling device.

178

The control device is programmed in several modes to permit speed control for cutting reinforcing bars, shaped steel, pipes, etc., embedded in concrete. (Patent pending.)

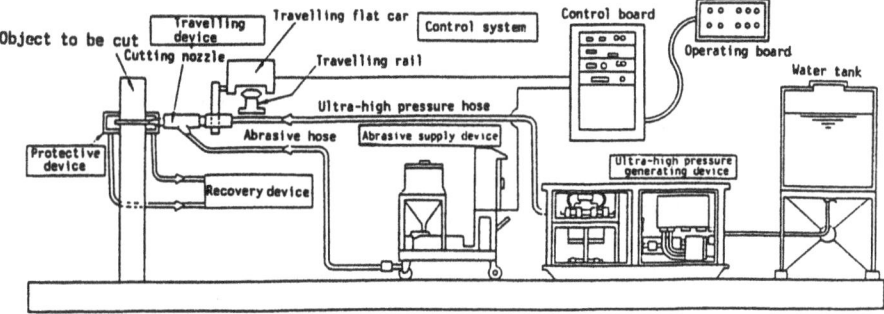

Fig. 1.   Composition of abrasive waterjet device.

4. Research and development for practical use

Basic and verifying experiments to find out the optimum cutting conditions and the cutting method that is essential to put this method to practical use have been repeatedly performed. Discussion on a part of the results thereof and the cutting ability follows.

4.1 Jet stream, abrasive

1) Jet stream - To cut deeper and at higher speeds using the AWJ method, it is an absolute condition that the best suited abrasive is efficiently mixed with the jet stream of the maximum power. Therefore, experiments to determine the maximum capacity (jet stream power) of the ultra-high water pressure generating device (type owned by Taisei: 40 EQ), were performed. In these experiments it was decided to select the optimum orifice diameter. Assuming water volume as Q (liters/min), pressure as P (kgf/cm^2), orifice diameter as d (cm) and orifice efficiency as η, the following formula was derived:

$$Q = 66 \times d^2 \times \eta \times \sqrt{P}$$

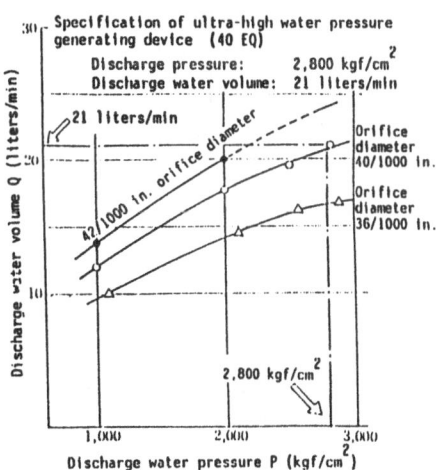

Fig. 2.   Discharge water pressure and discharge water volume curves by different orifice diameters.

179

By substituting orifice efficiency with 0.6-0.8, the orifice diameter d becomes d=0.1-0.08 cm at water volume Q=21 liters/min (max) and pressure P=2,800 kgf/cm^2 (max). From the above results of examination, the experiments were performed using three types of diameters, i.e. 42/1000 in., 40/1000 in. and 36/100 in., the results of which are shown in Fig. 2.

The only orifice that obtained water volume of 21 liters/min when the water pressure was 2,800 kgf/cm^2 was the one 40/1000 in diameter. Therefore, the optimum orifice for this type of equipment was decided to be the orifice 40/1000 in diameter.

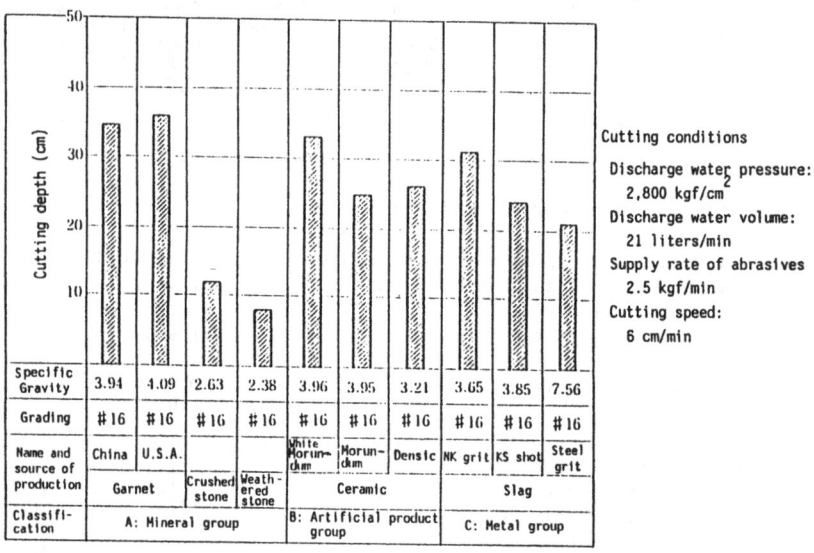

Fig. 3. Cutting depth of concrete by various abrasives.

2) Abrasives - In general, abrasives are classified into three groups, i.e. mineral, metal, and artificial. Although these abrasives differ in hardness, specific gravity, diameter, shape, etc., experiments on cutting ability were performed for several abrasives selected of the same diameter (#16) to find the abrasive with the best cutting ability irrespective of the physical properties. The results thereof are shown in Fig. 3. It can be seen from the figure that the garnet shows the best cutting ability.

Subsequently, experiments were performed to find out differences in cutting ability among the different grades of garnet.

The experiments were performed using #16 garnet from China, #16 garnet from the U.S.A., and #36 garnet from the U.S.A. at three cutting speeds: i.e. 6, 8 and 10 cm/min. Fig. 4 shows the results thereof. It can be seen from the figure that #36 garnet had the best cutting ability at respective cutting speeds.

180

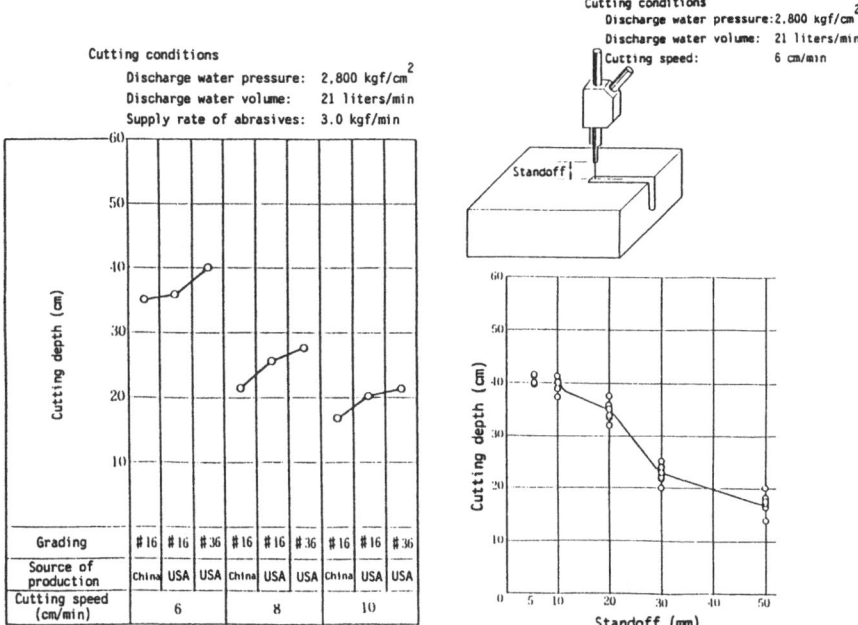

Fig. 4. Grading of garnet and the cutting depth.

Fig. 5. Standoff and cutting depth.

4.2 Cutting technique

Although jet stream power, selection of abrasives, etc. determine the cutting ability, the following two items become important in actual cutting:

(1) The spacing between the nozzle and the object to be cut (standoff).

(2) the angle of the nozzles and the object of cutting (nozzle angle).

Therefore, experiments to select the optimum standoff and experiments to verify the effect of the nozzle angle were performed, the results of which are shown in Fig. 5. It can be seen from the figure that although cutting depth does not vary between standoff from 5 mm and 20 mm, it decreases very much, by more than 10 mm. As shown in Fig. 6, no difference can be seen in cutting depth due to differences in nozzle angle. Another important element in cutting technique is whether the cutting of structures of the same thickness is to be accomplished by one cut at a low cutting speed or by several traverse cuts at a high cutting speed. In other words, how deep a cut can be accomplished by how many traverse cuts with the same charging energy in the same cutting time.

From the results shown in Fig. 7, it can be seen that up to structures of about 30 cm in thickness, the case of cutting at the traverse speed of 5 cm/min showed better results than the case of cutting twice at the traverse speed of 10 cm/min.

181

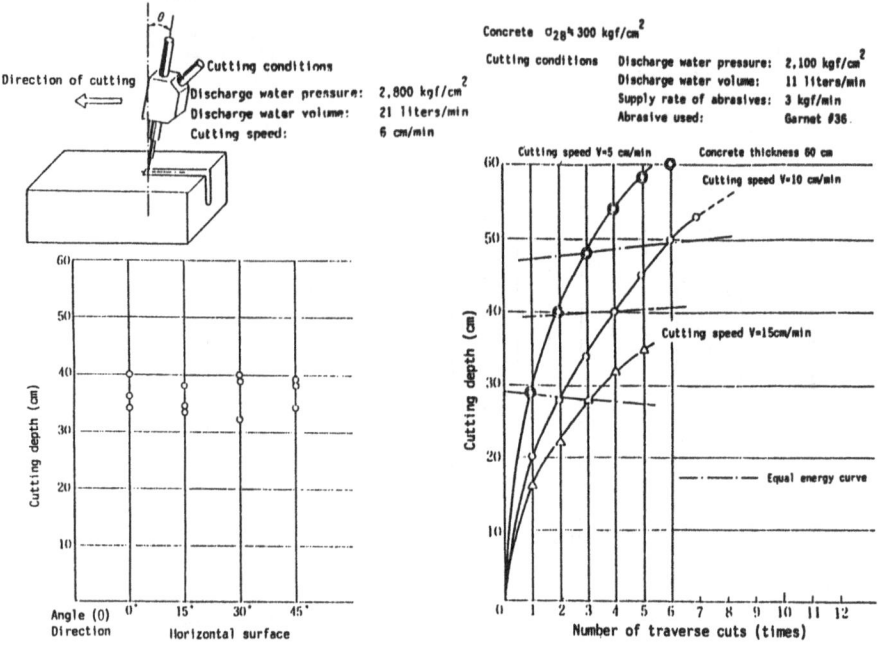

Fig. 6.　Nozzle angle and
　　　　　cutting depth.

Fig. 7.　Number of traverse cuts
　　　　　and cutting depth.

　Therefore, it is clear that structures up to 30 cm thick, one traverse can cut deeper than several repeated cuts within the same cutting time.　However, for structures thicker than 30 cm thick, it can be seen from the figure that better results were obtained from repeated traverse cuts.

　The optimum cutting conditions up to the present stage of research and development are summarized in Table 2.

Table 2.　Optimum cutting conditions.

Ultra-high water pressure generating device	Water pressure 2,800 kgf/cm^2, water volume 21 liters/min
Orifice diameter	40/1000 in.
Abrasive	Garnet, #36
Standoff	10 mm
Nozzle length	To be determined depending on the object
Number of traverse cuts	To be determined depending on the thickness of the object

182

4.3 Cutting ability (representative example)

Experiments to verify cutting ability were performed at our institute prior to the actual execution of the work.

In the experiments, reinforced concrete specimens of 210 kgf/cm^2 in unconfined compression strength having two layers of steel bars 16 mm in diameter were cut at varying cutting speeds.

Fig. 8 shows the results thereof.

Cutting conditions

Orifice diameter	40/1000 in.		
Nozzle diameter	3.5 mm	Type of abrasive	Garnet #36
Discharge water pressure	2,800 kgf/cm^2	Q'ty rate of abrasive used 3 kgf/cm^2	
Discharge water volume	21 liters/min	Number of traverse cuts	1
Cutting speed	2-10 cm/min	Direction of cutting	Downward, vertical
Standoff	10 mm		

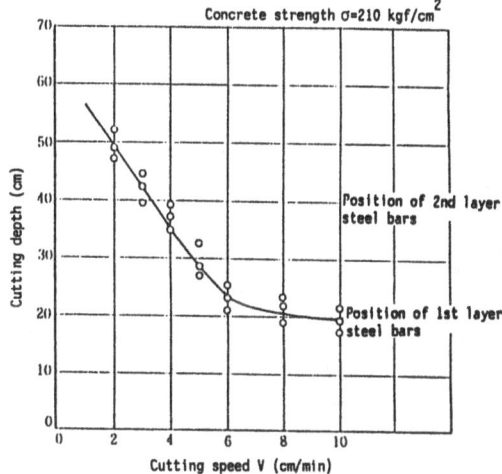

Fig. 8. Cutting depth and cutting speed.

It can be seen from Fig. 8 that at the cutting speed of 2 cm/min, cutting can be accomplished to the depth of 50 cm, while at the cutting speed of 10 cm/min, the 1st layer steel bars could not be cut and the cutting depth remained at 20 cm.

Fig. 9 shows a cross-section of cutting of levelling concrete at the cutting speed of 2 cm/min. It can be seen that although there is a small amount of levelling concrete yet to be cut, all the steel bars were completely cut.

Figs. 10 and 11 show a cross-section of cutting of levelling concrete at the cutting speed of 6 cm/min and 10 cm/min, respectively. From these figures, it can be seen that actual results are similar to the results obtained at our research institute. Walls with a thickness of about 15 cm (in the case of general structures) can be cut at the cutting speed of 8 cm/min.

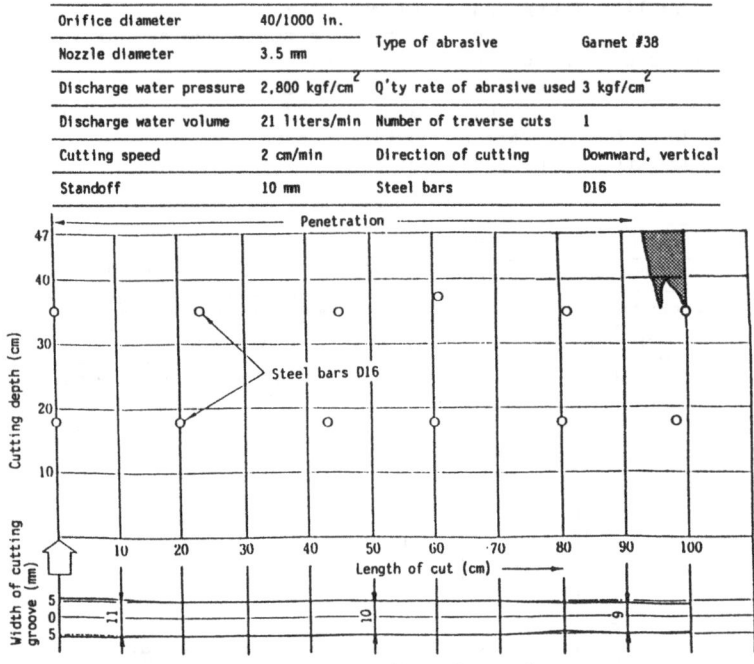

Orifice diameter	40/1000 in.		
Nozzle diameter	3.5 mm	Type of abrasive	Garnet #38
Discharge water pressure	2,800 kgf/cm²	Q'ty rate of abrasive used	3 kgf/cm²
Discharge water volume	21 liters/min	Number of traverse cuts	1
Cutting speed	2 cm/min	Direction of cutting	Downward, vertical
Standoff	10 mm	Steel bars	D16

Fig. 9. Cross-section of cutting.

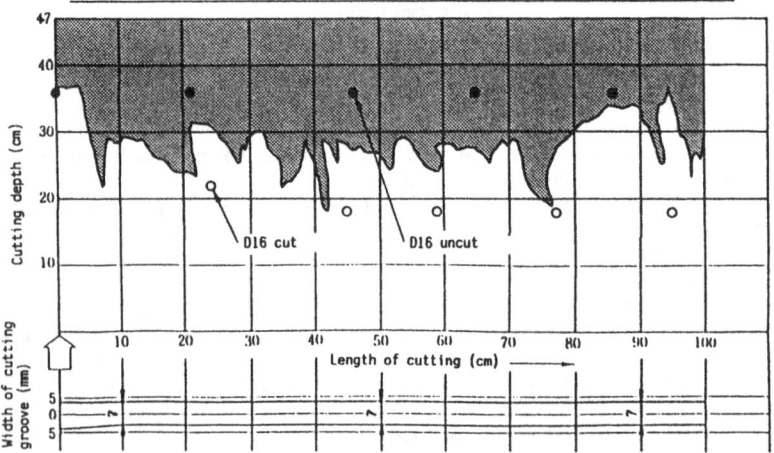

Orifice diameter	40/1000 in.		
Nozzle diameter	3.5 mm	Type of abrasive	Garnet #36
Discharge water pressure	2,800 kgf/cm²	Q'ty rate of abrasive used	3 kgf/cm²
Discharge water volume	21 liters/min	Number of traverse cuts	1
Cutting speed	6 cm/min	Direction of cutting	Downward, vertical
Standoff	10 mm	Steel bars	D16

Fig. 10. Cross-section of cutting.

184

Orifice diameter	40/1000 in.		
Nozzle diameter	3.5 mm	Type of abrasive	Garnet #36
Discharge water pressure	2,800 kgf/cm^2	Q'ty rate of abrasive used	3 kgf/cm^2
Discharge water volume	21 liters/min	Number of traverse cuts	1
Cutting speed	10 cm/min	Direction of cutting	Downward, vertical
Standoff	10 mm	Steel bars	D16

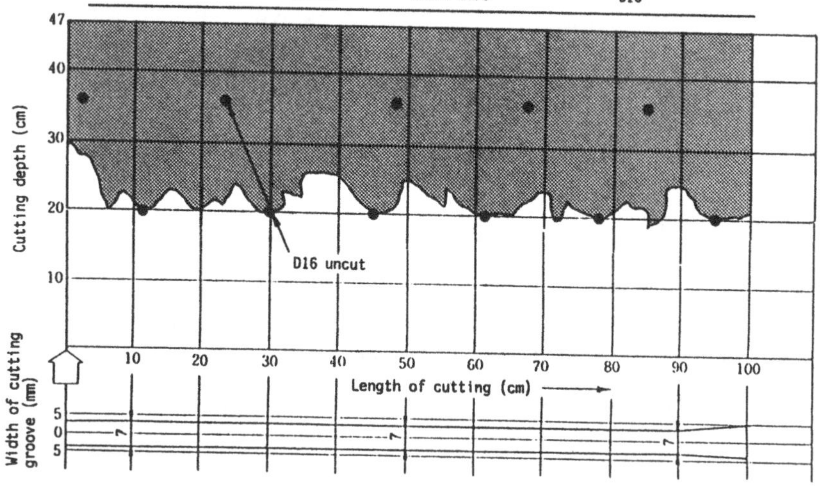

Fig. 11. Cross-section of cutting.

5. Work experience

Table 3 lists the examples of work using the AWJ method. Photo 1 shows the cutting condition of an 150 mm thick wall from bottom to top. Photo 2 shows the cutting condition of a levelling concrete 470 mm thick.

Table 3. Example of work using the AWJ method.

Location of work	Cutting location	Reason for adopting AWJ
Hotel	Wall at lobby, floor stairs, beam, etc.	Low noise, no vibration
Hospital	Partition of operating room	Low noise, no vibration, no dust
Iron works	Floor of precision machine room	No vibration, no dust
Newspaper company	Partition wall	Low noise, no vibration
Important structures	External wall, etc.	Cutting of curved section of thick wall

185

Photo 1.   Cutting condition of   Photo 2.   Cutting condition of
           150 mm thick wall   levelling concrete
           from bottom to top.   470 mm thick.

6. Conclusion

Thus far research and development for putting the AWJ method to
practical use have been described. Although research and development
of robotization to lower the cost of dismantling is required, we
believe it has already reached the technological level where it can be
used relatively easily.

In the future, research will center on cutting devices for
large-scale structures that make dismantling easier.

PARTIAL DISMANTLING OF A HOSPITAL BY WATER JET SYSTEM

YUSUKE MATSUSHITA
KUMAGAI GUMI CO., LTD.

Abstract
"Medical services are normally continued in some rooms, while the area
adjacent to them is being dismantled." The most recent dismantling
technology has enabled to achieve such incredible scheme.
This paper describes why and how the K-Jet system was used in the reconstruction of Tokyo Kosei Nenkin Hospital (Welfare Annuity Hospital
constructed at central Tokyo in 1952), showing also detail of the
partial dismantling work.
Key Words: abrasive water jet, dismantling, cutting

Introduction

Needless to say, the easiest solution in reconstruction is to build a
new building in a vacant land. But now, it is difficult to find such
a space, due to the outrageous land prices. Accordingly there is increasing need for solutions which enable reconstruction in the initial
site without ceasing medical activities. The Tokyo Kosei Nenkin Hospital, the first phase reconstruction of which is now under way, was
designed by Mr. Mamoru YAMADA and built in 1952, famous especially for
its long history of the orthopedic surgery.
    Recently, the space per bed in this hospital has been reduced be-
cause of the introduction of more advanced medical instruments. On
the other hand, the equipments such as air conditioning, pipings for
water feed and drainage, pipings for medical services are deteriorated.
These are two main reasons for the reconstructing project. The method
selected for this project was, taking into account the requisition mentioned above, to execute dismantling and new construction alternately,
displacing for a certain period inpatients to a zone in the hospital
and continueing the normal medical services.

Outline of the work

The work discussed here is the first phase work in the whole reconstruction project. This phase plans dismantling of the South Block of
the 5 storied main buildings constituted of three wings (South, West
and East Blocks). Table 1 shows the outline of the work.

187

Photo 1

Aeroview of
 Tokyo Kosei Nenkin
 Hospital

Table 1. Outline of the work

Work name	Reconstruction of Tokyo Kosei Nenkin Hospital
Work site	23 Tsukudo-cho, Shinjuku-ku, Tokyo
Client	Social Insurance Agency
Architect	Mamoru YAMADA Architectural Office
Supervisor	ditto
Execution	Joint venture of Kumagai-gumi, Mitsui, Matsui and Tatsumura-gumi Co., Ltd.
Work period	December 1986 to March 1989 (First phase)
Building area	approx. 10,000 m2
Total floor area	approx. 50,000 m2
Structure	composite structure of reinforced concrete and structural steel, 4 floors underground 9 floors above ground

Problems in planning of the dismantling work

Severe constraints are imposed on the work, as the destruction of about 1/3 of the existing building should be performed in such a condition that there is only a temporary partition panel between the area where the hospital staff and inpatients live and the portion to be dismantled. Accordingly, the client, the hospital and the architect required that extreme caution should be exercised not to give disagreeableness or uneasiness to the hospital staff and inpatients. Causes of disagreeableness and uneasiness may be:

(1) Vibrations and noises
(2) Production of dust
(3) Influences on the structure, such as cracking in the existing building (The portion remaining after dismantling must be durable enough to withstand the elements for 3 to 7 years after the reconstruction.)

Selection of the work method

Main problems to be solved in selecting the work method are:

- the three problems mentioned above;
- removal of about 5000 m2 (about 1500 m3) of concrete during the work period (3 months); and
- sufficient safety.

Considering the reasons described later, we established a plan that the dismantling shall be performed totally above the ground, using a hydraulic crusher equipped with a extra-long boom.
Moreover, we studied methods for isolating the portion to be dismantled from the portion to be remained, as the work projected is a partial dismantling of a building where the normal activities are continued. The five alternatives chosen as available methods are:

(1) hand breaker,
(2) cutter drum,
(3) core-drill continuous drilling,
(4) water jet, and
(5) flame jet.

These methods were assessed globally from the viewpoint of environmental problem, workability, influence upon the remaining building, cutting performance. As shown in Table 2, summarizing the result of such assessment, the breaker was rejected (symbol in the table: D) as it exerts undesirable influence upon the environment and upon reinforced concrete around the cutting part; the cutter drum is less desirable (C) in the problem of noise and work performance; the core drill presents no problem except in the working efficiency but requires higher cost than the water jet. Consequently, we judged the water jet method, using a hydraulic crusher also, is optimum for the projected dismantling.

189

Table 2. Assessing dismantling methods

Methods	Dismantling principle	Environmental problem			Workability			Influence upon the reinforced concrete	Execution performance			Global evaluation
		Noise	Vibration	Dust	System size	Reaction	Safety		Cutting depth	Cutting speed	Working ratio	
1 Hand breaker	Crushing by shock using oil hydraulic force or compressed air	D	D	D	C	C	D	D	C	C	C	D
2 Cutter drum	Cutting by rotating diamond bit	D	A	A	C	B	B	A	B	D	D	C
3 Core-drill continuous drilling	Continuous boring holes using core drill	B	A	A	C	C	B	A	A	D	D	B
4 Water jet	Cutting by ejecting under extra-high-pressure the mixture of water and abrasive	B	A	A	B	A	B	A	A	B	C	A
5 Flame jet	Fusion with burning rod of iron and aluminum alloy	D	A	D	B	A	D	D	B	C	D	D

A : Excellent   B : Good   C : Less desirable   D : Undesirable

Water jet method
- Outline of the K-Jet system -

"Cutting solids with water" is now the focus of attention in various fields. The water jet is not only attractive for its innovative process but also advantageous in many points.

For example, it ensures effective cutting of concretes, rocks and metals, as well as excellent linearity and accuracy free from thermal deformation in cutting. Considering the excellent cutting performance inherent in the water jet, we have developed the K-Jet system (KUMAGAI-Water Jet Cutting System). In this system, the cutting performance is, due to abrasive mixed in water, three or four times higher than the ordinary water jet. The K-Jet system is composed of, as depicted in Fig. 1, extra-high-pressure water jet pump, abrasive feed device and cutting device with jet nozzle.

The jet pump incorporating a booster produces extra-high-pressure water, i.e., 3000 kgf/cm2 of discharge pressure at maximum. The extra-high-pressure water produced is led to the cutting device through an extra-high-pressure hose (actually, the maximum length of the hose usable is about 100 m, taking into account the pressure drop due to resistance on the tube wall and to the joint metals). The cutting is remotely controlled by oil hydraulic system or electric motor, allowing speed regulation as desired. Abrasive is fed automatically and constantly in quantity to the nozzle, as shown in Fig. 2, and mixed there with water, then ejected at a speed of Mach 2 through the abrasive nozzle.

The cutting performance of the K-Jet system is given in Fig. 3: ejecting pressure 2500 kgf/cm2, discharge rate 14 lit./min, cutting speed 5 cm/min, maximum cutting depth 300mm. In comparison with the conventional cutters and the breakers, this system features these merits:

(1) cutting without vibration, exerting no influence, crack for example, upon the part other than the portion subject to cutting
(2) accurate cut corresponding to the cutting scheme, regardless of the thickness of the object
(3) protection against scattering of jet water and sufficient noise absorption (55 dB in the adjacent room), using a special cover and vacuum system
(4) capable of cutting reinforcing bars and structural steels
(5) no dust produced in execution

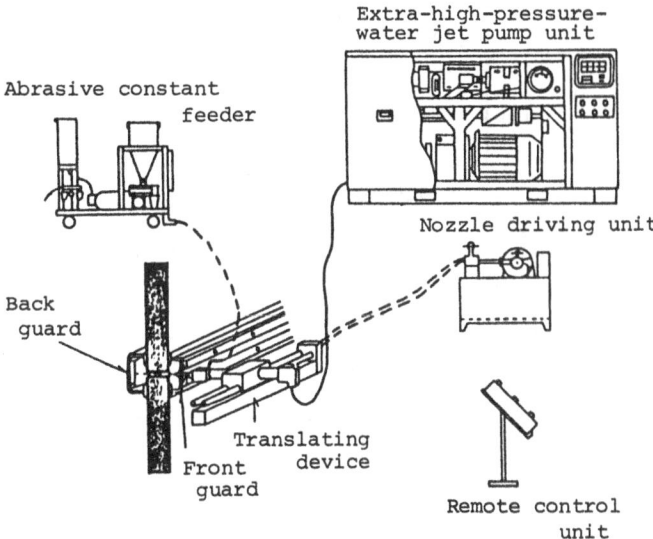

Fig. 1   K-Jet system

191

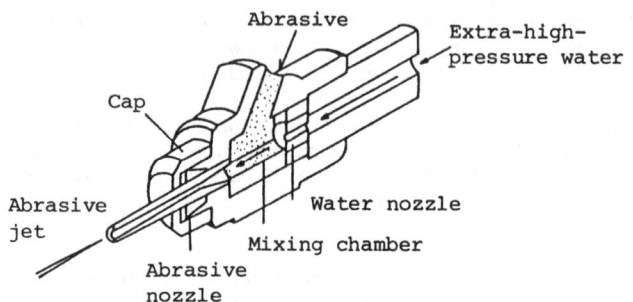

Fig. 2   Section of the nozzle

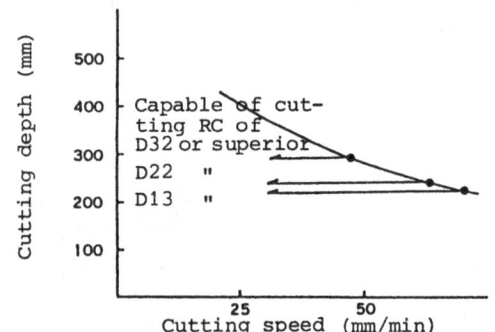

Pressure	: 2500 kgf/cm2
Discharge rate	: 14 lit./min
Nozzle distance	: 10 mm
Abrasive	: 4 kg/min
Compressive strength of concrete	: 270 kgf/cm2
Covering depth of Re. bar	: 100 mm

Fig. 3   Concrete cutting performance

Work process

Prior to the cutting with the water jet system, ALC panels (t=80) for exterior wall were installed on the portions which will be open after dismantling. After setting the ALC panels which serve as shield when dismantling, the remaining portion and the portion to be dismantled were, from the roof to the ground floor, separated completely. Then, the dismantling was executed by the hydraulic crusher.

Dismantling by the K-Jet system

The building to be dismantled was of doubled flooring structure having beams in the lower part in the 2nd and 5th floors. The upper floors were removed before beginning dismantling by means of the K-Jet cutting system, and cutting of the floors and the beams was made in such a manner as shown in Fig. 4. When cutting normal walls, we adopt a cutting device and guards shown in Fig. 5, the latter of which are destined for treating cutting water and insulating noise completely.

The part of the building which has been cut through this project is partially shown in Fig. 6, and the materials hereof are provided with the specification which follows:

Floor	: 13 cm in thickness
Reinforcing bar	: 9 mm in diameter
Girder	: 30 cm in width, 78 cm in height
Reinforcement	: doubled type with reinforcing bars of 13 in diameter

Taking into consideration the above, the K-Jet system was operated with the following conditions:

Cutting speed	: 5 cm/min
Injection pressure	: 2500 kgf/cm2
Added abrasive (garnet)	: 3 kg/min

The cutting speed is adjusted according to the thickness of the cut part and quantity of the reinforcing bars, but 400 m in total was cut with an average speed of 5m/min. The period required for the execution of the K-Jet was 2 months from the beginning of January to the end of February 1987. The virtual working ratio of the K-Jet system was not so high, that is, 40%, due to the hospital-related specific conditions such as long waiting time and frequent change of preparation procedures.

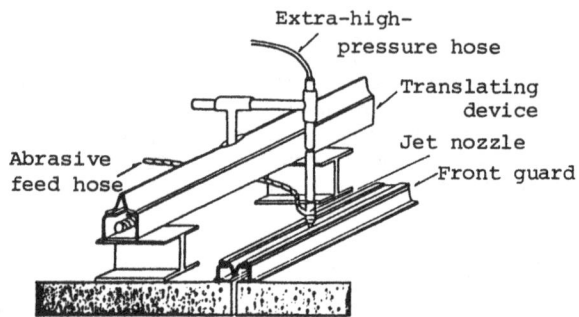

Fig. 4 Schematic view of the K-Jet system (1)

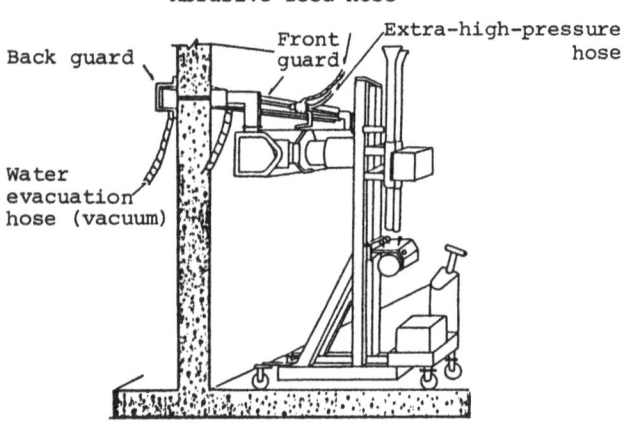

Fig. 5 Schematic view of the K-Jet system (2)

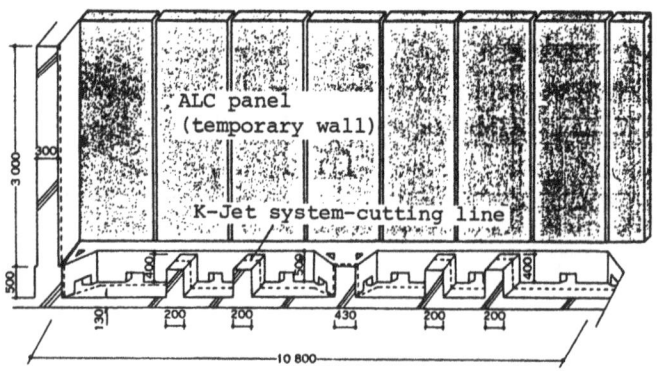

Fig. 6 Sectional view of the shear wall

194

Photo 3   View of the site under the cutting
works by the K-Jet System

Photo 4   Cutting of beams, wall and floor
by K-Jet System

Conclusion

This development of a new technology, K-Jet system, is intended for performing, as a trial, a large scale dismantling for the hospital building with inpatients in it, ensuring at the same time environmental requirements (no production of environmental disconfort, uneasiness) for the benefit of workers or inpatients concerned (Photos 3 and 4).

Concerning vibration generated from dismantling works, almost no vibration was produced because of an insulation established through K-Jet system. Such objects, as far as they concern the working technology, could be achieved successfully by fulfilling the initial requirements, but it remains problems to be studied and solved from the cost-related view point; for example, shortening of time needed for change of preparation steps, reduction of consumables such as abrasives and nozzles. Nevertheless, when reviewing all the processes have been satisfactorily completed without any accidents, we are convinced that our experiment has attained the desired targets, with hope of further improvement of the K-Jet system and its enhanced feasibility into many fields.

DISMANTLING TECHNIQUE FOR REACTOR BIOLOGICAL SHIELD
- ABRASIVE-WATERJET

M. YOKOTA, H. NAKAMURA and T. KONNO   Department of JPDR
                                  Japan Atomic Energy Research Institute
H. YOSHIDA, M. MIURA   Kajima Institute of Construction Technology
and Y. MIYAZAKI   Kajima Corporation

Abstract
The decommissioning program of the Japan Power Demonstration Reactor is being performed. For the purpose of dismantling the biological shield of JPDR, an abrasive-waterjet cutting system has been developed taking into consideration both minimization of personnel radiation exposure and prevention of the spread of the radioactive materials. This paper describes not only the features and the components of the system but also the way of dismantling the biological shield by this system.
Key words: Decommissioning, JPDR, Biological shield, Abrasive-waterjet

1. Introduction

Based on the national policy of the decommissioning way for the nuclear facility, the decommissioning program of Japan Power Demonstration Reactor (JPDR) has been performed by Japan Atomic Energy Research Institute (JAERI) under a contract with the Science and Technology Agency. The objectives of the program are to develop the decommissioning technology in preparation for the decommissioning of large nuclear power plant in the near future and to get the knowledges and actual experiences concerning the reactor decommissioning. Therefore, this decommissioning program covers two phases, which consist of development of the decommissioning technology (phase I) and the physical dismantlement of JPDR (phase II). The most important feature in the dismantlement of JPDR is to cut the reactor vessel in small pieces rather than to remove that in one piece. Technology development for the decommissioning of JPDR is almost completed and its physical dismantlement was commenced from December, 1986.

In general, concrete structures of the nuclear power plant could be roughly devided into the biological shield surrounding a reactor and the buildings themselves. Of these structures much attention must be paid in dismantling the biological shield. Because the biological shield is radioactivated with irradiation of neutron during the operation of a reactor, it is difficult for workers to dismantle that directly in proximity. Furthermore, with regard to the structure, the biological shield is a massive and strong structure made of the reinforced concrete to satisfy the requirements for radiation shield and the resistivity against earthquakes. Therefore, remote dismantling technique having a high performance compared with the existing tech-

197

niques is needed in dismantling the biological shield. At the same time, special measures for preventing spread of radioactive contamination are needed in order to maintain radioactive waste and the influence on the environment as low as possible.

Under the background described above, a total of four techniques have been developed by JAERI since 1981 as the dismantling methods for the biological shield. The abrasive-waterjet technique, which is planned to be used to the highly-activated region, will be focused in this paper. In particalar, the features and components of the cutting system and its dismantling procedure are described.

2. Fundamental testing on abrasive-waterjet cutting

For the purpose of applying the abrasive-waterjet cutting technique to the dismantlement of activated reactor biological shield, fundamental data to be used in the design of the cutting system have been accumulated through the testing on characteristics of abrasive-waterjet cutting. A part of the testing results is introduced in the following.

2.1 Experimental setup and procedure

The equipment used in the cutting testing was composed of a nozzle which jet a mixture of water and abrasive, a nozzle holder, a high-pressure pump, an abrasive supplying unit, a nozzle holder drive unit, and a slurry collector. Of these the high-pressure pump had the capacity to supply 50 l/min of water with a pressure of 196 MPa.

Cutting testing was performed using reinforced concrete specimens simulating the biological shield. In this testing, cutting characteristics and generation of secondary products were investigated. The parameters involved in the investigation were the abrasive material, the abrasive size, the abrasive flow rate and the number of cutting passes, and the standoff distance were held constant in all the testing. The depth of cut was measured after each test at 1 cm intervals.

2.2 Testing results

(1) Effect of number of cutting passes on the depth of cut

In the case of dismantling radioactivated biological shield, whether the cutting conditions are good or not exerts a great influence on the working time of the dismantlement as well as generation of waste. In order to select the optimal cutting conditions for reinforced concrete, the effect of the number of cutting passes on the depth of cut was investigated.

Figure 1 shows the test results with the cumulative depth of cut in the concrete region at three different nozzle traverse rates. The consecutive figures by the curves stand for the number of cutting passes. It is found in the figure that the depth of cut per unit energy is increased by increasing the number of cutting passes with the nozzle traverse rate made faster.

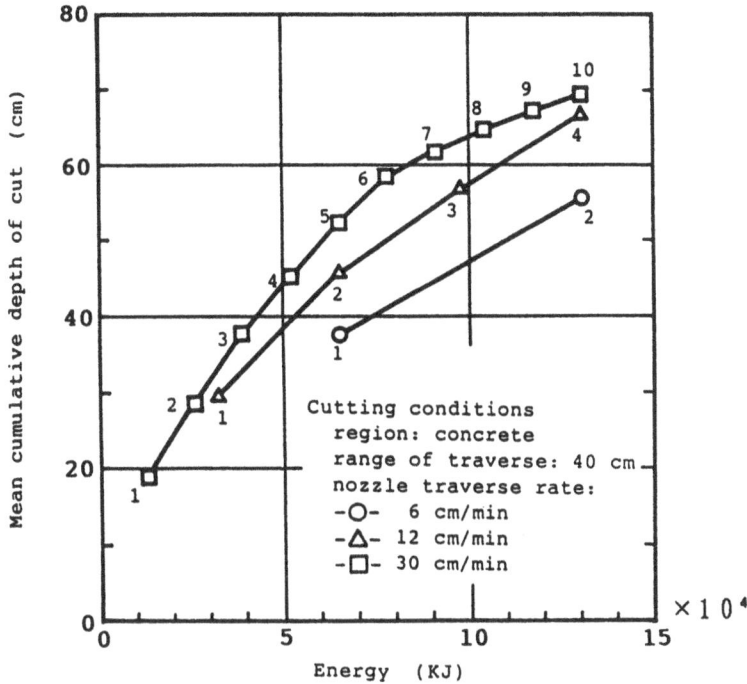

Fig. 1.   Effect of number of cutting passes
on the depth of cut

(2) Reinforcing bar cutting as a function of the embeded depth

In order to investigate the cutting performance of abrasive-waterjet
to the steel bars having different embeded depth, reinforcing bar cutting was carried out as a function of the nozzle traverse rate and the
number of cutting passes.
   The influence of the embeded depth on the real cutting rate of the
steel bars having different embeded depth is shown in Fig. 2.   The
word "real cutting rate" stands for the traverse rate devided by the
number of cutting passes which took for the cutting of the steel bar
at a certain position from the surface of the reinforced concrete
specimen without steel plate, and means the rate capable for cutting a
steel bar by one pass.   For the bar having shallow embeded depth, as
the figure shows, the real cutting rate becomes large with increasing
the nozzle traverse rate.   However, the real cutting rate decreases
with an increment of the embeded depth owing to the increased number
of cutting passes necessary for cutting the bar.   The real cutting
rate at the embeded depth of 80 cm shows little difference attributed
to the nozzle traverse rates, and converges to about 0.5 cm/min.

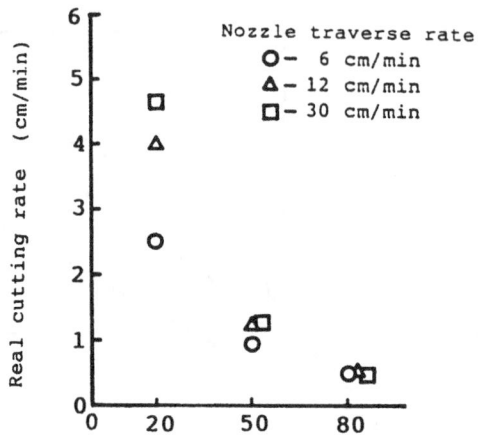

Fig. 2. Relation between embeded depth and real cutting rates of steel bars having different embeded depth

(3) Cutting volume

The width of the kerf made in a specimen has a tendency of expanding from the surface into the inside as shown in Fig. 3. Therefore, the apparent cutting volume, which is derived from the multiplication of the kerf width on the surface, the kerf length and the depth of cut, differs from the inside cavity volume in the specimen. The inside cavity volume was measured with white cement to know the real cutting volume.

Fig. 3. Cross section of reinforced concrete specimen after cutting

Table 1 shows the comparison between the white cement used and the apparent cutting volume. The test results indicate that the real cutting volume is 1.5 to 2 times the apparent cutting volume.

Table 1.   Real cutting volume

Case	Abrasive	White cement (cm^3)	Apparent volume (cm^3)	Volume ratio (real/apparent)
1	Garnet mesh 20	1,190	813	1.5
2	Steel grit mesh 40	6,350	3,614	1.8
3	Steel grit mesh 40	1,640	837	2.0

3. Abrasive-waterjet cutting system

A dismantling system for the biological shield with abrasive-waterjet cutting technique has been designed and fabricated since 1986 on the basis of the results of fundamental testing so far and the design policy such as radiation protection of personnel, minimization of radioactivity release to the environment and prevention of spread of the radioactive contamination. As shown in Fig. 4, the system consists of a water generating unit, an abrasive supplying unit, a nozzle drive assembly, a lift assembly, a block bucket, a slurry treatment unit, a ventilation unit, and a control unit. Of these components the waterjet generating unit, the abrasive supplying unit, the nozzle drive assembly, the lift assembly, and the control unit have been completely fabricated, but the others are to be fabricated by March, 1988. Outline of each component fabricated to date are described about its functions and features in the following.

3.1 Waterjet generating unit

The waterjet generating unit, which is composed of two high-pressure pumps, high pressure-resistant pipes, a high pressure-resistant hose, and a nozzle part, has a function of jetting the water generated by the pumps from the nozzle together with abrasives.

The high-pressure pump adopts a method of sending the oil pressured by a hydraulic pump to the low-pressure cylinder of a booster and compressing the water in the high-pressure cylinder with a piston drived right and left. The high-pressure water raised by a factor of ten through the booster is suppressed in the pulsatory motion by an accumulator, and is introduced to the outlet of the pump. The maximum pressure at the outlet of the pump is approximately 200 MPa and the water flow rate a pump is about 25 l/min. With regard to the feeding

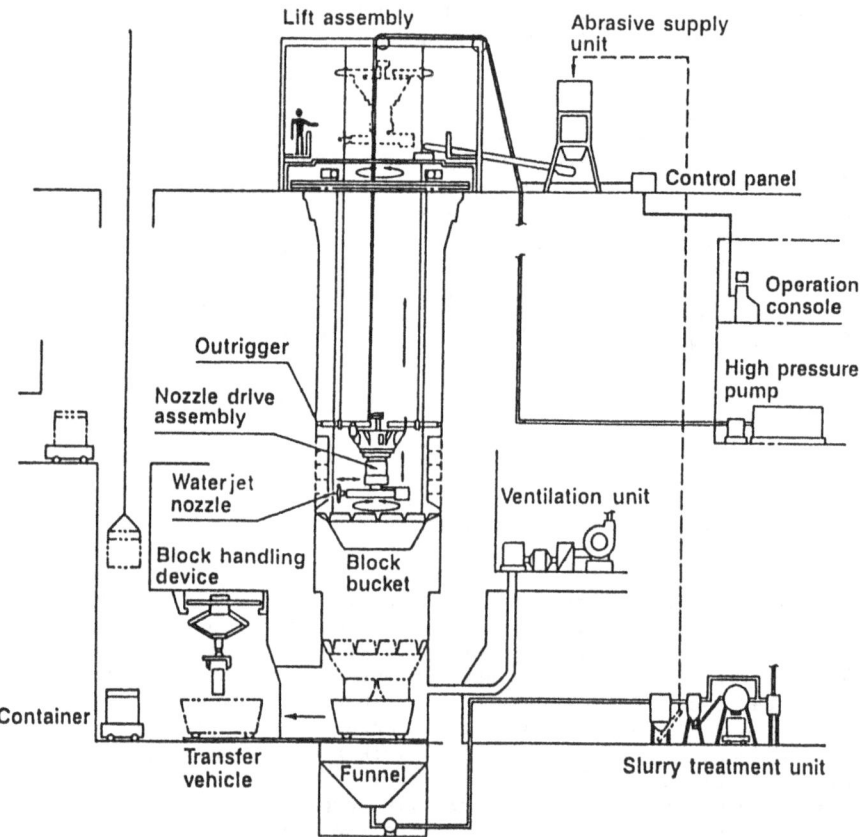

Fig. 4.   Abrasive-waterjet cutting system

line of high-pressure water, high pressure-resistant pipes are used at the immovable portion to reduce the pressure drop in the pipe and high pressure-resistant hose is used at the movable portion to give flexibility.

The nozzle part consists of a nozzle and a nozzle holder. The nozzle gathers mixed turbulent flow of water and abrasives. The nozzle holder converges the high-pressure water introduced by the hose at the orifice and mixes abrasives with waterjet at the mixing chamber by means of the vacuum produced by the jet. Concerning the nozzle holder , two types of nozzle holders, a holder for vertical and lateral cuttings and a holder for rear cutting, are ready for removing the biological shield concrete in the form of block.

3.2 Abrasive supplying unit

The abrasive supplying unit has a function for supplying a certain abrasive to the mixing chamber of the nozzle holder successively and is composed of a primary abrasive tank, an weighing hopper, a mini-conveyor, a second abrasive tank, and an abrasive supplying hose.

The primary abrasive tank is a tank for keeping abrasives to be used for the work in one day in store and has 600 l capacity capable of supplying abrasives for about five hours at the rate of 5 kg/min. The weighing hopper has roles of adjusting and measuring the abrasive flow rate as well as a saucer for the abrasives from the primary tank. The quantity of abrasives in both the primary tank and the weighing hopper is weighed with load cells and is maintained in good conditions based on the results. The mini-conveyor has a role of conveying abrasives up to the funnel of the abrasive supplying hose and is composed of two series of conveyors. Adjustment of the abrasive flow rate is carried out by changing the clearance between the conveyor and the outlet pipe of weighing hopper and the belt speed of the mini-conveyors. The abrasive flow rate can be adjusted within the range of 3 to 7 kg/min. The second abrasive tank is a tank for detecting flow and blockade of abrasives. The flow and the accumulation of abrasives in the tank are checked for possible trouble with the sensors attached to the tank. The abrasives dropped by the gravity are introduced through the abrasive supplying hose and are finally supplied into the mixing chamber of the nozzle holder with the suction in the chamber.

3.3 Nozzle drive assembly

The nozzle drive assembly has a ability to move the nozzle freely to three directions and allows the choice of the lateral, vertical and rear cutting actions, which are needed for cutting off the biological shield concrete in the form of block. That is, the nozzle drive assembly consists of three mechanisms for turning, up and down, and back and forth movements of the nozzle in order to set the nozzle at arbitrary positions. While both the mechanism for driving the nozzle back and forth and the mechanism for driving the nozzle up and down use the hydraulic-pressure-drived stepping cylinders, the turning mechanism uses a motor as its driving source. The movement of the nozzle is controlled with ranges of ± 190° in rotational movement, 700 mm in vertical movement, 750 mm in radial movement, respectively.

The body of the nozzle drive assembly is suspended with three steel wires, and is fixed by stretching three outriggers against the bio-logical shield wall so as not to cause the swinging of the body owing to the reaction force of waterjet and the weight imbalance of the body. With respect to the accessories, an industrial television (ITV) camera for watching the tip of the nozzle and a touch sensor are pro-vided to position the nozzle at an appointed point in cutting and to prevent the contact of the nozzle with the biological shield wall, respectively. A level, a reflector for laser beams and a ITV camera for watching the spot of laser beams on the reflector are also attach-ed to level the nozzle and to maintain the body of the nozzle drive assembly in the center of the cavity of the biological shield. Total weight of the assembly is about 2.8 tons.

3.4 Lift assembly

The lift assembly is so designed as to suspend the nozzle drive assembly at arbitrary positions in the cavity of the biological shield. Furthermore, the lift assembly is formed by the box-shaped structure equiped with working stage and winch, and allows the maintenance of the nozzle drive assembly. The movement of the nozzle drive assembly using the lift assembly is performed by means of three wires. With respect to the accessories, a laser beam emitter and an ITV camera for watching the inside of the biological shield cavity is attached.

3.5 Control unit

The control unit consists of a operation console and a control panel. The operation console has the functions for giving instructions to each component of the system in order to operate by remote control and for watching the instantanious state of work by means of an ITV monitor or the graphic display on a CRT. Therefore, various kinds of buttons, levers and meters which are needed in the remote operation are provided on the console. On the other hand, the control panel transmits signals into each component of the system on the basis of the command from the operation console and returns the signals from the sensor attached to each component to the operation console as information to the contrary.

4. Dismantlement of biological shield

4.1 Biological shield of JPDR

The biological shield of JPDR is a reinforced concrete structure built into the reactor enclosure building. It has the cylindrical shape having a cavity for installing the reactor pressure vessel in the center. The shield wall is thick (about 3 m at maximum) to satisfy both the requirements for a radiation shield and the resistivity for the earthquakes, and the inner surface is lined with the 13 mm thick carbon steel plate. With reference to the wall structure, latticed main reinforcing bars 29 mm in diameter are located in the inside at intervals of 150 mm and radially stretched reinforcing bars for shear force at intervals of 300 mm in axial direction and at intervals of 500 mm in circumferential direction. In addition, a number of cooling pipes and seven ion chamber guide tubes are embeded. A cross section of the JPDR biological shield is shown in Fig. 5 including the contour line of equi-radioactivity and the cutting techniques to be used. The abrasive-waterjet cutting technique will be applied to the lower half of the projecting part of the biological shield because of the radiation dose rate (approximately 200 mR/h) in the cavity after the pressure vessel is removed.

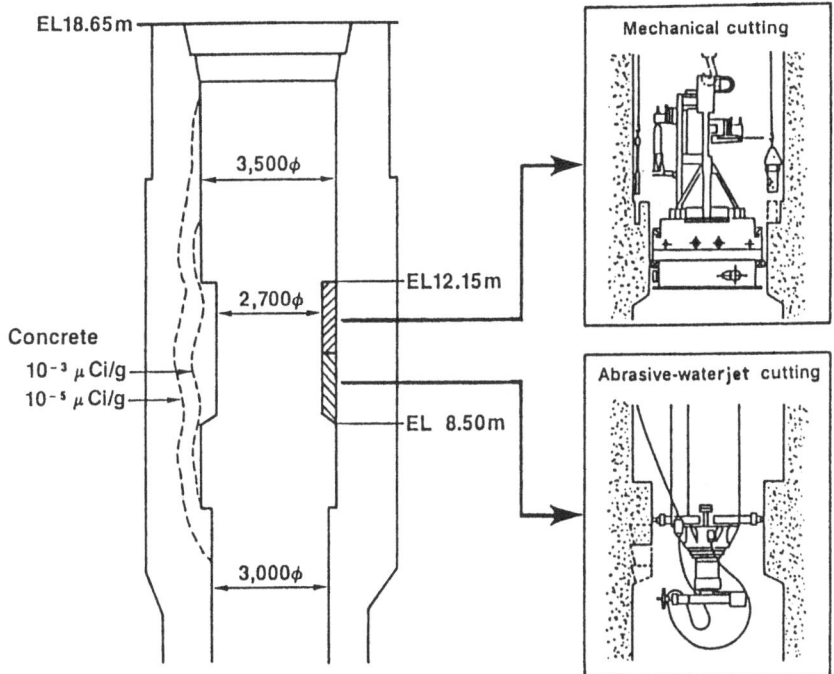

EL18.65m

3,500φ

2,700φ

EL12.15m

Concrete

10⁻³ μCi/g

10⁻⁵ μCi/g

EL 8.50m

3,000φ

Mechanical cutting

Abrasive-waterjet cutting

Fig. 5. Cross section of JPDR biological shield

4.2 Dismantling procedure

Dismantlement of the biological shield will be performed after the
pressure vessel is removed. The following is main dismantling procedure with the abrasive-waterjet cutting system.

   (a) The nozzle drive assembly is hung down in the cavity of the
biological shield by the lift assembly installed on the service floor,
   (b) Its inclination and deviation from the cavity center line are
adjusted,
   (c) The nozzle drive assembly is fixed with its three outriggers,
   (d) Nozzle trajectory is traced without jetting water before actual
cutting,
   (e) Biological shield concrete is cut into block by repeating rear,
vertical and lateral cuttings in this order,
   (f) Dismantled block is dropped into the block bucket and is removed from the bottom of the biological shield.

Considering the capacity of the using steel container and block handling, block size was decided to be 40 cm x 40 cm x 80 cm. In this
dismantling work, reinforced concrete of 7 tons will be removed. Personnel radiation exposure during the work is estimated at 0.6 man-rem.

205

5. Concluding remarks

The abrasive-waterjet cutting system described here is scheduled to be put into practical use of dismantlement of the biological shield in 1991 after its systematic functioning is adequately confirmed in the mock-up tests. It is expected that this application results in a precious experience and brings various kinds of useful reference data to the reactor decommissioning in the future.

References

Ishikawa, M. and Kikuyama, T. (1985) Decommissioning plan and present status of technical development in JPDR. _Proceedings of International Nuclear Reactor Decommissioning Planning Conference,_ Bethesda, 450-468.

Ishikawa, M., et al., (1986) Decommissioning program of the Japan Power Demonstration Reactor by JAERI. _Proceedings of International Low-, Intermediate-, and High-Level Waste Management and Decontamination and Decommissioning Meeting,_ Niagara Falls.

Ishikawa, M., et al., (1987) Present status of JPDR decommissioning program. _Proceedings of 1987 International Decommissioning Symposium,_ III18-III30.

EXPERIMENTAL STUDY FOR REALIZATION OF THERMIC LANCE CUTTING METHOD

T. UCHIKOSHI   Japan Atomic Energy Research Institute
N. MACHIDA   Kumagai Gumi Co., Ltd.
Y. KATANO   Kumagai Gumi Co., Ltd.
Y. KAMIYA   Kumagai Gumi Co., Ltd.

Abstract
A series of experiments on thermic lance cutting method was carried out to obtain useful data for the purpose of making practical application of this method to the dismantling of reinforced concrete.
As the first step of these experiments, basic performance experiment was executed to study basic cutting performance concerning oxygen consumption, extent of bar loss and cutting speed, and also by-products generated during cutting work such as powdered dust, gas, fumes and slag. An automated and remote-controlled cutting machine was then developed by utilizing automated bar supply and ignition.
This paper describes the result of these experiments.
Key words: Thermic lance cutting method, Lance bar, Remote-control, By-products, Cutting into blocks, Dismantling of biological shielding walls of nuclear reactors.

1. Introduction

The authors paid attention to the thermic cutting method, which has been already proven to be usable for cutting steel plate or concrete with high cutting speed. Under an agreement with Science and Technology Agency, we carried out a series of experiments for the purpose of making practical application of this method to the system of dismantling nuclear reactors.

In a lance bar cutting method, one of thermic methods, the lance bar itself is burnt by heating the tip of the bar with oxygen gas supplied in the lance bar. The resultant reaction heat of oxidation of iron and aluminum melts, bores, and cuts irons and concretes.

The conventional thermic lance cutting is carried out manually and the cutting result and performance are much influenced by workers' skill. To reduce the influence of workers' skill on the method, an experimental cutting machine has been developed in two steps and then cutting tests are carried out.

As the first step, an experimental machine was made to perforate and cut the objects automatically without workers' skill. The basic cutting test was carried out by the experimental machine to obtain various cutting data such as those of cutting performance and by-products generated during cutting work.

207

In the second step, on the basis of the cutting performance obtained at the first step, we have prepared a cutting machine with function of forwarding and igniting bars automatically, and its peripheral system as well, and a remote-controlled cutting test was carried out to pursue the possibility of continuous cutting.

In this report, based on fundamental data on the cutting performance (cutting into blocks) and the by-products generated during cutting work obtained from the above two tests, the authors study the possibility of making practical application of this method to dismantling of biological shielding walls of nuclear reactors and problems in applications.

2. Basic cutting test

A basic cutting test was carried out to obtain the basic cutting performance of the lance bar method (oxygen consumption, bar consumption and cutting speed) and the characteristics of by-products generated during cutting work (dust, gas, fumes and slag).

2. Test outline
The used cutting test machine and test conditions are as follows:

(1) Cutting test machine
The cutting test machine consists of a bar forwarding device, shifting device, electric equipment, oxygen feed pipe and others. The bar forwarding device delivers the bar toward the object to be cut. The shifting device moves the bar together with the bar forwarding device itself horizontally. The dust generated during cutting work is exhausted by a dust collector. Fig. 1 shows a schematic drawing of the lance bar cutting test machine.

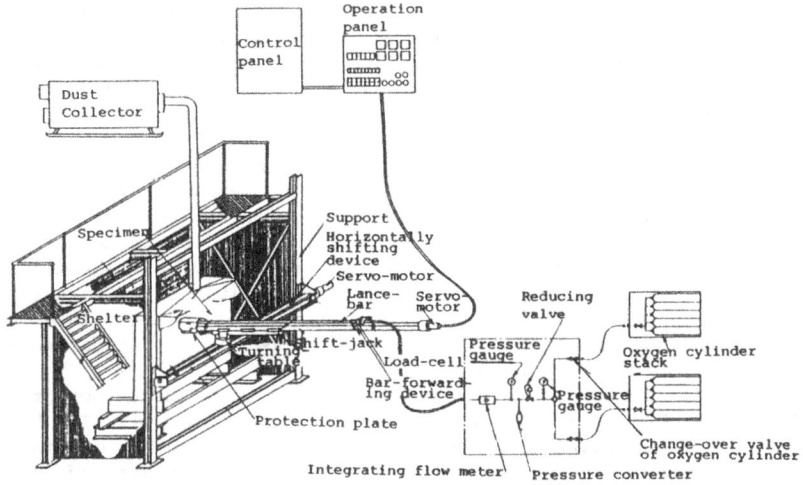

Fig. 1 Schematic Drawing of Test Machine for Lance Bar Cutting

(2)　Shape, dimension and arrangement of reinforcement of specimen
As a rule, a specimen simulated a part of JPDR biological shielding
objects, having a rectangular parallelepiped shape of 1.0 m x 1.0 m x
0.5 m.　Both horizontal and vertical pitches of reinforcement were
150 mm.　A steel pipe was embedded in some of the testpieces and an
iron liner plate was set on the surface of some testpieces.　Eight
testpieces were prepared.

(3)　Tested items
(a)　Cutting performance test
The cutting performance of the lance bar method was tested with oxygen
gas pressure, bar forwarding speed, specimen materials, cutting direction and other factors taken as a parameter.　Set values of test conditions are shown below in Table 1.

　　Two modes for testing the cutting performance were employed: horizontal cutting and continuous perforation cutting.　In the horizontal
cutting mode, the objective is cut with the burning bar, being fed forward and horizontally at the same time.　In the continuous perforation
mode, the objective is bored (spot perforation) and then cut by repeating perforation.

　　The test was carried out changing the parameters such as oxygen
pressure, perforation pitch, horizontal shifting speed, bar forwarding
speed, perforation depth, cutting direction, cutting mode, cutting section, specimen material, liner thickness and bar angle.

Table 1　Set Values of Test Conditions

No.	Item	N. of levels	Level		
1	Preset oxygen pressure	3	8, 11, 14		kg/cm2
2	Preset perforation pitch	3	P-5, P, P+5 ✕1		mm
3	Preset horizontal shifting speed	3	5, 10, 15		cm/min
4	Preset bar forwarding speed in horizontal shifting	at each oxygen pressure 1	Pressure (kg/cm2) 8 11 14	Bar-forwarding when shifting (cm/min) 70, 74 80 87	
5	Preset bar forwarding speed	at each oxygen pressure 3 (4)	Pressure (kg/cm2) 8 11 14	Bar forwarding speed (cm/min) 100, 110, 120 110, 120, 130 120, 140, 150, 160	
6	Preset perforation depth	4	10, 20, 30, 40		cm
7	Cutting direction	4	Front Front, Horizontal, vertical, (downward and upward), Upper side		
8	Cutting mode	4	Spot perforation, Spot horizontal shifting, Continuous horizontal shift- ing, Continuous perforation		
9	Cutting section	2	Middle parts, Corners		
10	Specimen materials	4	Concrete, Reinforcement, Steel pipe, Liner plate		
11	Liner thickness	3	6, 13, 45		mm
12	Horizontal bar angle	3	0, 15, 30　(to right) °		
	Vertical bar angle	4	0, 5, 15, 30 (downward) °		

✕1　P is a horizontal hole diameter at spot perforation.

(b) Mechanical characteristic test
A load cell was installed in contact with the bar catching holder.

(c) Cut block removing test
The dimensions of a block to be cut out were determined in such a manner that it could be removed after cutting a liner plate 13 mm thick, reinforcement D29 and steel pipes arranged in 3 rows and on two stages.

After the specified cutting was over, a wedge was set and inserted into the horizontal frontal-cutting-line with a jack to remove the block.

When determining the wedge-penetration force, the maximum value was selected among those recorded through the block-removing works. Removing work of the cut block is described in Fig. 2.

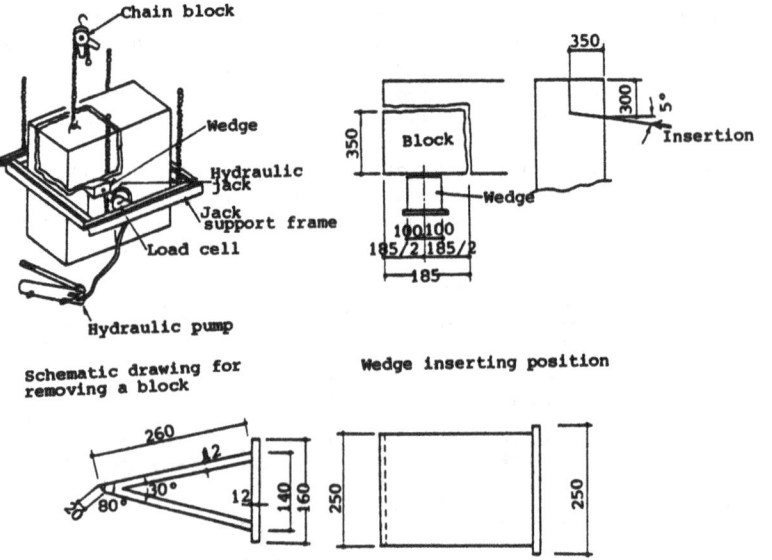

Fig. 2   Removing of Cut Block

(d) By-products and environmental test
The burning bar cutting method will produce gas and dust. To obtain gas density, gas chemical composition, amount of dust, concentration change of dust, dust grain size and dust chemical composition, we planned to collect all amount of gas and dust. In this test, the cutting section of the specimen was covered with a shelter as shown in Fig. 3. To prevent leakage of gas and dust produced during cutting work from the opening of the shelter, air suction speed was adjusted with the dust collector, so that the air inflow speed might become more than 1.0 m/sec at the opening. The air suction speed was set to 18.0 m/sec (corresponding to 1.0 m/sec of the air speed at the front of the shelter) with the damper of the dust collector, while the air speed in the duct was monitored with the Pitot tube.

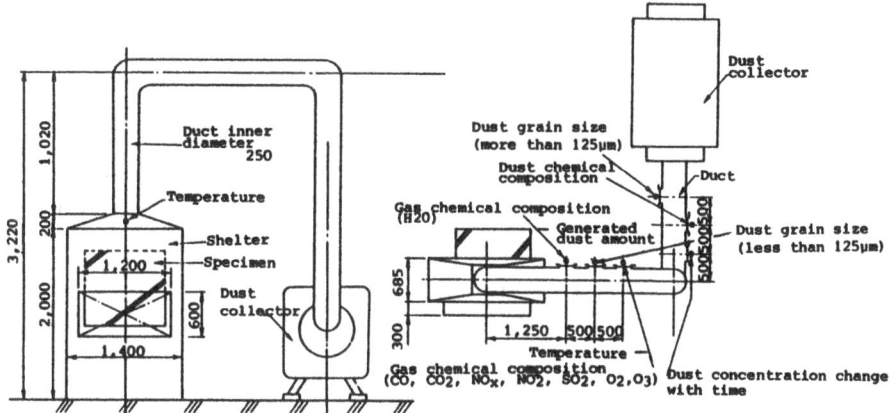

Fig. 3   By-products (gas and dust) Collecting Machine

2.2   Cutting test results

Cutting test on lance-bar process : at first, judging from all of data
concerning cutting efficiency and cutting conditions, the items such
as;
- oxygen pressure
- bar forwarding speed
- horizontal shifting speed (for horizontal cutting only)
- perforation pitch (for continuous perforation only)

are determined as parameters to be preset and given as optimum values
for horizontal cutting and the continuous perforation. By executing
cutting under the conditions thus determined, we obtained the basic
data of the cutting performance mechanical properties, by-products,
environmental properties.

(1)   For horizontal cutting test, the optimum testing conditions are
established as follows:
    Oxygen pressure : 11 kgf/cm2, Bar-forwarding speed : 80 cm/min.;
    Horizontal shifting speed : 15 cm/min.
The cutting area was 56.4 to 83.4 cm2/min. for concrete, 86.4 to 87.6
cm2/min. for lined concrete and 136.2 cm2/min. for steel plate. The
maximum cutting depth in lined concrete was 9.6 cm which is sufficiently deep to cut reinforcement. It is known that sufficient cut depth
can be obtained by repeating twice horizontal cutting at the same
place.

(2)   The optimum conditions for a spot perforation test are set with
oxygen pressure of 14 kgf/cm2 and bar-forwarding speed of 150 cm/min.
But if a continuous perforation cutting is executed under these conditions, it is found that the bar being shifted too fast, cutting tends
to deviate towards the previous perforation-cut face.

    To avoid this, the bar-forwarding speed is changed to 130 cm/min.
(oxygen pressure : 14 kgf/cm2). The pitch between holes is set at 47
mm. The cutting area is set at 120 cm2/min. for all materials, except
for steel plate with a larger value of 138.0 to 160.2 cm2/min.

(3) In continuous perforation cutting, the thrust exerted through perforation is 184 kgf at maximum on frontal and vertical cutting. The thrust force generated when detecting a specimen is 359 kgf at maximum, larger than the above values. When backing the bar at high speed after perforation cutting, the extructing force needed is 291 kgf at maximum, because of the bar sticking to solidified dross. At this moment, the bar was not detached from the holder, but kept in it safely.

(4) The amount of dust generating during cutting was about 0.7 to 1.9g and dust density was known to be 3.2 g/m3 at maximum. Dust grain size distribution was as follows;
    500 µm or less   :   99 % or more in weight
    125 µm or less   :   50 % approx.
      1 µm or less   :   5 % approx.
    0.3 µm or less :   0.2 % or less.
The dust thus generated was subjected to chemical analysis and found to have the following composition: the elements Fe and Al, as they are constituents of the bar, are dominant in content, 72 to 82 % and 4.1 to 7.6 % respectively. Si, contained mainly in cement, was 2.1 to 13.7 %. The other elements such as Mg, Ca, Mn, C, S and P are 2 % or less respectively.

(5) When reinforced concrete is subjected to continuous perforation cutting, dross is generated at a rate of 282 kg/m2 of section. It is composed of dropped dross with a weight of 214 kg and that sticking to the specimen surface with a weight of 15 kg. Dross remaining on the cutting sections weighs 53 kg. As in the case of dust, the most dominant element is Fe which occupies 69.4 % of the dropped dross, 63.9 % of the sticking dross. Dropped dross can be recovered by an electric magnet. The secondly dominant element was Si, occupying 25 to 27 %. And Al, one of the alloy elements of the bar, was 7.1 to 7.3 %.

(6) Observing the oxygen density along with time, it is found that it remains almost always in the range of 20.8 to 21.6 %, without making its supply excessive nor insufficient.
    The density of the other gases (CO, CO2, SO2, NO2, O3) was measured and all of them but for O3 were in the tolerances specified in labour hygienics.

(7) Noise was 108 dB at maximum (back ground noise : 71.5 dB) when measured at the point 1.2 m high above the grade, being 3 m apart horizontally from the cutting place.

(8) The burning temperature of the bar was 2,250°C and the fused dross generated thereby was 1,627°C.
    A protection plate, which is provided in the cutting unit for receiving dispersed fusing dross, was 479.3°C. The temperature of an inspecimen spot 5 cm apart from the cutting face rising to 911.7°C at maximum, the concrete surrounding the cutting surface is credited with brittle property. As for a place in the specimen which was located 15 cm from the cutting surface, its maximum temperature was limited to 65°C, and this means heat is considerably attenuated during being transmitted through concrete.

212

(9) With a test verifying if a block can be cut out from specimen,
a block of 103.0 kg (81.9 %) could be cut off, while it had been originally planned to be 125.7 kg. Wedge penetration force required for
peeling off a block was 2,085 kgf. (when finishing cutting, part of
concrete remained, but was completely cut off by using a wedge.)

3. Remote-controlled operation test

For performing continuous cutting by the lance bar method, the system
should ensure automatically spare bar feeding and bar ignition. Based
on the results of the basic cutting test described in the preceding
article, an experimental machine having such functions was developed
and its continuous cutting performance was tested.

3.1 Outline of the test
The configuration of the cutting machine and the test conditions are
described below.

(1) Configuration of the experimental cutting machine
The experimental machine (Fig. 4, 5) is composed of :
    a cutting unit provided with functions of automatic bar feeding and
    bar-tip automatic ignition, and
    a control system of the cutting unit and industrial television.
By remounting the cutting unit, perforation cutting by this system is
practicable for both vertical and upper faces. Programmed sequential-
ly, the serial cutting steps are carried out automatically, the flow
of which is shown in Fig. 6.

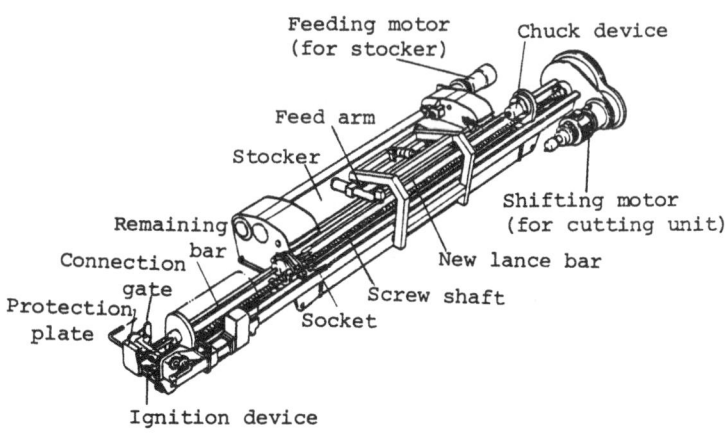

Fig. 4 View of the Cutting Unit

213

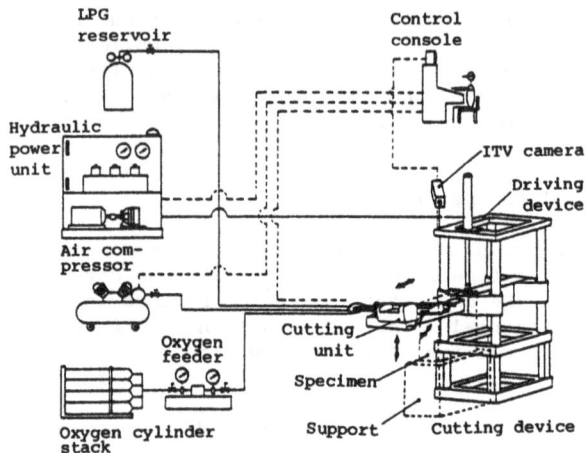

Fig. 5 Configuration of the Experimental System

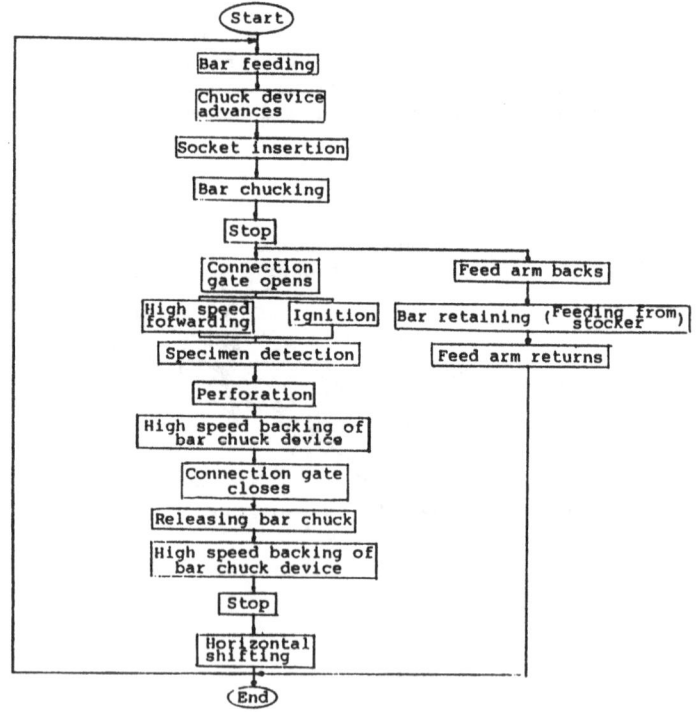

Fig. 6 Flow of Continuous Cutting Steps

(2) Cutting conditions and specimen shape
The continuous cutting test was conducted, using the experimental machine and the specimen specified in the preceding section to obtain data of cutting performance.

3.2 Test results and consideration
In the following are described the test results of the cutting performance.

① Cutting efficiency
Table 2 summarizes the obtained data of cutting efficiency and Fig. 7 the time division indicating time required for each step in the whole work.

Table 2 Result of Cutting Performance Test

	Number of bars	A Bar consumption length (cm)	B Oxygen feed volume (Nm3)	C Total work time (sec)	D Cutting surface (cm2)	A / D (cm/ cm2)	B / D x10⁻³ Nm3/cm2	D / C cm²/ min.)
Continuous perforation cutting test	6	1000	9.2	897 (1144)	1200 (1212)	0.8	7.7	80.3 (63.6)

Note) The values in parentheses represent the data of work by the conventional machine.

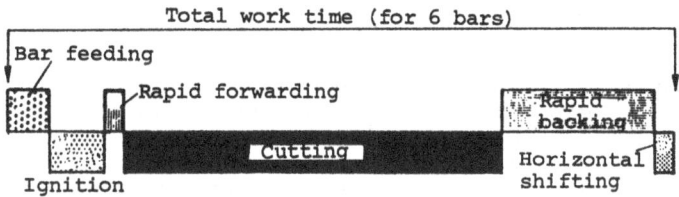

Fig. 7 Time Division of the Whole Work Process

The cutting area per unit time recorded in the test with the experimental machine is 80.3 cm2/min. It is 21 % larger than the speed (63.6 cm2/min.) of the conventional machine by which bar feeding and ignition are done manually. This is because, with the automatic device, the bar feeding time is reduced to less than 1/3.

② Bar consumption length and perforation depth
Fig. 8 shows the variation in bar consumption length and perforation depth.
    The bar consumption length is around 1,650 mm, the variation range being 122 mm. As this variation is within the tolerance (230 mm) for the length of bars set in the cutting unit, the continuous cutting is judged satisfactory. The perforation depth agrees well with the preset value, varying within the preset value ± 10 %.

215

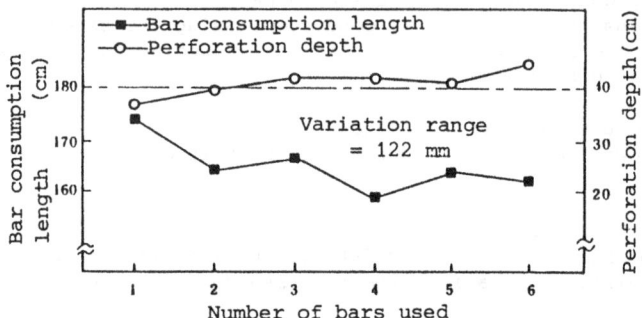

Fig. 8 Variation in bar consumption length and perforation depth

The test as described above has attested that :
    a. automatic bar feeding and ignition can be done smoothly in the
       continuous cutting work, and
    b. there is no remarkable difference in cutting performance between
       the automated exeperimental machine and the conventional machine.

4. Conclusions

The lance bar works, which have been executed manually, are successfully automated. Here, the applicability of the automatic cutting method
to dismantling works is assessed and problems to be solved by further
studies are enumerated.

Applicability :
1)   The method is also applicable to reinforced concrete structures
    with steel bars of large diameter or thick steel plates.
2)   Satisfactory cutting in block is done by the semi-remote-controlled
    method as shown by the test results. Now, we have a perspective
    that this method could be applied successfully to dismantling,
    through remote control, concrete structures containing many metal-
    lic parts, such as biological shielding structures used in nuclear
    reactors.
For ensuring higher dismantling performance of the method, it is necessary;
1)   to enhance the thermal resistance of the machine, so as to endure
    the heat generating during the cutting work, and
2)   to develop a highly efficient system of disposing and treating
    fumes and slag generated during the work.

DEMOLITION OF CONCRETE STRUCTURES BY HEAT - A PRELIMINARY STUDY II

JOHN M. McFARLAND
McFarland Wrecking Corporation

Abstract

Examination of concrete buildings involved in fires shows that the concrete most directly exposed is weakened by heat and visibly altered. This paper explores the feasibility of demolition and/or decontamination of massive concrete structures by heat. It also considers the expansion of heated reinforcing steel and its bonding to concrete. It proposes the use of electrical heating units to decompose radioactive contaminated surfaces of reactor pits to facilitate their removal. It also suggests the removal by melting of vertical reinforcing bars by electrical resistance heating, and the use of the resulting voids to place explosives or expansion agents.

Key Words: Demolition, Heat, Concrete, Decontamination, Melting, Reactor pits, Oxidation.

Introduction

The background for this study includes an analysis in 1976 of a Pressurized Water Reactor (PWR) of the Westinghouse design for the purpose of establishing the methods and cost parameters of its removal. Later a similar study was done of a Boiling Water Reactor (BWR), WPPS Plant No. 2, Hanford, Wash. Both studies were done for Battelle's Pacific Northwest Laboratories (PNL). The procedures and costs reported were based on conventional demolition methods. A primary assumption is that all decontamination will have been done by others prior to demolition by heat.

In 1979, under the sponsorship of PNL, a paper was presented at the American Nuclear Society Conference on Decontamination and Decommissioning of Nuclear Facilities, at Sun Valley, Idaho. That paper suggested the use of heat to demolish massive concrete structures, using the Reactor Containment Vessel of the Trojan (Westinghouse) PWR nuclear power plant as its model.

This paper reviews that earlier suggestion and offers related possibilities for your consideration, including:

1. Technical background data on the reactions of concrete and steel to heat.
2. Application of that data to the demolition of a PWR containment vessel.
3. Application of that data to the decontamination of a reactor pit.

4. Consideration of the use of electrical resistance heating to demolish a BWR and other reinforced concrete structures.

Technical Background Data

Concrete is composed of cement paste, CaO + clay oxides + sand + gypsum, (20% to 30%) and rock and sand aggregate (70% to 80%). The cement glues the aggregate together. Concrete has some 5% to 6% water by weight, even when dry. In general, chemically bound water equals some 20% of the weight of the cement paste. The remainder of the water in the concrete is free water.

When the temperature of concrete is raised to 100°C, the free water is driven off as steam. At about 500°C, the chemically bound water is driven off. $Ca(OH)_2$ in the cement paste decomposes to produce CaO and H_2O. This dehydration causes the cement paste to shrink and lose much of its adhesion.

When the temperature is raised to 573°C, there is a change in the crystalline structure of quartz (from quartz alpha to quartz beta) which results in swelling and internal cracking. Concrete with quartzitic aggregate has lost 50% to 75% of its strength at this approximate temperature. At 870°C, such concrete has less than 20° of its strength.

When the temperature is raised to between 700°C and 870°C, the $CaCO_3$ in calcarous aggregate converts to CaO + Co_2. Driving out the CO_2 in this manner and driving the H_2O out of the cement paste $Ca(OH)_2$ is referred to as calcining. At 870°C the strength of calcarous aggregate concrete drops to about 10%.

This decomposition of the concrete enables it to be readily demolished. There could be spontaneous disintegration of the structure because of the CaO absorbing moisture from the atmosphere and converting to CaOH. CaOH is considerably less adhesive than CaO, allowing rocks and sand to fall away.

The specific heat of reinforced concrete is a nominal 0.20 kilocalories per kilogram. It takes obout one-fifth as much heat to raise one kilogram of concrete 1°C as it takes to raise the same weight of water 1°C. It is noted that the specific heat of concrete generally decreases with increasing temperature while the specific heat of steel (0.111 KC/Kg) rises. Since specific heat varies with temperature as well as with changes of composition of the concrete, the rounded figure of 0.20 is used as a nominal value.

Application of Data

There are two separate but concurrent actions which destroy a concrete structure by heat:
1. Internal alteration of the concrete. Included in this action is the dehydration of the cement paste, differential expansion within the aggregate caused by alteration of the crystalline form of quartzitic components, and/or calcination of the calcarous aggregate.
2. Differential expansion between the concrete and the steel rein-

forcement. This attempts to break the bonds which concrete has
to steel. These bonds are chemical adhesion, friction, and
mechanical interaction.

Internal Alteration of the Concrete

A principal problem of heating the concrete to 870°C, aside from the
amount of the heat input, is that heat transmission through concrete
is so slow that a fast fire will have burned itself out before it
could penetrate the concrete to any appreciable depth. The heat would
simply be wasted out the exhaust. Over a period of time, heat from a
continuous source will penetrate a massive concrete wall. The temperature of a cross section of wall cannot exceed a linear relationship
between the heated inside surface and the unheated outside surface.
The unheated outside surface would experience uneconomical heat loss
above about 150°C due to radiation. With an assumed 1093°C temperature
in the inner face of the chamber and 150°C outer wall temperature,
the midpoint could only reach 650°C and the three-quarter point could
not exceed 427°C.
There are four devices or procedures which could increase the temperature of the wall:
1. Insulation added to the exterior could raise the economically
   achievable exterior temperature.
2. Oxygen added to the combustion air intake could raise the interior temperature above the 1093°C.
3. Petroleum base coke could provide a combustion temperature of
   1510°C.
4. Combustion heat applied to the underground post-tensioning duct-
   work gallery of the Westinghouse design containment vessel could
   heat the inside of the wall through the ducts, thereby raising
   the cross wall temperature gradient.
Radiation heat losses increase in proportion to the fourth power of
the absolute temperature which tends to limit the outer temperture
of the wall. Radiation heat losses can be reduced and the steady-
state temperature of the wall increased by covering the outer surface
with insulation. Five centimeters of rock wool insulation could cost
in the neighborhood of 50¢ per square foot, and would raise the allowable temperature of the outer wall to some 427°C. More expensive
high efficiency insulation is available. Insulation is valuable for
its conservation of fuel.
With the oxygen necessary for combustion, there would be four times
as much inert nitrogen. (Air is approximately 20% oxygen and 80%
nitrogen.) The nitrogen acts as a dampener on the heat of combustion.
Natural gas, for instance, burns at 1150°C with air, while pure oxygen
produces a combustion temperature of 2315°C.
Preheaters for incoming combustion gases as well as for fuel are
available. By preheating the air used for combustion, for example, to
260°C by using the 538°C stack exhaust gases for heat, much of the
stack heat loss would be eliminated and the combustion efficiency
increased.
If an oxygen supplement is added after the structure has been heated
to its limits by ordinary air combustion, it would raise the temper-

219

ature inside the RCV depending upon its percentage mix with ambient
air. A wood or oil fire using ambient air burns at a nominal 1093°C;
while using 100% oxygen, it burns at 2204°C. Assuming a straight
line relationship, 12½% added oxygen (net 10%) would result in combustion temperature of 1204°C while increasing the reactivity of the
combustion air by 50% (Oxygen increased from 20% to 30%).

An advantage of added oxygen is that its higher temperature and
increased reactivity could oxidize the inner steel liner of the RCV.
In effect the steel would become fuel for the fire. In this reaction
two atoms of iron combine with three atoms of oxygen to form ferric
oxide (2 Fe + 3 O combine to form Fe_2O_3). In so doing they liberate
a heat of conversion of 199 kilocalories per mole. Adding the combined
atomic weights plus a allowance of 10% excess oxygen, the reaction
would require approximately 50% by weight oxygen to that of iron.
At $55 per metric ton for oxygen, it would cost $22.50 per metric
ton to dispose of the steel liner if this reaction can be made to
occur.

A unique feature of the Westinghouse RCV design is that it has
a basement post-tensioning gallery underneath the circular wall.
From here steel cables pass up and over the dome and back to the
gallery on the opposite side through hollow ducts. If these cables
are removed, heat through these ducts could help increase the speed
of heat transmission through the concrete. It would also raise the
thermal gradient of the wall. (These ducts could also be used to
place explosives or expanding agents.)

Differential Expansion

Reinforcing bar in slabs or outer courses of walls will, in general,
break the bond of the concrete cement when its temperature exceeds
that of the concrete by 150°C. Mark's Mechanical Engineering Handbook
states that "When the deformation arising from change of temperature
is prevented, temperature stresses arise that are proportional to
the amount of deformation that is prevented.... In the case of steel,
a change of temperature of 12°F will cause in general a unit stress
of 2,304 lb/in².". At 195 lb/in² per degree Farenheit, a change of
302°F would generate a stress of some 58,890 lb/in². This approximates
the tensile strength of carbon steel and could reasonably be expected
to cause a failure of the concrete's bond.

Substantiating this a Japanese patent, Itoh,et al., U.S. Patent No.
8,727,982, for heating of buried concrete reinforcement by electrical
induction heating, claims that when the reinforcement has been heated
150°C (302°F) above the heat of the concrete, the reinforcement will
break its bond and the concrete can then be readily removed.

The reservation must be made in the case of massive containment
vessels that the restraining effect of the heavy reinforcement and
thick walls would require substantially higher stresses to enable the
reinforcement to break its bonds. This could possibly be achieved by
a higher temperature differential; by delivering this stress to large
amounts of the reinforcing simultaneously; and by concurrent weakening
of the concrete itself by the heat.

The principal obstacle to causing a differential expansion between reinforcing steel and concrete is that their coefficients of expansion are so closely matched, 0.00084 for steel and 0.0008 for concrete. If the concrete and steel are heated together, a significant differential between the two could not be achieved below calcining temperatures.

The objective would be to get the heat directly to the steel and utilize the extreme difference in thermal conductivity, (the rate at which heat is passed through a substance) of the steel vs. the concrete, 26.2 to 0.54, a 48:1 ratio. The means of accomplishing this is presented by the existence of the steel liner covering the inner wall of the reactor containment structure. The liner is used during construction as the inner form for the concrete pour and supports the rebar placement. It has extensive ribbing on its backside and direct ties with the rebar network. It would be most difficult and costly to attempt to cut this liner away from the concrete without the heat process. It would be equally or more costly to leave this liner bonded to the concrete and rebar while atempting demolition. This disadvantage can be turned into an advantage by using the liner to transmit heat directly to the rebar network. The heating would free the liner and the adjoining network of reinforcing steel from the concrete. In addition, it would project the heat throughout the massive walls and perhaps achieve the heat distribution that the concrete's insulating qualities tend to prevent.

Fuels

Three fuels have been selected for consideration: wood, diesel oil, and petroleum base coke.

Wood has the advantages of easy availability, low cost, and familiarity. It burns relatively quickly but its coals continue to produce and hold heat. It burns down to a fine white ash which is generally harmless, but which can be annoying to neighbors. Unless continually fed, a wood fire would be of too short a duration for its heat to penetrate through four feet of concrete. A fire lasting three days with coals lasting another four days would not be sufficient to overcome the insulating qualities of the concrete.

An unknown factor here is the penetration of the heat along the steel reinforcing network with its induced stresses and contact alteration of the concrete. It remains possible that a simple loading of wood of the requisite heating energy capacity could so stress and alter an RCV that it could be easily demolished by conventional means of crane and ball demolition.

Diesel oil can be fed into the combustion area at a controlled rate for a long as the fire needs to be maintained. It costs about twice as much as wood. It can be regulated so that it does not have an exhaust gas problem, and it does not have the fly ash problem. An oil fire could be metered to gradually heat the entire mass to the limits of its theoretical heat gradient. This could be done without the severe heat loss that a fast burning fire in a well insulated RCV could encounter. This slower heat regulated to achieve optimum heating of the concrete could have the inherent disadvantage of not providing a maximum temperature differential between the concrete and the steel.

If the concrete heats as the steel heats (or to the extent that it does so), the differential expansion stresses will be reduced. If the oil input is increased to provide the maximum heat conduction in the steel, it might become uneconomical as regards altering the concrete. Coke has the properties of wood without the fly ash problem. In addition it has a high energy content, burns at a higher temperature than either wood or oil (1510°C), and it retains its heat for a longer period of time. Coke has the potential to quickly heat the steel to several hundred degrees higher than a wood or oil fire. It could ef- fectively and rapidly direct heat through the steel network. The High- er temperatures could melt the steel liner in a reducing atmosphere, thus eliminating the consideration of iron oxide being released to the atmosphere as a pollutant dust.

Coke's cost per kcal appears to be lower than either wood or diesel oil. The alternatives of loading the coke into the chamber in a single load, or using a feed mechanism to add fuel, are options to be con- sidered. Either would appear to be feasible.

The effectiveness of each of these fuels could be enhanced by ex- terior RCV insulation, preheating of combustion air, and/or supplement- ing the combustion air with pure oxygen, as discussed above. These enhancements can be measured for cost effectiveness by comparing their cost and benefits vs. the cost of the fuel saved.

Heat Input and Fuel Cost Calculation

To determine the heat input and cost of heating a reactor containment structure, the Trojan Pressurized Water Reactor has been selected as the reference case in the following calculations. The mass of its reinforced concrete in kilograms, times the specific heat of concrete, times the temperature difference in degrees Celsius, will determine the theoretical heat input in kilocalories. To the theoretical heat input, add a heat loss factor. This sum times the cost per kcal of the fuel selected will determine the total cost:

Volume of reinforced concrete	15,500 m³
Mass per cubic meter	X 2,000 kg/m³
Total mass	31,000,000 kg
(mass) (specific heat) (temperature change) =	kilocalories
31,000,000 kg x 0.20 kcal/kg x 838°C =	5.2 billion kcal
Heat loss @ 200%	10.4 billion kcal
Total heat input required	15.6 billion kcal

Fuel Alternatives:	Cost per million kcal	Total cost of 15.6 billion kcal
Wood	$9.90	$155,000.
Diesel heating oil	19.00	297,000.
Coke	6.67	104,000.

Decontamination by Heat

An extension of the concept of weakening concrete structures by heat is to apply it as a means to decontaminate radioactive reactor pits. Cooperheat of Somerset, New Jersey, manufactures electrical heating units up to 73 KVA capacity which can withstand temperatures up to 1093°C. These units can be used in multiples as needed. A critical advantage of electrical heat is that it does not add combustion gases to the radioactively contaminated space.

This heat source, suspended into a reactor pit, has the advantage that it can be thermostat regulated. By this it is possible (but not certain) that the problem of exploding concrete can be overcome. For example, if the the temperature is held at about 95°C, just below the boiling level, it could allow the "free" trapped water in the concrete to escape as vapor before it is converted into explosive steam and throws concrete "Shrapnel". Shielding of the heating units should be added to protect them from chemically bound water converting to steam. Raising the temperature to 870°C would calcine the surface of the reactor pit. The depth of penetration of the calcining should be able to be precisely regulated so that unnecessary amounts of uncontaminated material will not be removed. A technique for heating under a covered blanket with provision for trapping escaping gases has been demonstrated by Battelle's Pacific Northwest Laboratory in its process for vitrifying buried contaminated waste in place.

A caveat of this procedure concerns the respiration of vapor from and back into heated concrete. It is known that concrete will exhale its water content in the form of vapor as it is heated. This would have the benefit of exhaling finely divided free radioactive material if it were in a state in which it could be moved by the movement of vapor. If the cleanup of the radioactive inner crust of the reactor pit is not completed before the central portion of the reactor pit wall begins to cool, and reverses the exhalation process, there could be a drawing of atmospheric moisture back into the wall. If this returning moisture passed through radioactive material, it could conceivably draw that radioactivity further into the wall. It is known that calcined concrete does draw atmospheric moisture back into itself. It is not known if non-calcined cement draws atmospheric moisture back into itself after it has been heated: but I would suspect that it does, based on the known action of other porous construction material (wood), and the fact that concrete tends to retain a percentage of free moisture even when "dry".

223

This should not be a concern if the cleanup of the radioactivity proceeds when the heating is complete and the pit has cooled sufficiently to permit work. If the problem is anticipated, proper planning and execution of the work together with temperature monitoring should take care of the problem.

Electrical Resistance Heating

In my 1979 paper, I dismissed the practicality of using electrical resistance heating on buried rebar because of what I believed to be the difficulty of exposing it for electrical connections. Also, I had the mistaken belief that the massive rebar grid would dissipate the current and make the bars impossible to heat.

Professor Kasai's experiments with electrical resistance heating solved the access problem by saw cutting and inserting copper conductor bars on which to attach the electrodes. In addition he established that the electric current is unaffected by the steel rebar gridwork.

His work opens a further line of inquiry into the use of resistance heating, namely the use of electric current to completely melt the vertical rebar to where it pours out the bottom connection. (This lower connection would not be a sawcut bar connection because provision must be made for escape of the molten steel and its safe retention.) Possibly a water jet could be adjusted to remove concrete without damaging the steel.

The theory which supports the complete liquification of the steel reinforcing bar is two fold:
1. The insulating qualities of the concrete hold the heat in the steel rather than allowing it to be dissipated by radiation.
2. The endothermic phase of the heat of fusion phenomenon stores excess heat in the steel until melting occurs, then liberates this excess heat in the following exothermic phase.

This latter phase would tend to sustain and even accelerate the liquification. Since the excess heat should be stored throughout the steel, there should be no "cold" spots to impede a complete flushing.

This is consistent with the explanation of the sudden rise in specific heat of concrete at the 500°C point at which the chemically bound water of cement paste, $Ca(OH)_2$, is liberated in the calcining conversion to $CaO + H_2O$. An inorganic substance absorbs heat until saturated, then goes into a chemical change which releases the heat and the temperature increases.

A principal advantage of completely melting out the steel reinforcing is that the resulting voids could be subsequently filled with either explosives or expanding agents. Present trends are to use the expanding agents rather than the explosives. These expanding agents could be used in conjunction with the established sawcut bar placement technique applied to the horizontal reinforcing steel.

The melt-out method could be used for lengths of rebar 6 meters to 13 meters or longer for each melt. Its distance limitation would be the length of the continuous conductor. It would apply to all thick-walled structures. It would apply particularly to the BWR RCV. The latter has walls 152 cm thick with 4 grids of 5.7 cm rebar on 30 cm centers both horizontally and vertically. It is constructed vertically inside a 60 meter high building with service floors every 6m. The floors could be used as work stations to gain access to the steel, melt it out, and to place expanding agents in the voids.

Summary:

Massive heat applied to a reactor containment vessel can weaken it by:
1. Dehydrating the cement paste, thereby weakening the concrete as well as its chemical bond to the steel reinforcement.
2. Distorting the steel liner and reinforcement to loosen it from its bonds with the concrete.
3. Reducing the strength of both quartzitic and calcarous aggregate concrete to less than 20% of original, thereby weakening its bonds to the steel reinforcement as well as making it easier to demolish.

The fuel cost to accomplish the destruction of a representative RCV using petroleum base coke is estimated at $104,000. U.S.

The concept of weakening concrete by heat has support in theory, economics, and in the observation of fire damaged buildings. The procedure saves steel for recycling; should cost materially less than drilling and placing explosives or ball and chain; and it appears to be safer in that it minimizes human exposure.

Further application of the principles involved suggests that the selective addition of heat to RCV reactor pits could be used for radio-active decontamination.

The reaction of steel to electrical resistance heating has applications in the removal of vertical steel reinforcing bars and the destruction of heavy walls without explosives.

Conclusion:

It would be most interesting to model these theories by computer simulation and to field test them on a reactor containment vessel due for demolition. Until they have been subjected to study and testing, they remain in the field of theory and must continue to be labelled "preliminary".

I would appreciate your thoughts, suggestions and questions, now or later.

References

Smith, R.I., Konzek, G.J. and Kennedy, W.E. Jr. (1978) Technology, Safety and costs of Decommissioning a Reference Pressurized Water Reactor Power Station, NUREG/CR-0130, U.S. Nuclear Regulatory Commission Report by Pacific Northwest Laboratory, p. g-45

Oak, H.D., Holter, G.M., Kennedy, W.E. Jr., Konzek, G.J. (1980) Technology, Safety and Costs of Decommissioning a Reference Boiling Water Reactor Power Station, NUREG/CR-0672, U.S. Nuclear Regulatory Commission Report by Pacific Northwest Laboratory, Appendix L.

McFarland, J.M. (1979) Demolition of Concrete Structures by Heat, Proceedings of the American Nuclear Society Conference on Decontamination and Decommissioning of Nuclear Facilities, Sun Valley, Idaho.

American Society of Heating, Refrigeration, and Air Conditioning Engineers Handbook (1977 ed.) pp. 37.3 and 37.4.

Itoh (1973) Methods of Electrically Destroying Concrete and/or Mortar and Device Therefore, U.S. Patent No. 3,727,982

Vidosic, J.P. Mechanics of Materials, Mark's Mechanical Engineering Handbook, pp. 5-17, 8th Edition.

Kasai, Y. (1985) Demolition of Reinforced Concrete by Heating Reinforcing Bars with Electric Power, Nihon University

Abrams, M.S. (1978) Behavior of Inorganic Materials in Fire. ASTM Special Technical Publication 685, pp. 14-75.

Fintel, M. (1974) Handbook of Concrete Engineering, Van Nostrand Reinhold Co.

Harmathy, T.Z., (1970) Thermal Properties of Concrete at Eleveted Temperatures, ASTM Journal of Materials, Vol. 5, No. 1, pp. 47-75.

JET FLAME CUTTER METHOD

TAIITSU NAKAJIMA
Sumitomo Construction Co., Ltd.

Abstract
The jet flame cutter method, one of the thermal cutting methods used
for dismantling structures by cutting them into blocks, has been
developed, and its use has been studied.

In the jet flame cutter method, the cutting depth can be adjusted
by varying the cutting speed or the fuel flow rate. This method can
also be used for the simultaneous cutting of concrete structures with
a large quantity of steel members, and is particularly effective for
cutting underwater structures. Its utilization is being expanded,
making full use of these features.

The outline of this method and some of its applications are
described below.

Key words: Flame, High temperature, Thermal cutting, Simultaneous
            cutting, Composite member

1. Introduction

As a method for dismantling old structures or underwater stuructures,
the authors have been engaged in the development of the jet flame
cutter method, one of the thermal cutting methods. This is a method
of melting and cutting structures at a high temperature, and uses
oxygen as an oxidant for obtaining a high temperature. Its applications are being expanded to various fields such as the cutting of
concrete, the simultaneous cutting of concrete and steel, the cutting
of composite materials and refractories containing ceramics, and the
cutting and crushing of materials difficult to cut. This method is
also used in combination with other methods for economic reasons.
It is also used for cutting underwater structures because of its advantage that no noise is generated under water as well as its superior
cutting effect.

2. Outline of the method

The jet flame cutter consists of a cutting unit including a cutter for
generating a supersonic jet flame, and a controller for controlling
the flow rate and pressure of oxygen, kerosene, and cooling water

supplied to the cutter, and a drive unit for holding and remotely moving the cutter.

The jet flame cutter method is a method used to melt and cut concrete and other materials at a high temperature and energy, by ejecting kerosene and oxygen supplied from separate systems into a combustion chamber to form a mixture gas, and igniting and burning the gas to generate a flame jet which is ejected from the nozzle.

Since the external walls of the combustion chamber are exposed to high temperatures, they are cooled by cooling water from outside.

3. Structure of Equipment

Fig. 1 shows the schematic diagram of a jet flame cutter. The cutting unit consists of a "supply unit" for supplying kerosene, oxygen and cooling water to the cutter, a "cutter" and a "controller", as shown in Fig. 1.

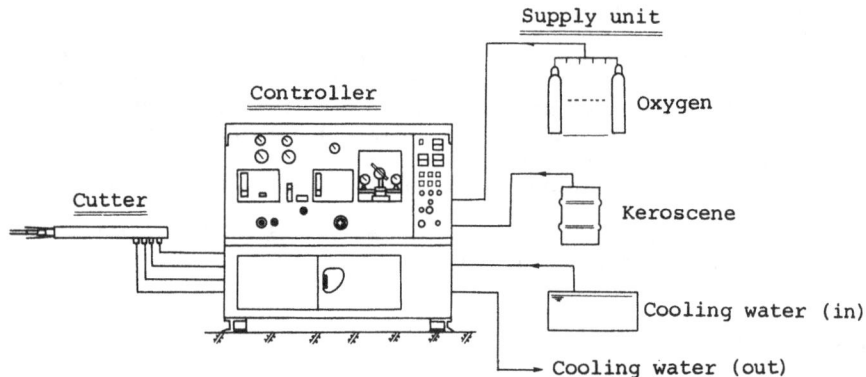

Fig. 1.   Schematic diagram of jet flame cutter

3.1 Supply unit

The supply unit supplies the cutter with kerosene, oxygen and cooling water. Kerosene and cooling water are supplied using a pump, and oxygen is supplied from an oxygen cylinder through a reducing valve except in special cases.

Table 1 shows the specifications of the supply unit.

Table 1.   Specifications of Supply Unit

Item	Specifications	
Oxygen reducing valve	Maximum flow rate (Nm^3/h)	: 500
	Range of secondary side (kgf/cm^2)	: 15 - 50
Kerosene pump	Flow rate (ℓ/min)	: 4.5 - 10.0
	Pressure range (kgf/cm^2)	: 7 - 50
Cooling water pump	Flow rate (ℓ/min)	: 130

3.2 Cutter

The cutter is a device for mixing and burning kerosene and oxygen to generate a jet flame of a high temperature and a high pressure.

The typical specifications of the cutter are shown in Table 2.

Table 2. Specifications of type 20 cutter

Item	Specifications	
Oxygen	Flow rate (Nm^3/min)	: 2.46
	Supply pressure (kgf/cm^2)	: 40.0
Kerosene	Flow rate (ℓ/min)	: 1.0
	Supply pressure (kgf/cm^2)	: 32.0
Burner tube	Throat diameter (mm)	: ϕ6.7
	Outlet diameter (mm)	: ϕ15.0

Fig. 2 shows a cross-sectional view of the burner tube, Photo 1 shows an example of the burner tube, and Photo 2 shows an example of the cutter.

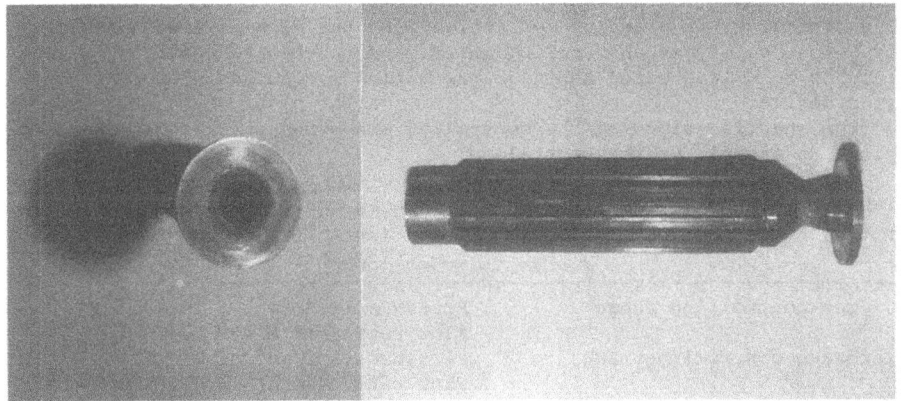

(Front view)   (Side view)

Photo 1. Burner tube

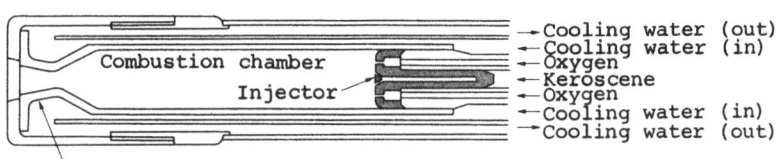

Fig. 2. Cross-sectional view of burner tube

229

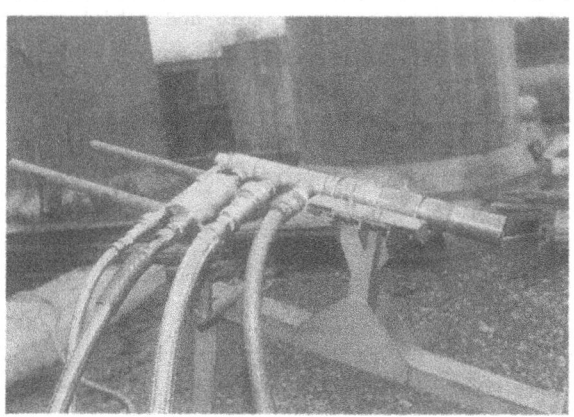

Photo 2.   Cutter

3.3 Controller
The controller is a device for controlling the flow rate and pressure of kerosene, oxygen and cooling water, and incorporates safety devices such as a cooling water shortage prevention mechanism and an emergency stop device.

The specifications of the controller are shown in Table 3.   Photo 3 shows an example of the controller.

Table 3.   Typical specifications of type 20 controller

Item	Specifications
Oxygen controlling range	Pressure (kgf/cm^2) : 15 - 50 Flow rate (Nm^3/min): 0.4 - 4.0
Kerosene controlling range	Pressure (kgf/cm^2) : 0 - 40 Flow rate (ℓ/min) : 0.2 - 2.0
Cooling water controlling range	Pressure (kgf/cm^2) : 0 - 10 Flow rate (ℓ/min) : 14 - 140
Safety device	Check valve, flow switch emergency stop device

Photo 3.   Controller for jet flame cutter

4. Drive unit etc.

The jet flame cutter uses a drive unit which can adjust cutting speeds
and cutting angles, together with an automatic cutting unit having a
remote control mechanism suitable for the materials to be cut.
Devices for the recovery and treatment of byproducts and heat, for
monitoring etc. are combined in the system as required.

5. Cutting performance

The cutting performance of the jet flame cutter depends on the materials to be cut, environment, etc. However, the performance of the
jet flame cutter is varied by the amount of kerosene supplied or the
content of steel members in concrete structures, and the cutting
capacity is greater for a higher steel member content.
Although molten dross formed during cutting affects cutting performance, it has been found that good results can be obtained in
underwater cutting. This method is also used for purposes other than
cutting, such as drilling. As an example of cutting performance, the
change in cutting depth for steel bars of different diameters is shown
in Fig. 3, and the change in cutting depth for different amounts of
kerosene supply is shown in Fig. 4.

231

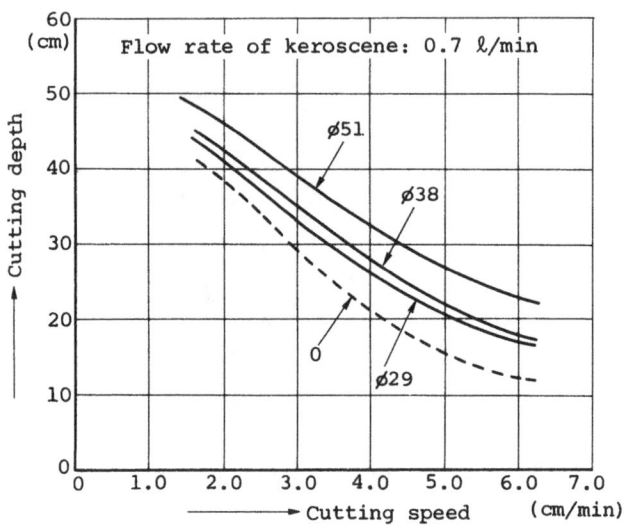

Fig. 3. Change in cutting depth for steel bars
of different diameter (in air)

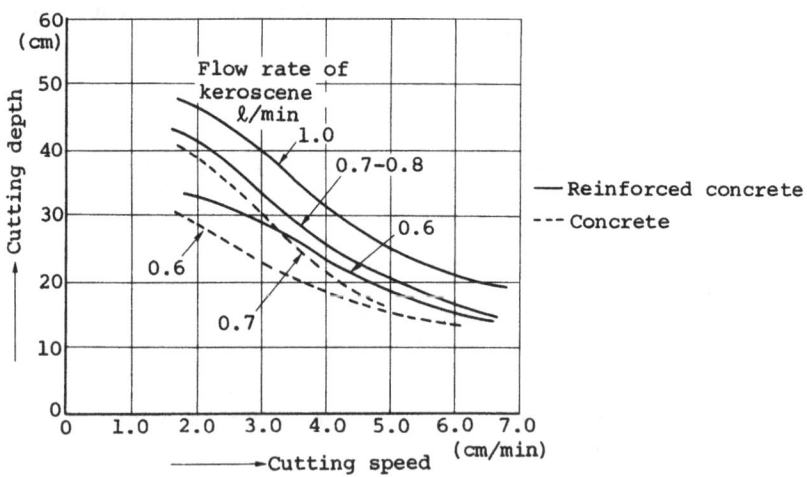

Fig. 4. Change in cutting depth for different
amounts of kerosene supplied (in air)

6. Application examples

Some of the application examples of the jet flame cutter method are described below.

6.1 Dismantlement of Underwater Structure (Double steel tube concrete piles)

Among the dismantlement of temporary structures for the Ohnaruto Bridge, the method was used for underwater cutting of double steel tube concrete piles. Photos 4 - 7 show the cutting conditions (cutting speed: 8 cm/min.).

Photo 4.   Part of temporary structures

Photo 5.   Temporary structures during dismantling

Photo 6. Hanging after cutting   Photo 7. Cut surface

6.2 Dismantlement of old bridge piers

The method was used for cutting a concrete structure with a wall thickness of 600 mm. Photo 8 shows a member cut and dismantled.

Photo 8. Concrete member cut and dismantled

6.3 Cutting of stainless steel

An example of cutting stainless steel (thickness: 100 mm, cutting speed: 100 mm/min) is shown in Photo 9.

Photo 9.   Cutting of stainless steel (Appearance)   (Cross-section)

6.4 Cutting of ceramics

From the results of cutting alumina-silica-zirconia, zirconia, and alumina ceramic materials, the result of cutting an alumina-silica-zirconia ceramic material of a thickness of 75 mm is shown in Photo 10.

Photo 10.   Ceramics after cutting

7. Conclusion

Efforts have been made for improving the cutting performance of the jet flame cutter, and for expanding the fields of its applications.

Various materials are being made available, and techniques are being accumulated.

THERMAL CUTTING BY EUTECTIC GENERATION : PRINCIPLE AND PERFORMANCE

Y. MALIER
Public Works Research Institute (L.C.P.C.), Ministry of Construction, France

Abstract
Constant improvements in the properties of materials and increasingly strict requirements with regard to quality and environmental factors are contributing significantly to the growing difficulties encountered during the modification or demolition of concrete structures.

The solution we have developed in the research laboratory and then implemented in the field in the past 10 years or so ivolves a methodology derived from prefabrication. It required the development of a thermal cutting process, called "eutectic generation", derived from the traditional thermal lance and improved by scientific research in the field of vulcanology.

The principle of this process will be reviewed.

The performance of the equipment will then be described (consumption, speed, depth of cuts, nature of materials, and so on) as well as some results from research relative to the conservation of the local properties of the material subjected to a high temperature gradient in the vicinity of the cut (distribution of temperature and strength, adherence, etc.).

Key words : Thermal cutting.

1. DEMOLITIONS IN THE FUTURE

In the last decade, there has been a growing awareness in all industrialized countries of the technical difficulty of demolishing structures, in particular structures made of reinforced and prestressed concrete.

This difficulty is, moreover, much greater when strict conditions of security, economy, and protection of the environment must be met.

Accordingly, demolition work, hitherto conducted in a manner that is often archaic, and dangerous, costly, and generator of pollution as well, will in the next few years have to become a sector marked by innovations.

The size of this sector, while difficult to ascertain precisely, can be illustrated by a few figures, given as examples :

It would seem that the mass of concrete and heavy masonry demolished in the countries of the European Community in 1980 may be estimated at 40 million tons. Forecasts suggest that this mass will have doubled by

1988 and quintupled by 1995 (a period during which many post-war buildings that waste energy and cannot be rehabilitated at reasonnable cost will be coming down).

In 1983, there were 33 known nuclear power plants worldwide that were more than 20 years old. This figure had doubled by 1987.

2. THE SPECIFIC CHARACTERISTICS OF DEMOLITION PROBLEMS

2.1 Three orders of technical problems :

The stability of the structure during all the intermediate stages of demolition (especially in the case of prestressed concrete structures). This first point in fact affects the satefy of the site and nearby. A look at accident statistics shows beyond a doubt that this is an area in which there is urgent need of improvement in the demolition sector.

Limiting nuisances (noise, dust, vibration) in the vicinity of the site ; we all know that this is increasingly important.

Avoiding damage, in particular mechanical damage, to the parts of the structure that are to be preserved ; this problem must be faced in all rehabilitation and conversion projects, which are becoming increasingly numerous.

2.2 Economic factors are also important
The time an operation will require and its cost, hitherto most often regarded as hard to predict, and almost aways problematical, will have to be estimated by precise, rational methods, because this is in most cases the basis of the decision made by the owner when a major project is concerned.

In addition, the possibilities of recycling materials and components will have to be completely re-analyzed. There is, to be sure, a long tradition of recuperation among demolition contractors, but it was for a long time limited to a few items for which there was a ready market (copper, lead, finishings, and so on). What is needed now is to initiate a vaster and more systematic recycling process accompagnied by some hard thinking about what to do with both materials and components. What are we to think, for example, in the long run, given the price of oil, of the convoys of trucks carrying old concrete from the certres of our larger cities to points 150 kilometres away and regularly passing other trucks carrying aggregates to be used in making new concrete to the same cities ?

3. OUR PROPOSAL OF A DEMOLITION METHOD

On the basis of the various factors we have just mentioned, we believe it possible to put forward our proposal concerning a demolition method, one we have worked to developed both in the research laboratory and industrially (from 1968 to 1980).

This method is in fact inspired by prefabrication, and includes :

(a) a stage in which the structure is cut into large pieces on the site, following a sequence worked out in advance in an engineering office that is based primarily on considerations of the structure's stability and of what can be transported conveniently ;

(b)   a transport stage (as short as possible) ;

(c)   a processing stage.

This approach presupposes :

(a)   the existence of processes for cutting reinforced and prestressed concrete structures that are sufficiently effective and reliable and keep the generation of nuisances to a minimum ;

(b)   the existence of "demolition plants" on the outskirts of large cities, where the waste materials would be systematically conditioned for disposal or recycling (here, one might have the same plants produce components made of recycled concrete).

4. CONSEQUENCES FOR RESEARCH OBJECTIVES

Three directions seem to us to merit priority, and as we shall see in what follows our own work has concerned two of them :

4.1 Cutting processes

Research and development work on these processes must be conducted with three aims :

economy : through a precise determination of the performances and energy consumptions of all processes, conventional or new, with a view to accurate prediction of both cutting rates and operating costs ;
safety : through equipment and machine design that takes the specific conditions at building sites into account ;
protection of the vicinity : through precise quantification both of the local alterations of the cut material in the vicinity of the cut and of the nuisances generated by the cutting operation.

4.2 "Plant processing" techniques

The concept of "demolition plants", permanently installed at fixed locations, would allow for greater use of powerful and efficient processes so far little used because they are hard to adapt to mobile facilities or difficult to amortize under such conditions. Technologies in which there have been major advances in recent years, but from which we have been unable to benefit fully in the context of conventional approaches - I am thinking in particular of explosives and of large crushers - would find considerable scope for action at such plants and would in the long run be economically advantageous, because the plants could be arranged, in particular as regards the abatement of nuisances, to allow the systematic use of these large-scale technologies.

4.3 Processes for recycling the materials

In the case of concrete, for example, the investigations we carried out in France in 1973, and studies conducted elsewhere, notably in the United States,, the Netherlands, Denmark, etc., have demonstrated the feasibility of recycled concretes.

Methodes of composition now becoming available to us make it possibleto prepare recycled concretes having precise strength and workability values that can be used without difficulty to produce high-quality components.

5. RESEARCH AND DEVELOPMENT WORK ON THERMAL CUTTING

5.1 The principale of thermal cutting by the generation of eutectics. "Thermal lance" processes, developed long ago in metallurgy, use the heat of combustion of a metal in an oxygen jet to reach the melting point of the material to be cut so transform it locally into a slag fluid enough to be removed, producing a hole.

In our process, to improve the kinetics of the reaction we combine the development of this exothermal reaction with optimization of the eutectic point of the slag constituted by the mixture ot the metallic oxides produced by the exothermal reaction and the components of the materials to be cut.

For example, taking the binary mixture SiO_2-CaO as approximately representative of the concretes we encounter, let us examine the ternary system SiO_2-CaO-FeO from the point of view of the melting points of the mixtures (fig. 1).

This diagram reveals the existence of a thermal valley that is interesting both because the melting point is relatively low (1100 to 1300°C) and because the valley is rather wide, meaning that it can be reached witout extremely precise proportioning of the three components.

We have consistently found these two features (low temperature level and broad valley) when we have examined systems based on iron oxides and the mineralogical components used in the production of concretes.

It is instructive to note, on the other hand, that aluminium and magnesium oxides, which are produced by reactions that are much more exothermal and would therefore seem to have greater potential (and they are in fact used in other sectors of industry), give us valleys that are often shallower and invairably much narrower.

Using them would therefore call for very precise proportioning, which we regard as incompatible with site equipment we wish above all to make simple and reliable.

5.2 The process
I shall not give a detailed description of the process, for which a number of configurations are possible, depending on the nature of the additives used, which may be in powder and/or wire form. We shall see the simplest of these configurations in the film.

5.3 Performance

(a) Cutting characteristics
The process produces a hole from 25 to 50 mm in diameter, depending on the configuration, in all materials, up to thicknesses of several metres.

The rates attained in silica-aggregate and limestone-aggregate concretes are of the order of 40 to 60 cm/mn (fig. 2).

It should be noted that the presence of reinforcements (in reinforced or

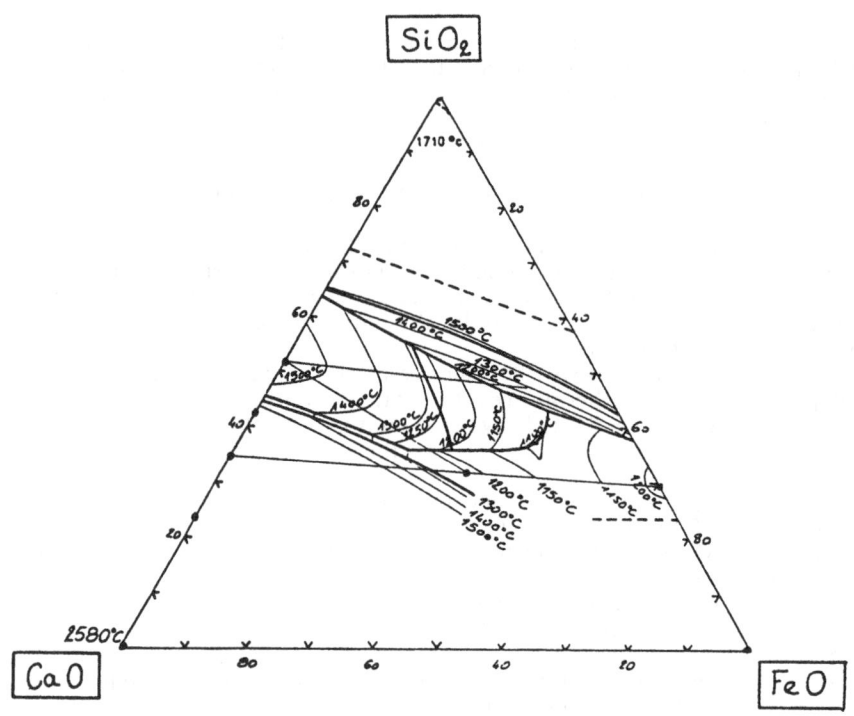

Fig. 1 SiO₂-CaO-FeO system

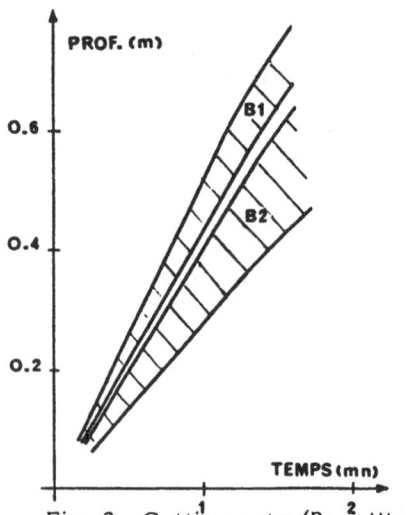

Fig. 2 Cutting rate (B₁, silica-aggregate conc. ; B₂, limestone-aggregate conc.)

Fig. 6 Protection of operator

prestressed concrete) substantially increases the cutting rate (since the iron adds to the heat of the reactions).

But for the engineer calculating cutting times and costs, it is more useful to think in terms not of cutting rate but of the area per unit time (e.g., square metres per hour) of the plane surface produced by the boring of tangent holes. When the work is properly organized, this ranges, in a known manner governed by various parameters, from 0.35 to 0.90 m²/h and may be regarded, for a given type of structure, as constant for the duration of the operation. This last porperty makes it possible to rationalize calculations of the time required for an operation, a feature that is still the exception in demolition work.

(b) Energy consumption

Without going into details, we shall state merely that, here again, the consumptions of oxygen, of metal, and of any additives have been precisely determined. As in the case of the cutting rates, ghese consumptions can be referred to the area cut, and the laws so established allow a rigorous calculation of the costs of an operation.

5.4 Alterations in the vicinity of the cut

In the case of delicate modifications to structures (work on prestressed concrete, on shells, near piping, etc.), the temperature rises in the vicinity of the cut must be known. These temperature rises have been the object of theoretical models and of experimental investigations in which all parameters related to the thermophysical properties of the concrete, the geometry of the element to be cut, the combustion characteristics, etc., were taken into account. We may mention, for example, the temperatures recorded during the cutting of a silica-limestone-aggregate concrete (fig. 3) or, more generally, the probable ranges of maximum temperature (fig. 4) in a concrete shell versus the distance from a plane cut (range A) or a cylindrical hole (range B) produced by thermal cutting.

Such curves may be used to predict with sufficient accuracy any alterations that thermal cutting might produce in the portion of a struture that is to be preserved. In most cases, except where very narrow restrictions are imposed, these alterations may be regarded as negligible at a distance of 5 cm or more from the cut.

The temperatures rises in reinforcements parallel and perpendicular to a cut have also been determined (fig. 5). These curves show that the distance affected by the temperature rise is very short (for example, 5 diameters for a temperature rise of 50°C). This point is important, because it means that the adhesion of cut reinforcements (in reinforced or prestressed concrete) is preserved - as adhesion tests also confirm.

5.5 Nuisances and safety

The only gas thermal cutting uses is oxygen. Unlike other processes, thermal cutting uses no water, no electric power, etc. It is completely silent and generates no vibrations, but it should be pointed out that the sparks produced (in large quantities at the beginning and end of a cut) make it necessary to protect the operator (fig. 6), as we shall see in the film. The quantity of fumes is generally small - negligible in outdoor work - but may become larger during the cutting of a concrete made with aggregates that are highly calcareous (or impregnated with oil, bitumen, and so on) ; if necessary, in the case of indoor work in premises that must remain in use

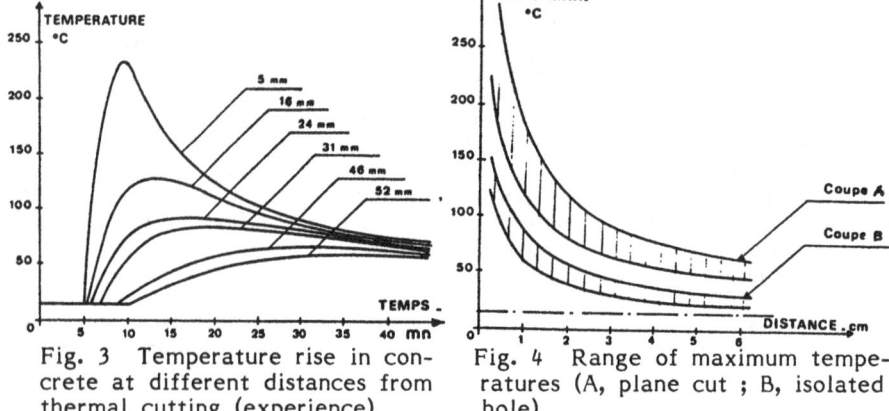

Fig. 3 Temperature rise in concrete at different distances from thermal cutting (experience)

Fig. 4 Range of maximum temperatures (A, plane cut ; B, isolated hole)

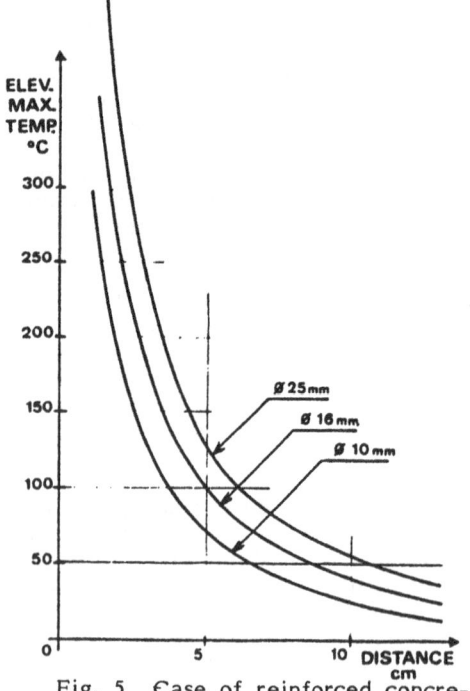

Fig. 5 Case of reinforced concrete : temperature of cut reinforcements

Fig. 7 Example of cutting

242

(hospital, computer centre, shop, etc.), they can if necessary be caught at the source.

5.6 Areas of application

Since the nature of the materials to be cut imposes few limits on its use, thermal cutting may be used extensively in demolition projects, and even more so for the modification of structures in cases where the work must be done quickly and precisely, without vibration or noise, and in strict compliance with safety conditions.

Its reliability and constant productivity are most often the decisive reasons for choosing it. Experience has even shown that, provided that a priliminary study is carried out to match the stages of cutting to the lifting equipment used, the costs of demolishing reinforced concrete structures can be made lower than with conventional techniques.

Finally, the modification and demolition of prestressed concrete structures are applications in which the advantages of this technique really count : by allowing "controlled relief" of the prestressing, it ensures the stability of the structure at all times and ensures safety in its vicinity. The film shows a good example.

CONCLUSIONS

My duties since 1978 have had more to do with the construction than with the demolition of large structures, but from 1968 to 1977 I was engaged in demolition work (industry and research), and I wish to address my conclusion to both research and the industry.

To research to encourage the discovery and development of new processes. There are many available principles, some of which are even used successfully in other sectors of industry, but little work has been done to adapt them to the needs of the construction industry. Others have been tested in the laboratory but have never undergone the full-scale testing essential for their adaptation.

To the industry, because safety, the protection of the environment, and the optimization of costs require a rationalized and structured professional organization like that of construction contractors, with engineering departments, even laboratories, to meet the needs of a growing market.

In the longer term, at least the broad outlines of a general waste recycling policy must be worked out, and its influence on demolition techniques will be so large that the development of the latter cannot be properly analyzed without taking it into account. Such at least is our conviction !

REFERENCES

Bowen N.L., Schairer J.F. and Posnjak E. (1935) The system CaO-FeO-SiO₂, Am. J. Sci, 9, 35
Malier Y. (1977) Le découpage thermique des bétons armés et précontraints, Annales de l'I.T.B.T.P., Paris, 353, 93-112
Malier Y. (1978) La démolition des structures de génie civil par découpage thermique, thèse doctorat ès Sciences, Paris, (E.N.S. Cachan)

Malier Y. (1979) The process of thermal and mechanical cutting of concrete offshore structures, <u>International Symposium on Offshore Structures,</u> F.I.P. - C.E.B. - R.I.L.E.M., Rio de Janeiro, proceed : 11 pages

Turner and Verhagen (1961) Ignesus and metamorphic petrology, <u>Mc grill B.C.,</u> N.Y.

THERMAL CUTTING BY EUTECTIC GENERATION : MONOGRAPHS OF APPLICATIONS TO REINFORCED CONCRETE STRUCTURES

Y. MALIER
Public Works Research Institute (L.C.P.C.), Ministry of Construction, France

Abstract
An analysis is presented of the most pertinent problems that we have encountered.
Then we propose the special examination of some difficult cases :

- demolition of a railway bridge without interrupting traffic,
- demolition of a reinforced concrete theatre between two hotels in service,
- demolition of temporary structures used in the construction of a tunnel under a river,
- demolition of the upper storey of the Conservatoire National de Musique de Paris without interrupting convervatory courses,
- providing large openings between a major computer centre and a telephone exchange, without noise or vibrations, through 60 cm of concrete less than 1 metre from the computers, which continued to operate,
- modification of the statically indeterminate structure of an eight storey building after ground settlement,
- demolition of temporary structures (public works),
- demolition of damaged structural parts.

A synthesis will make it possible to look into the economic aspects, giving some simple rules, and to show that, beyond these economic implications, the rationalization of demolition operations contributes significantly to the improvement of safety.
Key words : Thermal cutting

1. PRELIMINARY REMARKS

It is not our aim here to deal exhaustively with all the applications of thermal cutting. We have simply confined ourselves to an analysis of the main types of technical problems that may be solved advantageously by this technique.

For each of them, we shall examine more particularly an example we consider to be typical.

Thermal cutting is used in its most frequent industrial applications in the building and public works area as a means of meeting one or more of the following types of requirements :

(a) Stability of the structure

In particular, the need to ensure static equilibrium during the work and, for the future structure, to ensure the strength of the conserved part.

(b) Environment and site

Requirements here are almost always related to the limitation of nuisances and, more particularly, noise and vibrations.

(c) Scheduling

Scheduling and lead times are often decisive in selecting the methods and approaches used.

(d) Jobsite safety

Demolishing a structure into large pieces reduces the number of work stations and consequently enables them to be placed in perfectly stable and accessible locations where safety requirements are more easily fulfilled.

The result is that the main application areas are in fact the following :

1. Complete dismantling of reinforced or prestressed concrete structures, with the possibility of using thermal cutting alone or in combination with other techniques.

2. Demolition of temporary structures.

3. Modification of buildings or structures for aesthetic, functional or regulatory reasons (adaptation of premises to the extension of services, adaptation to meet new fire safety regulations, etc.).

4. Modification of structures for stability reasons (let us mention the case of structures subjected to accidental shifting of foundations).

5. Demolition of damaged sections of structures.

6. More special applications, such as underwater work, cutting of rock boulders, and the strengthening of reinforced concrete structures.

Let us now look into the main aspects of these applications.

2. COMPLETE DISMANTLING OF STRUCTURES

2.1 Public works : demolition of a highway bridge over railway tracks. The reinforced concrete bridge was of the type with girders under the pavement and two continuous spans (fig. 1) and had become unsuited for handling urban traffic. Another bridge with a prestressed concrete slab was built alongside it. It was consequently no longer used and had to be demolished without disturbing road traffic on the new bridge and without stopping or slowing railway traffic (in particular, the passage of high-speed turbotrains) coming out of two adjacent tunnels with limited visibility.

The operating procedure, controlled strictly by engineers of French Railways (SNCF) and of the Departmental Directorate of Equipment (DDE), was the following :

1) Bolting of slabs and transverse girders to steel beams resting on the longitudinal girders and on the deck support lines.

2) Cutting (to transport size) of the elements thus suspended.

3) Then, cutting of longitudinal girders and central pillars.

These operations were all carried out in 6 days, without any disturbance to

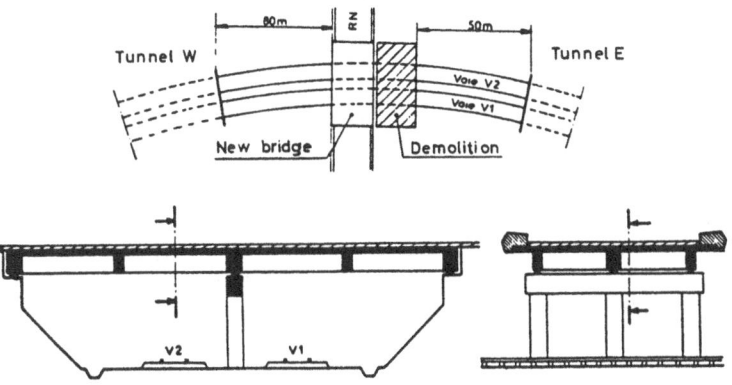

Fig. 1 Demolition of a highway bridge over railways tracks

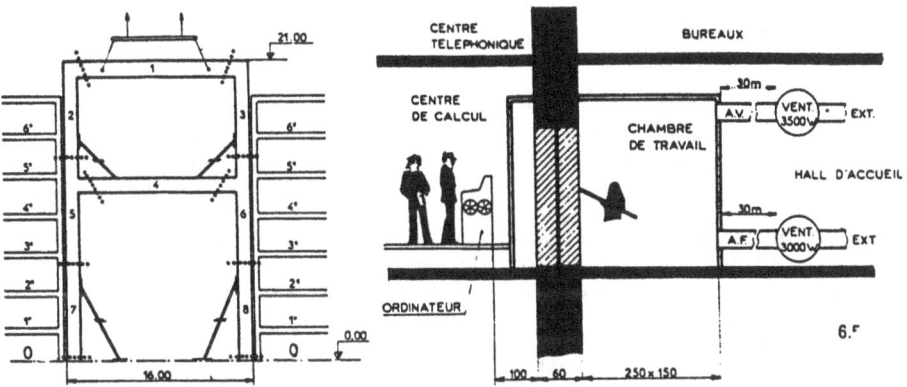

Fig. 2 Demolition of a theatre

Fig. 8 Modification work on a telephone exchange

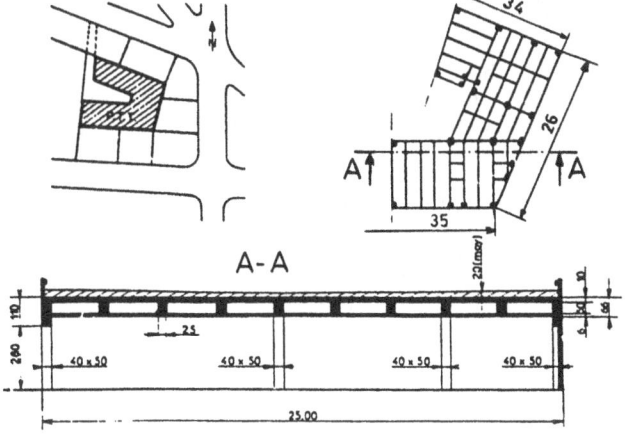

Fig. 9 ... Without
noise and vibration

railway traffic and complying strictly with safety conditions. Moreover, during the buidding, this solution was found to be less costly than the use of more traditional methods.

2.2 Building : demolition of a cinema-theatre

The entire hall (two balconies and an orchestra pit) were supported by seven reinforced concrete portal frames (21 m high, 16 m span) remaining after the rest of the building was demolished by traditional methods. The structure was located between two heavy traffic roads and was between a hotel and a residential and office building.

After installing stability struts, the operations for cutting, removal, recutting to transport size were carried out as indicated in figure 2 using an order compatible with jobsite safety and that of surrounding areas. The operations lasted 7 days.

2.3 Demolition of temporary structures

The removal of temporary structures always involves problems of scheduling, cost and sometimes also of envirnomental disturbance. This may be the case, for example, of diaphragm walls which have become partially unnecessary and which lend themselves well to the silent and fast process of thermal cutting.

Another example may be mentioned : the elimination of a gantry used for placing prestressed concrete caissons (weighing 2,000 t) in the construction of a metro tunnel under the Seine. With a height of 15 m and a span of 20 m, the reinforced concrete portal frame was previously braced in order to ensure stability during the intermediate phase, and cut into three blocks (fig. 3 to 7).

The upper girder was 2.7 m wide and 2.2 m high. Thicknesses at the location of the cuts varied from 40 to 80 cm. The entire operation was completed, by two cutters, in 4 days and in full safety.

2.4 Modification of structures for functional or aesthetic reasons

For these applications, where the structures are still in service, it is important to avoid disrupting the living conditions of the user and any damage to stored merchandise or equipment. These requirements will determine the procedure to be used.

Doing reconstruction work on a telephone exchange is particularly delicate because of the sensitivity of equipment and circuits to vibrations. One application involved the extension of an exchange by rebuilding, after demolishing, the top floor of the five-storey building.

The heavy structure with reinforced concrete beams and columns is shown in figure 8. The average sections are 80 x 30 cm for the beams and 40 x 60 cm for the columns.

In addition, the exchange was surrounded on all four sides by residential buildings and the National Conservatory of Music. It had no direct access to neighbouring streets.

As it was impossible to install a crane, all the reinforced concrete (beams, slabs and columns) had to be cut up into 800 blocks weighing from 300 to 500 kg to be removed by electric hoists.

The demolition required 3 months, owing especially to difficulties involved in getting materials out of the site.

Another important application area is in commercial buildings, hotels

Fig. 3 to 7
Phases of demolition (temporary structure)

and also hospitals. In all these structures, noise at any time is not tolerated, and is one reason why many projects, involving the demolishing of concrete, may have been abandoned. These operations have now become so routine that they hardly require a length description. We shall however cite one example.

This had to do with the provision of openings in a big computer centre, as well as in the lobby of a business building accessible to the public ("Club Méditerranée" in Paris. The project included complex operations, and difficulties were compounded by the fact that it was necessary to allow for communication between the two buildings and that the reinforced concrete separating them was 60 cm thick ! (fig. 9).

In spite of the proximity of computers (the closest was within a metre), cutting was carried out from a fireproof airlock within which the air was renewed constantly. A slight vacuum was maintained to prevent any leakage.

The operations went so smoothly that people in the computer centre did not even realise the work had been completed !

2.5 Modification of structures for stability reasons
In certain cases, thermal cutting can be used effectively to provide discontinuities in structures in the form of :

- expansion joints on elements subjected to temperature variations and not easily accessible for mechanical sawing ;
- break joints designed, in general, to allow the structure to adapt (without increasing stresses) to the effects of foundation changes resulting from differential settlement.

Different projects have been completed, for example on flat elements resting on the ground (open-air theatre gradins) or on beam-column frames in order to reduce the degres of redundancy (statically indeterminate frame).

However, it should be pointed out that such applications, in our view, are of a particular and exceptional nature. They obviously call for perfect knowledge of the mechanical behaviour of structure before and after the operations and very strict compliance with execution requirements.

2.6 Demolition of damaged structural parts
A second field of application we shall examine here has to do with structures subjected to damage.

Thermal cutting can (because it can be checked, is accurate, and does note generate vibrations) be a convenient means of dealing with a structure whose stability has been jeopardized by fire or explosion.

In this chapter, we shall have a tendency to include work carried out during the construction of a structure, either to correct installation errors or to meet belatedly indicated requirements of the owner, leading to partial demolition, or even to the fast demolition of an element having a manufacturing defect or having been subjected to damage during transport.

It is an "eraser" as it were, offered to the project and it should be noted that, in our opinion, such operations (as many others moreover) should not be carried out without the approval of inspection authorities.

It would be too lengthy to enumerate here all the applications in the building area, so we shall simply mention a small project as an illustration. During the construction of a viaduct in prestressed concrete, one of the 50 precast beams (40 m span, 2.20 m high, weighing 100 t) had to be demolished during the launching operations. Eight cuts were required. The work was done within a day without delaying the construction work.

3. ECONOMIC ASPECTS

Economic criteria must of course be borne in mind. It may be stated that there are very precise relationships between the section of the cut and factors on which cost will depend, namely thermal lance and oxygen consumption, on the one hand, and working time on the other. The result is that the cost and time requirements (for a medium or large project) can be established with very good accuracy : on an industrial project, the accuracy is of the order of a few percent.

However, as the unit cost applies to the cutting surfaces and not, as in conventional demolition, to volumes of materials, the total cost of the cutting-dismantling operation depends on the choice and use of handling equipment. This obviously calls for a prior study, which should be as rigorous as any study conducted for a construction project. Thus, among the examples we have given, mention may be made of a certain number for which the particular conditions of the project (environmental disturbances, and so on) were not of primary importance and the bid specifications moreover did not require the use of this technique. The deciding factor was consequently the cost. We cannot overemphasise, in this respect, how important it is for a concrete structure demolition project to be preceded by real "methods" procedures grouping the cutting, handling and removal of materials, combining them if necessary within the demolition-reconstruction operations as a whole. It nevertheless remains that many misconceptions are still currently held with regard to the real cost of such operations.

4. CONCLUSIONS

Concluding, we shall come back to our idea of demolition to state how important we feel it is to further develop new methods.

It was long possible in this area, and rightly so, to be content with a certain empiricism. This has now become insufficient.

In fact, developments taking place in construction, in the areas of masonry, reinforced concrete and prestressed concrete, each involve specific orientations with behaviours in structures that we feel are basically different during demolition.

Making direct use, in one of these areas, of the "guides of good practice" established and accepted conscientiously for any of the others can have serious consequences.

Throughout the years, construction has become rationalised. Demolition has to do the same, at the same rate and along the same lines.

This is the only way we shall be able to ensure that safe conditions are maintained on the jobsite as well as around the jobsite.

Going beyond this important problem, if man is to maintain control over the factors determining the quality of this own life and the wealth of the heritage he is to transmit to future generations, it is important for us, in these times of rapid urban development, for him to acquire the means of intercepting, modifying and even demolishing whatever, deep down inside, he is no longer satisfied with.

REFERENCES

Malier Y. (1977) Le découpage thermique des bétons armés et précontraints, Annales de l'I.T.B.T.P., 353, 93-113.

Malier Y. (1976) Le découpage thermique du béton, Symposium ACI (Chapter Quebec Ontario - G. Haddad).

Fig. 10 Underwater cutting operations on
reinforced concrete

STRIPPING DEMOLITION OF SEMI-CIRCULAR RC WALL BY APPLYING ELECTRIC
CURRENT THROUGH REINFORCING BARS

YOSHIO KASAI
College of Industrial Technology, Nihon University
WAHEI NAKAGAWA, KIWAMU NISHITA and TOSHIYUKI SUGAWARA
Nuclear Engineering Department, Maeda Construction Co., Ltd.

Abstract
When electric current is applied through reinforcing bars in a
concrete structure, cracks grow in the concrete. Therefore, the
cover concrete of the reinforced concrete structure is readily exfoliated.
    The present report describes an experimental study with a semi-
cylindrical mock-up simulating a part of the biological shield wall
in a small nuclear reactor. The following conclusions have been
obtained:
    (1) Exfoliation after applying current is achieved readilier than
that without current.
    (2) Most of demolition concrete chips, after current application,
are in lumps or plates with less quantity of fine chips.
Key words: Stripping demolition, Electric current through reinforcing
bars, Biological shield wall, Cracks, Concrete dust

1. Introduction

When electric current is applied through reiforcing bars in a
concrete structure, cracks grow in the concrete. Therefore, the
cover concrete of the reinforced concrete structure is readily exfoliated. In a previous report (Ref. 1), a basic experiment with
small-scale models is described. In another paper (Ref. 2), the
result of a demonstration test with a plane full-scale model is
described. The present report describes an experimental study with a
semi-cylindrical mock-up. Major subjects to be studied with the present experiment include the effect of shear reinforcing bars on
stripping property and the volume of produced concrete dust. In the
present experiment, dust was classified into "flying dust" which was
collected by the CPS high-volume method (also partially collected by
the optical scattering method), "precipitation dust" which fell or
was included in demolition chips (less than 53 μm) and "demolition
chip" whose grain size was 53 μm or above.

2. Method of experiment

2.1 Specimen
The specimen is a semi-cylindrical wall simulating a part of the biological shield wall in a small nuclear reactor. With cut grooves, the specimen was separated into four blocks of A, B, C and D. Two blocks A and C contained shear reinforcing bars while omitting the bars in the other two blocks. Fig. 1 and 2 show the shape and bar arrangement of the specimen. Steel bars conform to JIS SD35m D25 and D29 (deformed bar 25mmØ and 29mmØ). The mix proportion of used concrete is shown in Table 1.

2.2 Method of applying electric current for heating
Electric current was applied to bars by the following method. Contact electrodes, equipped with push-out mechanisms, were contacted with exposed bars in cut grooves to apply heating current, from a generator through an induction voltage regulator and a transformer. Bars ①, ② and ③ shown in Fig. 3 were heated up by electric current. With ① and ②, bars in the first layer were heated up with both ends not restricted by cutting bars in grooves b and c. With ③, bars in grooves were not cut leaving both ends restricted, thereby bars in the second layer were heated up.

In order to control the amount of current conducted in bars, conduction time was determined by another specimen for calibration with the same length of bars as ①, ② and ③, in which a constant current of 3000 A was conducted while measuring time to increase bar temperature to 400°C.

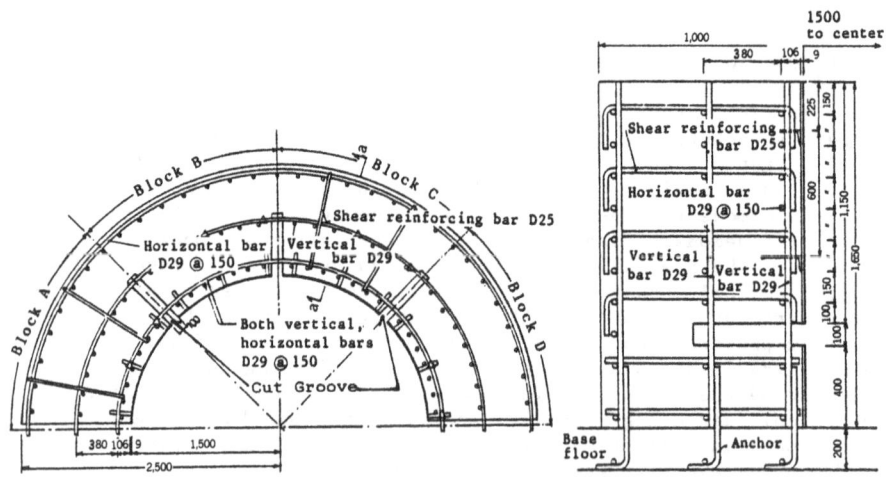

Fig.1 Horizontal section of the specimen   Fig.2 Bar arrangement view

Table-1 Mix proportion of concrete

Nominal strength kgf/cm²	Coarse aggregate mm	Slump cm	Air %	W/C %	S/a %	Unit quantity kg/m³				
						Water	Cement	Fine aggregate	Coarse aggregate	Admixture
240	25	12	4	530	44.3	157	297	813	1,035	0.743

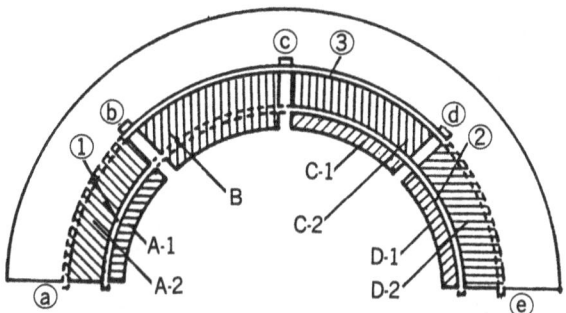

Fig. 3 Methods of current application and exfoliation

2.3 Measurement of bar temperature

The CA (chromel-alumel) thermocouples were embedded in bars after drilling holes, for measuring surface temperature of the bars. A temperature recorder was placed in a shield box to prevent the effect of magnetic field during current application.

2.4 Measurement of cracks

Cracks, produced on the top and side face of the specimen and in grooves during the current application were visually observed and recorded in photos and video tape.

2.5 Method of exfoliation

Exfoliation position was classified into seven blocks shown in Fig. 3, in each of which the wedge of a large hydraulic hammer was driven along steel bars in the first or the second layer after completion of heating, thereby exfoliating concrete in the surface layer. Time for working exfoliation and the number of hammer strokes were measured. Driving force of this hammer was about 40 tons. For exfoliating the second layer of Block D-2 was done by an exfoliating worker using a hand-held breaker to compare and examine dust data.

2.6 Methods of measuring demolition dust and chips

In order to measure demolition dust created during exfoliation and demolition work, a sealed chamber was constructed, whose wall and the ceiling were covered with PVC sheets while finishing the floor with epoxy lining. In order to prevent oil mist of the hammer, the base machine was placed outside the sealed chamber, while also covering the arm and hammer, located in the sealed chamber, being covered with PVC sheets.

Flying dust was collected by the CPS high-volume method and the optical scattering method. The former equipments and suction nozzles were moved for each exfoliation unit. The latter was measured in fixture. Fig. 4 shows the method of measurement.

Time for finishing measurement of dust was specified that the concentration of flying dust in the sealed chamber after exfoliation returned to the concentration before starting exfoliation. As soon as the dust concentration returned to that before exfoliation, the operators entered the sealed chamber while collecting fine particles with an evacuating device. Small exfoliation chips were collected in PVC bags and then measured for grain size and weight. Exfoliation lumps were treated to remove fine particles adhering on the surface and measured in dimensions and weight. Table 2 shows the devices and methods of measurement.

Table-2 Measuring devices and methods

Measuring item		Method	Device type	Remarks
Flying dust	Mass	Optical scattering method	Shibata Chemical Machinery Model AP-637	Suction rate; 30 /min Instant value(analogtype) Mean value (digital type)
		High-volume sampler method	Kimoto Denshi Model 121A	Suction rate; 60 /min
	Grain size	C.P.S. sampler method	Kimoto Denshi	1.6-18 um (4 steps) at 600 /min
Precipitation dust	Grain size	Standard sieve method	—	53, 74, 105, 149, 210, 250, 420, 500, 840, 1000, 1680, 5000 um
		Optical transmission method	Shimadzu SA-CP2-20	Less than 53 um
Demolition chip	Grain size	JIS A 1102	—	JIS Z 8801 Mesh Sieves 100, 50, 25, 20, 15, 10, 5 mm

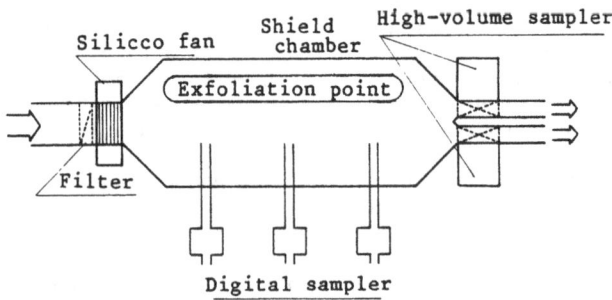

Fig. 4 Method of measurement

3. Results and discussions

3.1 Tests of concrete
Table 3 shows results of tests with concrete used for the specimen.

3.2 Heating with electric current
With the calibration specimen, a constant current value of 3000 A was maintained during current application at 50 Hz, resulting in total current application time up to 400°C of 4 minutes in the length of ①, and 4 minutes 30 seconds for lengths of ② and ③. However, it took 5-6 minutes until 400°C was achieved with the actual specimen. This difference resulted from unequal concrete temperatures because of different volumes of the actual specimen and calibration specimen. Consequently, current application time was controlled at 6 minutes with a constant current of 3000 A. Fig. 5 shows current application time vs. bar temperature for 7 bars in 3 (Blocks C-D). Fig. 6 indicates relationships between current application time and values of current, voltage and cumulative electric power.

Table-3 Results of concrete tests

Curing	Under water		In air	Test method
Age	28 days	35 days	35 days	
Mean compressive strength (kgf/cm²)	305	325	213	JIS A 1108
Mean tensile strength (kgf/cm²)	—	—	22.5	JIS A 1113 (Crack test)
Mean water content (%)	—	—	4.7	Weight method (Japan Cement Association)

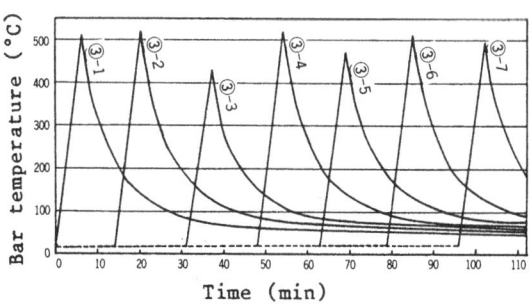

Fig. 5 Current application time-bar temperature chart

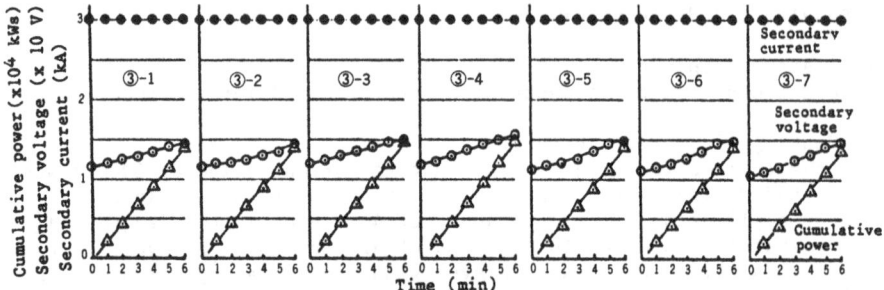

Fig.6 Current application time-current-voltage-cumulative power chart

3.3 Cracks in concrete

Fig. 7 shows the state of cracks on the top face of the specimen after current application. Results of observing cracks are shown in the following.

(1) ① (Block A): Cracks at the top face of the specimen occurred along the steel form. On the side face, cracks occurred radially to bars in addition to cracks between current conducting bars, because of expanded and pushed-out bars.

(2) ② (Blocks C-D): Cracks in Block C with shear reinforcing bars occurred between current conducting bars in grooves. At the top face, cracks, occurred only in a part of the left side where bars were not restrained. However, in Block D without shear reinforcing bars, cracks at the top face occurred along bars. In grooves, cracks occurred in straight lines along bars from the upper end to the lower end. Cracks on the side face are the same as those in ①.

(3) ③ (Blocks B-C): Cracks in the top face and grooves occurred along current-applied bars in both Blocks B and C. Widths of cracks were larger than those in ① and ② because current-applied bars in these blocks were not restrained at both ends. Widths of cracks in Block B without shear reinforcing bars were 11 mm at maximum, larger than those in Block C where there were shear reinforcing bars. Crack widths did not become narrower even after completion of applying current. Surface concrete was naturally exfoliated during current application.

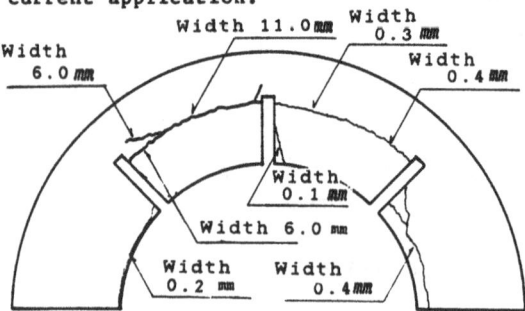

Fig. 7 Crack status

3.4 Surface exfoliation

Table 4 shows time required for exfoliating concrete using a large hydraulic hammer and a hand-held breaker and the number of strokes of the large hydraulic hammer. In the blocks without shear reinforcing bars, flake exfoliation occurred easily in 0.3 minute in Block B with the largest cracks and in 1 minute in Block D-1 with the next largest cracks. In the blocks with shear reinforcing bars, however, exfoliation occurred while leaving shear reinforcing bars both with and without current application. By comparing exfoliation in the second layer between each case above, time of exfoliation in Block A-2, without current applied, is about 2.5 times as large as exfoliation time of Block C-2 with current applied. The maximum exfoliation time, occurred in Block D-2 using the hand-held breaker, was about two times as large as in Block A-2 where no current was applied with shear reinforcing bars.

Table-4   Exfoliation time and number of strokes

Block	Tool	Current	Shear reinfor-cing bar	Exfoliating time (min)	No. of strokes (times)
A-1	Large hydrau-lic hammer	Yes	Yes	7	1,084
A-2		No	Yes	25	4,290
B		Yes	No	0.3	21
C-1		Yes	Yes	12	1,201
C-2		Yes	Yes	10	2,510
D-1		Yes	No	1	99
D-2	Hand-held breaker	No	No	53	-

3.5 Demolition dust and chips
  (1) Flying dust

A total of dust collected by the CPS high-volume method and the optical scattering method was defined as flying dust. The volume of dust collected by the optical scattering method was obtained using the accumulation method from the time variation chart of flying dust concentration in the sealed chamber. The volume of flying dust, produced during exfoliation, was greatly affected by the presence of shear reinforcing bars, application of current, exfoliating method (large hydraulic hammer or hand-held breaker) produced the largest volume of dust per unit exfoliation volume (g/m^3), namely 85 times as large as in Block B. The volume of dust collected in Block A-2 (without current) was twice as much as in Block C-2 (with current). Fig. 8 shows the volume of produced flying dust.

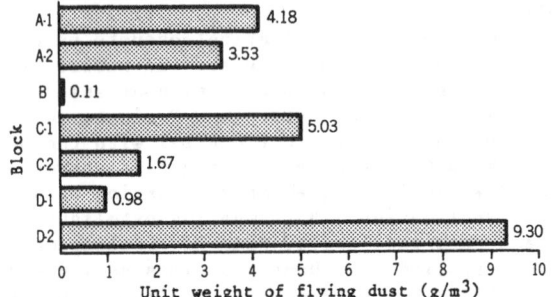

Fig. 8 Comparison of unit weights of flying dust

(2) Time variation of concentration of flying dust
Fig. 9 shows results of measured flying dust concentration using the optical scattering method. The pattern of dust is classified into the following three types.
  1 Pattern I: Blocks A-2, D-2
No current is applied in this case for both blocks. From the beginning of exfoliation, the concentration of flying dust suddenly increases arriving at about 6 mg/m^3 but quickly decreases upon comletion of exfoliating.
  2 Pattern II: Blocks A-1, C-1, C-2
There are shear reinforcing bars and current applied in this case for all three blocks. The concentration of dust reaches about 3 mg/m^3, then attenuates more slowly than in Pattern I.
  3 Pattern III: Blocks B, D-1
This pattern represents exfoliation where there are currents applied in both blocks without shear reinforcing bars. The concentration remains extremely low because the exfoliation was completed in a very short time (less than 1 minute). In addition, concentration damping speed is extremely low after completion of exfoliating.

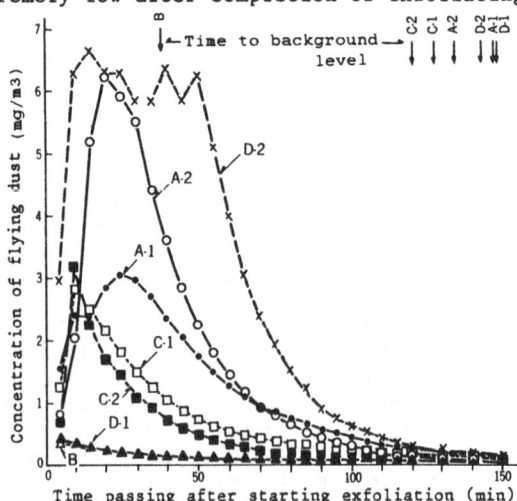

Fig.9 Time variation of concentration of flying dust in sealed chamber

(3) Precipitation dust
Fig. 10 shows results of measured dust smaller than 53 um, using the
optical transmission method. This volume is within a range of
0.01-0.10% of the volume of demolition concrete.
(4) Demolition chips
Fig. 11 shows the distribution of grain sizes for demolition chips.
In Blocks B and D-1 without shear reinforcing bars, flake exfoliation
occurred along and including reinforcing bars. Therefore, the volume
of demolition chips was extremely small. The distribution of grain
sizes for demolition chips in other blocks was substantially the same
in tendency.

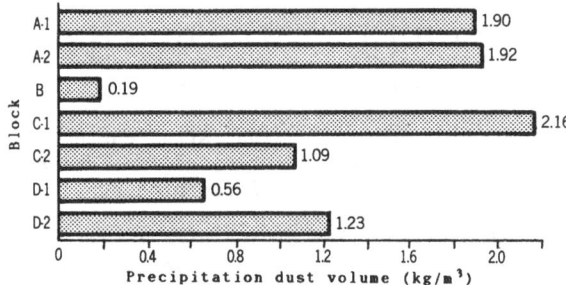

Fig. 10 Precipitation dust volume

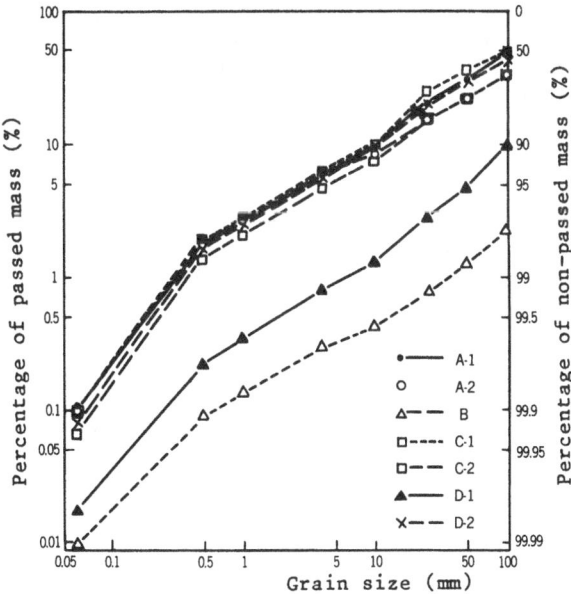

Fig.11 Distribution of grain sizes of demolition chips

4. Conclusions

The following conclusions have been obtained as a result of the present experiment using the semi-cylindrical specimen.

(1) Cracks produced during current application for heating occur more advantageously when both ends of current conducting bars are not cut, than cases where they are cut.

(2) Exfoliation after applying current, without shear reinforcing bars, has been proved to be excellent, like the result with the previous experiment. With shear reinforcing bars, time for exfoliation was reduced to less than 1/2 with current applied than the result without current.

(3) Of the volume of flying dust in demolition after current conduction, absence of shear reinforcing bars brings about extremely small dust volume. Even with shear reinforcing bars, the volume was only 1/2 of cases without current applied.

(4) Most of demolition concrete chips, after current application, were in lumps or plates with small quantity of fine chips. This means that the apparent volume of demolition products is so small as being much advantageous for disposing of the waste. The present experiment has proved applicability of the direct current application to steel bars for concrete exfoliation and demolition with RC shield wall in full scale and actual specifications. The authors would like to proceed with developing more advanced system including remote application.

Acknowledgement: The authors would like to express their sincere thanks to Prof. M. Kawamura, College of Science and Engineering, Nihon University for his informative guidance in carrying out the present experiment. The authors also sincerely thank Yamato Alloy Ltd. for their cooperation in the experiment.

References

Kasai, Y. et al., (1983) Experiments on Stripping Demolition of Concrete Cover by Applying Electric Current through Reinforcing Bars. Transactions of the 5th Japan Concrete Institute, Vol. 5, pp 97-100.

Kasai, Y. et al., (1985) Stripping Demolition of Sealed Wall in Nuclear Power Plant by Applying Electric Current through Reinforcing Bars. Transactions of the Japan Concrete Institute, Vol. 7, pp 97-104.

Kasai, Y. et al., (1984) Experiments on Stripping Demolition of Concrete Cover by Applying Electric Current through Reinforcing Bars - Part 2. Proceedings of the Annual Meeting of the Architectural Institute of Japan, pp 553-554.

DEMOLITION WORK FOR OPENING OF DIAPHRAGM-WALL SHAFT BY APPLYING ELECTRIC CURRENT THROUGH REINFORCING BARS

YOSHIO KASAI
College of Industrial Technology, Nihon University
WAHEI NAKAGAWA, KIWAMU NISHITA, TOSHIYUKI SUGAWARA
Nuclear Engineering Dept., Maeda Construction Co., Ltd.

Abstract
Heating method by applying electric current can be employed to the
partial demolition of reinforced concrete structures, such as cast-
in-site reinforced concrete diaphragm-wall shafts. Therefore, it
is necessary to prove the stability of demolition boundary.
    After a series of fundamental experiments and mock-up experiment
it was proven that the present method is effectively employable to
opening work in the diaphragm-wall shaft. This report describes an
example of actual demolition work creating shield-machine driving
opening in the cast-in-site reinforced concrete diaphragm-wall shaft.
Key words: Heating by applying electric current, Diaphragm-wall
shaft, Partial Demolition, shield driving port

1. Introduction

The heating method by applying electric current through reinforcing
bars has been developed as an exfoliation and demolition technology
for radiated and contaminated concrete surfaces upon the decommissioning of nuclear power facilities. This method can be employed
to a variety of demolitions. Especially, it is effective for the
case that demolition is decided in the stage of design. The example
is a partial demolition work in the diaphragm-wall shaft.
    The demolition of diaphragm-wall shaft is awfully difficult,
because the wall is thick including large quantity of reinforcing
bars and its concrete strength is high. In addition, working
environment is bad, because places of demolition are deep under the
ground and narrow. To solve these problems, application of the present method was decided.
    In the stage of assembling cages with reinforcing bars for
diaphragm-wall, reinforcing bars for current-application are arranged
at opening portions beforehand. Then immediately before demolition
work, these reinforcing bars are heated by applying electric current.
As a result, demolition efficiency is improved while especially fine
chips and dusts are reduced and therefore working environment can be
improved. For application of the present method to the partial demolition of reinforced concrete, a series of fundamental experiments
was carried out. Furthermore, a mock-up experiment was performed
with full-scale models on the ground. The mock-up experiment and a
actually executed work are shown below.

263

2. Mock-up experiment

2.1 Object of the experiment
The object of the experiment is to grasp thermal influence upon concrete at main body side, demolition efficiency and problems on demolition.

2.2 Outline of the experiment
The present specimen of full-scale model was placed on the ground simulating the diaphragm-wall shaft. The diameter of the opening was 3 m and the wall thickness was 70 cm. Using this specimen, temperatures of reinforcing bars during application of electric current and concrete and reinforcing bars at demolition boundary were measured to check thermal influence upon the main body. In addition, results of current-application were comparatively studied by applying or not applying electric current or changing devices of demolition.

2.3 Specimen
The specimen has three blocks, A, B, C as shown in Fig. 1. Construction arrangements of structural bars were same in each block, i.e., double arrangement including D32 @150 for vertical bars and D22 @200 for horizontal bars was employed. For the bars obtaining current-application, horizontal bars at demolition portion were used as shown in Fig. 2. A vertical layer of reinforcing bars (D22, @200) was additionally set up at the middle of double bar arrangement for applying electric current.
　　The mix proportion of concrete is shown in Table 1.

2.4 Equipment for applying electric current
Equipment for applying electric current was prepared with consideration of following items.
　　(1) The maximum output voltage was 50V for safety.
　　(2) Output current was large to obtain efficient heat generation.
　　(3) Simultaneous application of electric current to three reinforcing bars was enabled to have efficient applying work of electric current.
　　Table 2 shows specifications of devices for applying electric current.

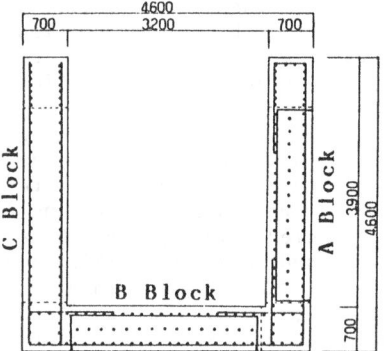

Fig.1　Plan of the Specimen

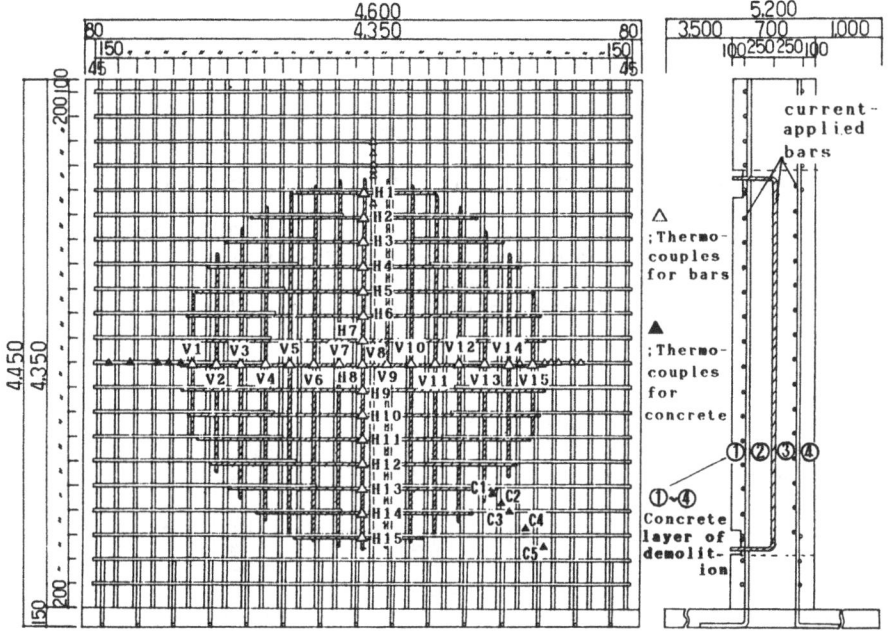

Fig.2 Arrangement of Bars and Current-applied Bars

Table-1 Mix proportion of concrete

Nominal strength kgf/cm^2	Coarse aggregate mm	Slump cm	Air %	W/C %	S/a %	Unit quantity kg/m^3				
						Water	Cement	Fine aggregate	Coarse aggregate	Admixture
350	25	18	4	42.5	41.6	174	409	704	1,014	1.022

Table-2 Specifications of equipment for applying electric current

Device	Output /input	Qty of phase	Voltage (V)	Current (A)	Frequency (Hz)
Controller (100% of continuous rating)	Input	3 phases, 3 wires	200	Max 500	50
	Output	3 phases, 3 wires	40-400 variable	Max 250	400
Transformer (100% ccontinuous rating)	Input	3 phases, 3 wires	400	Max 250	400
	Output	1 phase, 3 circuits	50/25 Max	1150/2300	400

2.5 Procedure of experiment
(1) Applying procedure of electric current
Foamed styrenes covering edge portions of reinforcing bars to be
current-applied were removed, then insulators such as iron rust and
mortars were removed by grinding. Electric current was applied
following procedure to Blocks A and B shown in Fig. 3. Block C was
without current-application.

Temperatures of reinforcing bars and concrete were measured using
a thermo-couples (cromel-alumel type) shown in Fig. 2. These data
were entered and recorded in a personal computer. These temperatures
were simultaneously monitored using a cathode-ray tube display to
control temperatures. These temperature measuring devices were set
up in a steel-plated chamber to prevent influence of magnetic field.

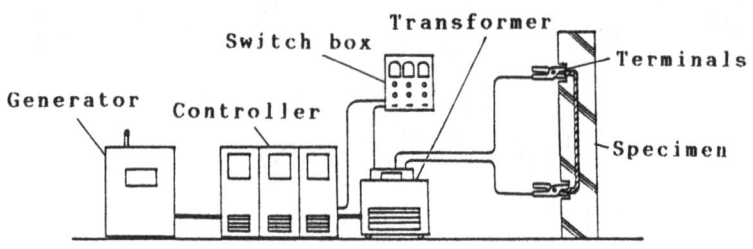

Fig.3   Layout of Electrical Equipments

2.6 Procedure of demolition
Demolition was carried out using machines shown in Table 3.
Demolition efficiency was recorded using photos and sketches. The
following shows the outline.
(1) Block A
Two workers demolished concrete with hand-held breakers until reinforcing bars appeared. Both ends of an exposed reinforcing bar were
cut with acetylene gas to remove the reinforcing bar. These works
were carried out for each layer of current-applied bars.
(2) Block B
Demolition and cutting work were carried out for each layer by a
large hydraulic hammer and a worker who cut reinforcing bars. For
the demolition boundary, two workers carried out demoliton with hand-
held breakers.
(3) Block C
Three workers carried out demolishing and cutting work for each layer
following same procedure described in Block A.

Table-3   Devices for demolition

Block A	Hand-held breaker #10 (weight 10 kg), 1 unit
	Hand-held breaKER #30 (weight 30 kg), 2 units
Block B	Large hydraulic hammer 0.4 m³, 1 unit
	Hand-held breaker #30, 2 units
Block C	Hand-held breaker #30, 3 units

Table-4 Results of concrete tests

Curing	Under water		In air	Test method
Age	7 days	28 days	28 days	
Mean compressive strength (kgf/cm^2)	277	384	328	JIS A 1108
Mean tensile strength (kgf/cm^2)	—	29.4	28.2	JIS A 1113 (Crack test)

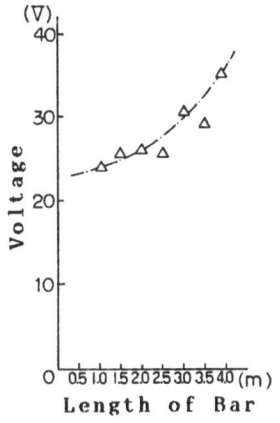

Length of Bar

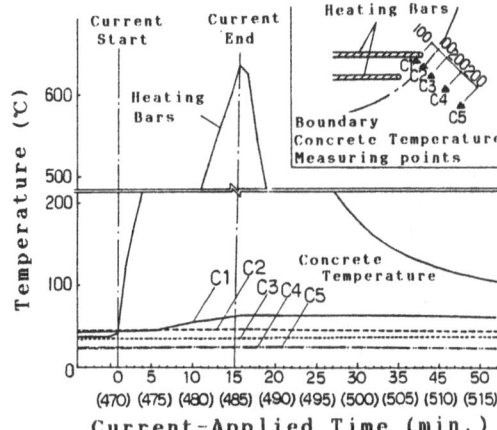

Current-Applied Time (min.)

Fig.4 Voltage Changing   Fig.5 Temperatures in the Concrete

2.7 Result of the experiment
    (1) Test of concrete
Table 4 shows test result of concrete properties used for specimens.
    (2) Heating by applying electric current
Constant electric current of about 1,000A was appplied. Regarding of
reinforcing bar length (max. length 3.8 m), the time until temperature of a reinforcing bar reached 400°C was approximately 10
minutes. On the other hand, voltage was increased as a reinforcing
bar became longer as shown in Fig. 4 showing increase of input
energy.
    Fig. 5 shows temperature influence of concrete at a main body when
reinforcing bars were heated by applying electric current. In this
figure, the concrete temperatures (C1 - C5) upon start of electric
current application were increased higher than the initial temperature, because they were influenced by reinforcing bars of the
first and second layers where electric current had been already
applied. At the beginning of experiment, the initial temperatures
concrete were 11°C in average. Temperatures were kept being recorded

for about 1 hour after all electric current applications were completed. From this figure, the following items were revealed.

1) C1 was located at the demolition portion 10 cm inner from the demolition boundary. Reinforcing bars in the second layer (V10, V11 V12) and those in the third layer (H13, H14, H15) which were located nearest to C1 were continuously current-applied. Therefore, C1 received the most severe thermal influence. However, only about 54°C was increased from the initial temperature (11°C).

2) C2 was located at the demolition boundary 10 cm apart from the curretn-applied bar end. Temperature was not increased after application of electric current to the nearest H13, H14 and H15. The temperature influenced by the reinforcing bars of the first and the second layers was maintained. Temperature of C2 was increased by only about 35°C from the initial temperature.

3) C3, C4 and C5 were located at the main body side. They were not directly influenced by H13, H14 and H15 similarly to C2.

(3) Demolition
Fine chips and dusts produced by demolition at Blocks A and B where electric currents were applied were extremely fewer than Block C where electric current was not applied.
In addition, demolition efficiency at Block A with handheld breakers was approximatley 1.6 times of Block C.
Table 5 shows demolition results. Demolition conditions of each block are described below.

1) Block A
Concrete of the first layer was exfoliated in shapes of flat plates as shown in Photo 1. The demolition was completed after 1 hour. The other concretes of the second, third and fourth layers were demolished in much larger lump than that of Block C. However, chipping and cutting efficiencies for removing reinforcing bars were same as those of Block C.

2) Block B
The first layer was exfoliated only after 10 minutes. The remaining layers were also demolished in a very short time. The demolished lumps were big as shown in Photo 2. Some of them needed to be crushed for scrapping.

3) Block C
There were remarkably lots of fine chips and dusts during demolition. Demolished pieces were small and even as shown in Photo 3.

Table-5 Demolition results

Block	Days for current-application	Days for demolition	Number of demolishing workers	Efficiency of demolition works
A	1	4	9	0.62 (m^3/man-day)
B	1	1	2	11.20 (m^3/unit)
C	-	5	14	0.40 (m^3/man-day)

Photo.1   A Block

Photo.2   B Block

Photo.3   C Block

3. Example of executed work

3.1 General
The present work was executed for shield-machine driving opening in the diaphgragm-wall shaft for constructing a sewerage structure. As shown in Fig. 6, the opening was located at approximately 20 m under the ground. Its diameter was 7.3 m.

3.2 Configuration
The diaphragm-wall shaft had double arrangement with main reinforcing bars (vertical bars) D32 @125 and distribution reinforcing bars (horizontal bars) D22 @250. The concrete design strength was 350 kg/cm^2. Wall thickness was 1.1 m. The opening with 7.3 m diameter was set up on three elements of the wall.

3.3 Reinforcing bars for applying electric current
Reinforcing bars for applying electric current were basically placed as shown in Fig. 7 using D22 @250. As shown in Fig. 8, four layers of reinforcing bars for applying electric current were set up. Number of the current-applied bars was 189.

3.4 Devices for applying electric current
Devices for applying electric current were those used for the full-scale experiment (shown in Table 2). Those devices were placed as shown in Fig. 9.

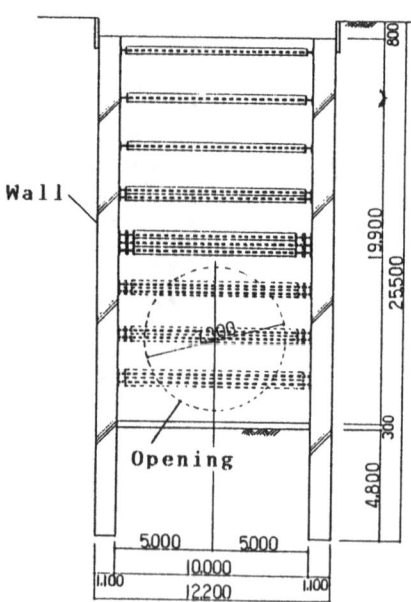

Fig.6   Section of Diaphragm-Wall Shaft

270

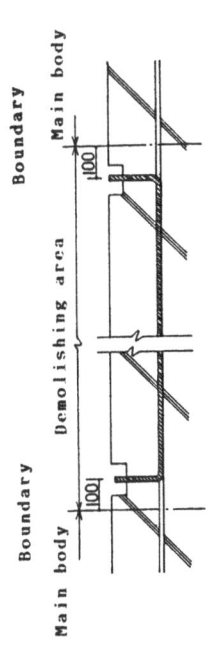

Fig.7 Concept of Setting
Current-applied
Bar

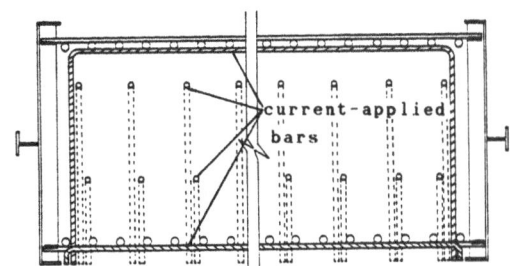

Fig.8 Arrengement of
Current-applied Bars

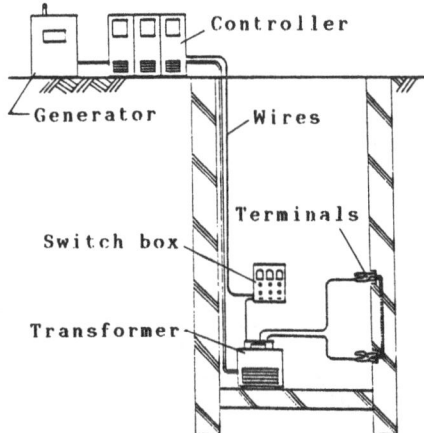

Fig.9 Layout of Electrical
Equipments

3.5 Procedure of work
(1) Preparation work
Foamed styrenes covering edge portions of reinforcing bars were removed, then the ends of bars to be clipped by terminals were cleaned.
(2) Current-applying work
Electric currents were simultaneously applied to three reinforcing bars in general. Temperatures of reinforcing bars were controlled by checking current-applying time, voltage and current.
(3) Demolishing work
A hydraulic breaker (0.7 m^3) was used for demolition.
The demolition boundary was demolished by hand-held breakers.

3.6 Result of work
(1) Current-applying work
Current-applying work was carried out at night (8.00 p.m. - 5.00 a.m.) because it made no noise. Current-applying work was executed

271

from the upper part proceding together with progress of excavating work in the shaft performed in the daytime. A hundred and eighty-nine reinforcing bars were current-applied during five nights.

(2) Demolishing work

A hydraulic breaker operator, a worker cutting reinforcing bars and a worker sprinkling water carried out demolishing work of four layers. Four demolishing workers carried out demolitions of the fifth layer and the demolition boundary using hand-held breakers after assembly of the shield machine. Table-6 shows actual results of demolition.

Table-6 Actual results of demolitions

Layer	Thickness (mm)	Demolishing time (min)	Days for demolition (day)	Discharging of demolished concrete (day)	Demolished quantity (m^3)
1	100	78			
2	250	23			
3	400	350	3	1	41
4	250	217			
5	100	-	2	0.5	5

4. Conclusion

In the fundamental experiments by applying electric current through reinforcing bars for RC structures, temperatures of reinforcing bars and concrete transmitted to the main body side, cracks and bond force with temperature rise of reinforcing bars were measured. In addition, with the full-scale experiment, temperature increase caused by stored thermal energy was measured when electric currents were applied to three layers. Through these results of measurements, it was proven that the main body side was not adversely influenced by current-application.

After the full-scale experiment and actually executed work, it was revealed that further rationalized constructions would be required. For example, improvement of electrode terminals, decreasing weight of devices and standardization of bar arranging method are necessary.

References

Kasai, Y. et al., (1983) Experiments on Stripping Demolition of Concrete Cover by Applying Electric Current through Reinforcing Bars. Transactions of the 5th Japan Concrete Institute, Vol. 5, pp 97-100.
Kasai, Y. et al., (1985) Stripping Demolition of Sealed Wall in Nuclear Power Plant by Applying Electric Current through Rein-forcing Bars. Transactions of the Japan Concrete Institute, Vol. 7, pp 97-104.

STUDY ON DISMANTLING METHOD OF REINFORCED CONCRETE BY INDUCTIVE HEATING

M. MASHIMO, K. OMATSUZAWA and Y. NISHIZAWA
Architectural & Structural Engineering Division, Tokyo Electric Power Services Co., Ltd.

Abstract
The paper describes an experimental and analytical investigation of the demolition of surface concrete utilized electrical inductive heating to reinforcing steel bars. The objective was to examine the effects of input power and frequency of electric current applied to an inductor on the demolition of surface concrete. The specimens used in the experiment were reinforced concrete with a size of $1.1 \times 1.1 \times 0.4 \, m^3$, and steel bars bundled in lattice for a preparatory experiment. In order to evaluate the distributions of temperature and stress within the specimen, heat conduction and thermal stress analyses were performed by using a two dimensional finite element model, and the analytical results were compared with those of the experiments.
Key words : Inductive heating, Reinforced concrete, Input power, Frequency, Heat conduction, Thermal stress.

1. Introduction

When nuclear facilities are dismantled, it is desirable to separate radiated concrete from non-radiated part wherever possible. Normally only surface concrete of the structures, such as the primary shield wall in the reactor building, is radiated, so a method for demolishing only surface concrete is required. Many methods for demolishing surface concrete have been developed. The method developed by Kasai et al.(1983) is to demolish surface concrete by applying electric current to reinforcing steel bars directly. The authors (Mashimo et al. 1987) also proposed use of electrical heating. Used, however, was not direct heating but inductive heating of steel bars. The merit of the inductive heating is that the steel bars can be heated at a some distance from themselves without any contact, and need not be exposed.

This paper describes the effects of power and frequency of electric current on the demolition of surface concrete by inductive heating based on experimental and analytical investigations.

2. Experimental study

2.1 Conditions
Conductive steel is heated in an alternating magnetic field induced by applying high frequency electric current to a coil as an inductor. This heating method is called inductive heating.

Figure 1 shows an experimental condition. That is, a rectangular coil having a side of square of 65 cm is placed on the reinforced concrete specimen illustrated in Figure 2. And, a high frequency current is applied to the coil. Heating time was 5 minutes to 10 minutes. Before the experiments of the reinforced concrete specimens, the specimens of steel bars bundled in lattice were examined to find the thermal characteristics of the steel bars by inductive heating.

The experimental cases are summarized in Table 1. The parameters of the experiment are input power and frequency of electric current applied to the coil. Table 2 shows the properties of steel bar and concrete. The steel bars, having a diameter of 38 mm, were welded at cross points of the lattice so as to constantly induce electric current.

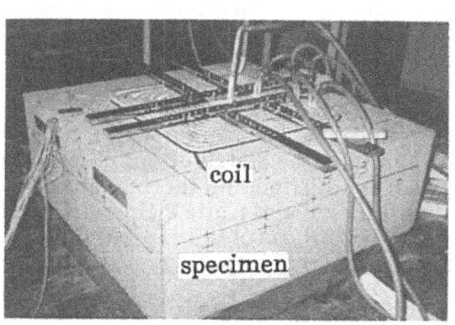

Figure 1. Experimental condition.

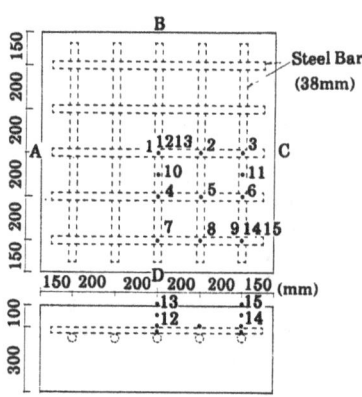

Figure 2. Reinforced concrete specimen.

Table 1. Experimental case.

Case No.	Power (kW)	Frequency (kHz)	Note
1	100	3	
2	100	32	
3	100	200	
4	200	3	
5	200	32	
6	200	200	
7	200	3	Coil movement

Table 2. Material properties.

Concrete	Compressive Strength (kg/cm²)	312
	Tensile Strength (kg/cm²)	28
Steel Bar	SD35 , D38 (JIS)	

2.2 Results

Figure 3 shows the relationship between time and temperature for the steel bar specimens against various input powers and frequencies. Note that the temperature is shown as the average temperature rise of each measured point. According to this figure, the frequency effect is seen to be small.

The temperature characteristics of reinforced concrete specimens for each frequency at the input power of 200 kW are shown in Figures 4, 5, and 6, while the vertical axis is a temperature rise of each measured point (see Figure 2). These figures show that the temperature of the concrete at points 12, 13, 14 and 15 increases only slightly at the early stage, and shows much difference in the temperature of the steel bars among each measured point. Moreover, it is found that the frequency effect on the reinforced concrete specimen is as small as on the steel bar specimen. At 20~30 seconds of heating time when cracks appeared on the concrete surface, the highest temperature of steel bars reached about 200°C.

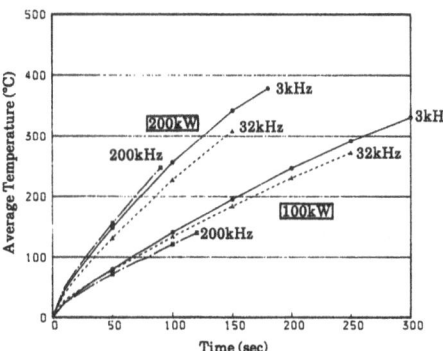

Figure 3. Temperature characteristics of steel bar.

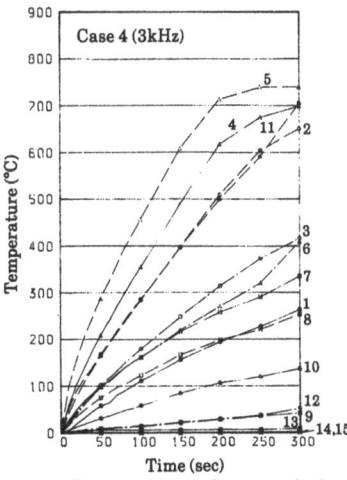

Figure 4. Temperature characteristics of reinforced concrete (3kHz, 200kW).

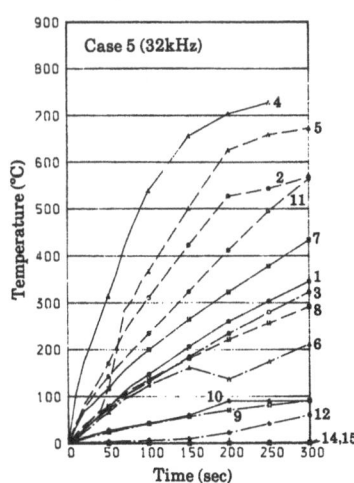

Figure 5. Temperature characteristics of reinforced concrete (32kHz, 200kW).

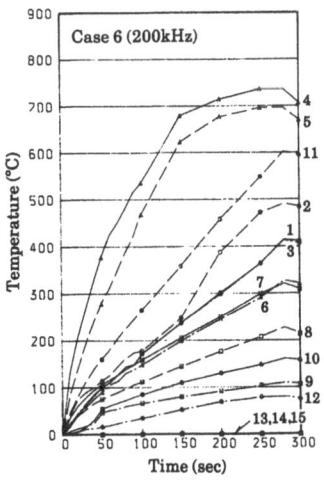

Figure 6. Temperature characteristics of reinforced concrete (200kHz, 200kW).

The demolishing load of surface concrete was measured by using the equipment shown in Figure 7 to confirm the reduction of the load by the cracks (see Figure 8). The values of the demolishing load are summarized in Table 3. It is found that the load was small when input power was large. Also, particularly in Case 7, in which the coil was moved on the concrete surface, the load value is noticeably small. The effect of frequency can not be seen in this table.

However, it is considered that the difference in frequency causes different distributions of temperature within the bar due to a skin effect of electricity, so the temperatures around the bar were measured. Figure 9 shows the relationships between time and temperature at each point around the bar for three frequencies. In the case of higher frequency, the temperature of the upper part was higher and that of the lower part lower when compared in the case of low frequency.

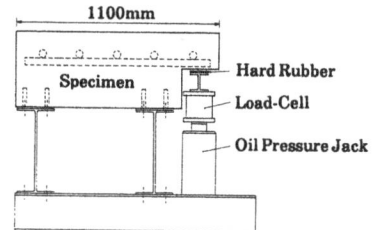

Figure 7. Demolishing equipment.

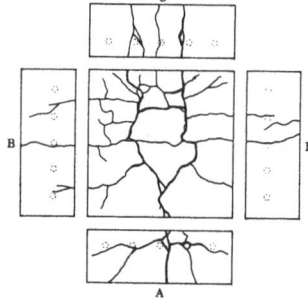

Figure 8. Crack pattern.
(3kHz, 200kW)

Table 3. Demolishing load.

Case No.	Power (kW)	Frequency (kHz)	Load (ton)	Note
1	100	3	13.5	
2	100	32	12.9	
3	100	200	13.0	
4	200	3	11.5	
5	200	32	11.5	
6	200	200	11.0	
7	200	3	10.3	Coil movement
8	–	–	19.0	Non-heated

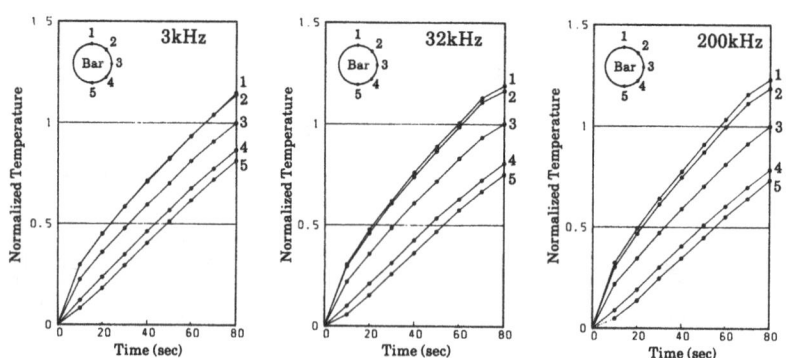

Figure 9. Surrounding temperature of bar by experiment.

276

3. Numerical Analysis

It is supposed that the distribution of temperature in the bar differs according to the frequency. As such, the distribution of stress in concrete was estimated by numerical analysis for two types of thermal load that show different distribution of temperature in the bar. This analysis comprises heat conduction analysis and thermal stress analysis which were performed by a two dimensional finite element method assuming constant material properties. The model is shown in Figure 10.

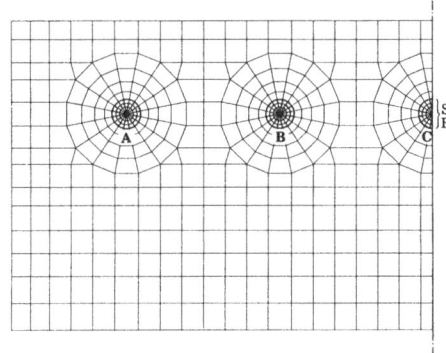

Figure 10. Analytical model (FEM).

(a) Load Case 1   (b) Load Case 2

Figure 11. Thermal load.
Dark colored elements indicate heat source. Total thermal quantities are equal in both cases.

3.1 Heat conduction analysis
As shown in Figure 11, two load cases, one being a low frequency model and the other a high frequency model, were considered. Heat was assumed to generate from bar elements which are colored dark in the figure. The value of total thermal quantity per bar was assumed to be 15000 kcal/h·m for both load cases, that is, the different thermal quantity per element for both cases. The material properties used in the analysis are shown in Table 4.

Figure 12 shows the temperature rise around bar B. Figure 13 shows the distribution of temperature at 30 seconds of heating time. The temperatures of concrete elements are almost zero degree except for the concrete surrounding the bars. A remarkable difference is noted in the detailed values of temperature between Load Case 1 and 2.

Table 4. Material properties for analysis.

	Density ρ (kg/m³)	Specific Heat C (kcal/kg·°C)	Thermal Conductivity λ (kcal/m·h·°C)	Modulus of Elasticity E (kg/cm²)	Poisson's Ratio ν	Coefficient of Linear Expansion α
Steel	7800	0.11	46	2.1×10^6	0.333	1.1×10^{-5}
Concrete	2200	0.21	1.4	2.1×10^5	0.167	1.0×10^{-5}

277

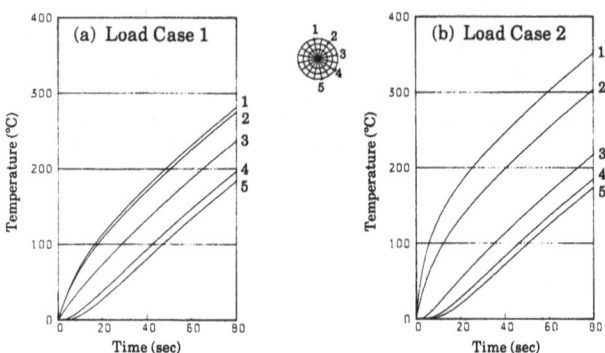

Figure 12. Surrounding temperature of Bar B by analysis.

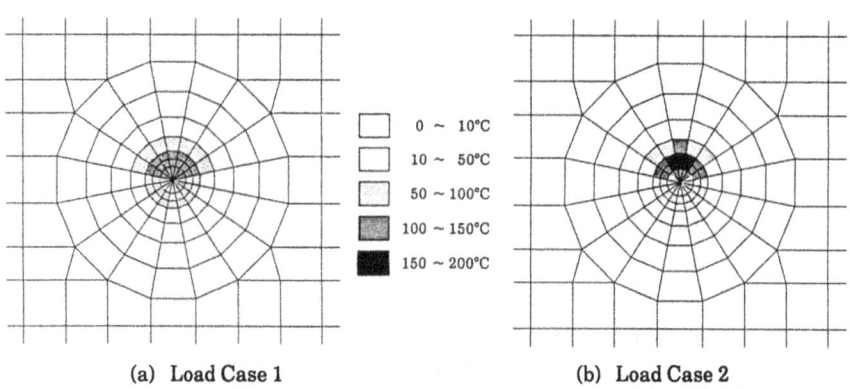

(a) Load Case 1

(b) Load Case 2

Figure 13. Distribution of temperature at 30 seconds
of heating time by analysis.

3.2 Thermal stress analysis

The thermal stress in the concrete was calculated on the basis of the results
of the heat conduction analysis. The distributions of tensile stress at 30
seconds of heating time are shown in Figure 14, while the stress value of
each element is marked in different color thickness corresponding to the
stress. For elements between bars or bar and the concrete surface, the
tensile stress of concrete exceeded 25 kg/cm^2, as shown in this figure. It is
noteworthy that the distributions of tensile stress of both cases indicated the
same tendency in spite of the different distributions of temperature as shown
in Figures 12 and 13. This result agrees with the experimental result
showing that the frequency effect was negligible.

From the above results, it is considered that high frequency, such as a 200
kHz and low frequency, such as a 3 kHz, do not differ much in total energy to
demolish the concrete around bar if the total thermal quantity generated
from the bar is equal.

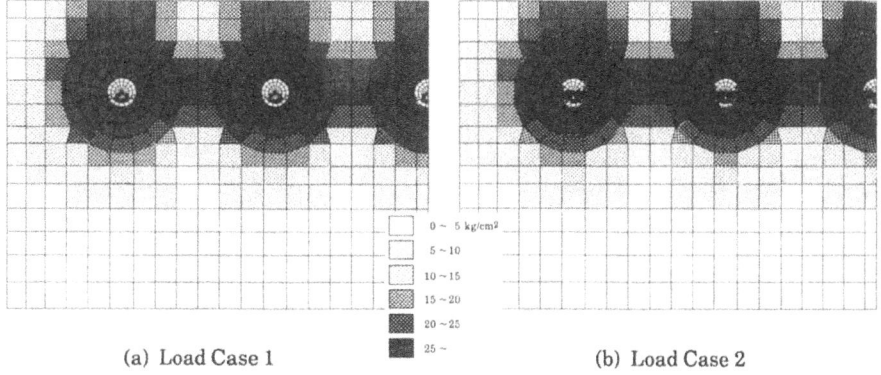

	$0 \sim 5$ kg/cm^2
	$5 \sim 10$
	$10 \sim 15$
	$15 \sim 20$
	$20 \sim 25$
	$25 \sim$

(a) Load Case 1   (b) Load Case 2

Figure 14. Distribution of tensile stress at 30 seconds
of heating time by analysis.

4. Conclusions

To examine the effects of input power and frequency of electric current
applied to an inductor in the case of applying electric inductive heating for
demolition of surface concrete, an experimental and analytical investigation
were performed. By heating the reinforcing steel bars inductively, cracks
appeared in the concrete surface, thereby easy demolition of surface concrete
was feasible. That is, the demolishing load of surface concrete for a heated
specimen was reduced to half as compared with that for a non-heated specimen. According to the results of experiments and analyses, frequency of
electric current had only a small effect on the demolition of surface concrete.
In regard to the effect of input power, it is found that the larger the input
power, the smaller the demolishing load.

5. Acknowledgments

The authors would like to express their gratitude to Professor Masatada
Kawamura of Nihon University of Japan for his valuable advice and
suggestions. Denki Kogyo Co.,Ltd. and Nippon Concrete Industries Co.,Ltd.
of Japan are also gratefully acknowledged for the cooperation of this
experimental study.

References

Kasai, Y. and Akiyama, N. (1983) Stripping demolition of concrete cover by
    applying electric current through reinforcing bars. 5th Annu. Conv.,
    J. C. I., pp97-100. (in Japanese)
Mashimo, M., Omatsuzawa, K. and Nishizawa, Y. (1987) Fundamental
    study on dismantling method of reinforced concrete by inductive
    heating. Trans. 9th SMiRT, Vol. H, pp603-608.

MICROWAVE IRRADIATION TECHNOLOGY FOR CONTAMINATED CONCRETE SURFACE REMOVAL

H. YASUNAKA, Y. IWASAKI, K. MATSUTANI, T. YAMATE, M. SHIBAMOTO,
M. HATAKEYAMA, T. MOMMA and E. TACHIKAWA
Department of Japan Power Demonstration Reactor,
Japan Atomic Energy Research Institute

Abstract
The paper briefly describes the microwave irradiation technology for
contaminated concrete surface removal and its necessity in reactor
decommissioning procedure. Japan Atomic Energy Research Institute
(JAERI) developed a prototype microwave concrete surface remover with
its performance test. The objectives of the work are to develop
effective equipment to remove a few centimeters of radioactively contaminated concrete surface in the Japan Power Demonstration Reactor
(JPDR) and to confirm its usefullness compared with the other conventional mechanical equipment. As a result of the test, some of the
operational data and prospect of its application to the actual dismantlement of the JPDR were obtained.
Key Words: Radioactive contamination, Decommissioning, Dismantlement,
Nuclear facilities, Radiation exposure, Concrete surface contamination, Microwave concrete surface remover.

1. Introduction

Nuclear power facilities such as a reactor building, a turbine building and a radioactive waste treatment building are to be dismantled
and demolished in a safe and rational manner. As a part of the
development of demolishing technology, the JAERI has been developing
a microwave irradiation technology for contaminated concrete surface
removal in the JPDR since 1983. Its usefullness for concrete surface
decontamination was confirmed through performance test. This paper
describes the performance of the microwave concrete surface remover
and its applicability.

1.1 Contaminated concrete surface removal in nuclear facilities

The decommissioning of nuclear facilities produces a large quantity
of concrete waste which is contaminated by spills and leaks of radioactive liquid and steam during operation. The radioactive concrete
waste must be treated and disposed in a controlled manner. If the
whole concrete structure is regarded as radioactive waste, the amount

This work was performed by the Japan Atomic Energy Research Institute
under contract with the Science and Technology Agency of Japan.

of contaminated waste will be about 15,000m³ in case of a large
commercial nuclear power plant. Usually, however, most of such contamination does not penetrate the concrete structures more than a few
centimeters. Therefore, the removal of the contaminated concrete
surface layer prior to demolition of buildings is very important to
reduce the volume of radioactive waste. It reduces the total quantity of radioactive concrete waste to 1/50 - 1/200 of those of the
whole concrete material used for the facilities. After the removal
of this thin layer, the other part of concrete structure can be
treated as a non-radioactive building. Conventional demolishing
techniques and machines can be applied to the building and no special
treatment for radiation exposure of workers is necessary.

1.2 Concrete contamination in JPDR

JAERI surveyed concrete surface contamination as a part of the
planned schedule for the dismantlement of the JPDR. Loose contamination was measured by the smear method, and penetrating contamination was measured by analyzing drilled samples. Figure 1 shows
distribution of contamination on the first basement of the turbine
building. Contamination was widely distributed on floors and partly
found on walls. Penetration was usually limited within 2cm from the
surface of the concrete. Figure 2 shows contamination area as a
function of penetration-depth. Eighty three percent of contamination
was within 2cm from the surface in the JPDR.

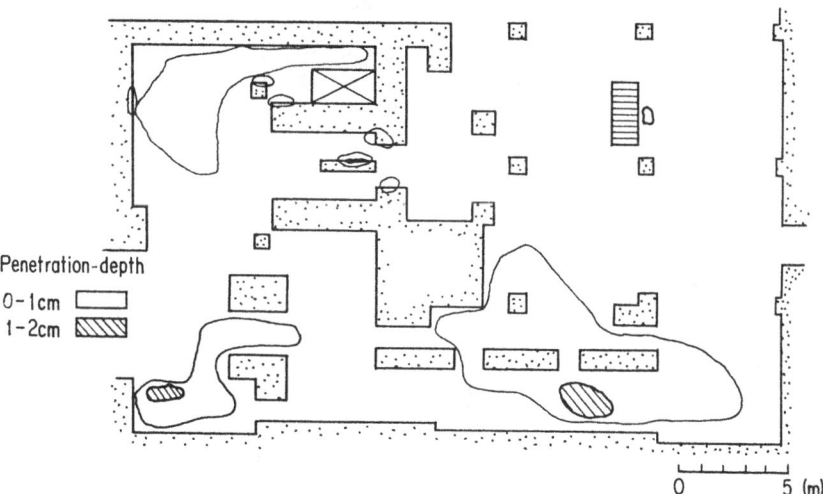

Fig.1 Concrete surface contamination in the first basement of JPDR
turbine building

281

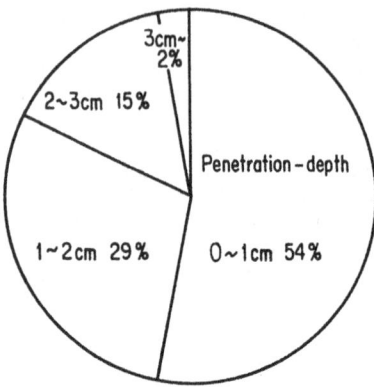

Fig.2 Contaminated area by penetration depth

2. Microwave concrete surface remover

2.1 Mechanism

Concrete contamination limited to surface coating can be removed by
conventional equipments such as scabblers and plainers. However,
they hardly provide sufficient performance to remove penetrating contamination. A microwave concrete surface remover has been developed
for removing this type of contamination. In exposing concrete to
microwave, energy is absorbed within a few centimeters from the
surface and heats water therein. This results in establishing high
steam pressure followed by concrete surface removal by a spalling
process as shown in Fig. 3. The curves shown in Fig. 4 are plots of
transition of the temperature inside irradiated concrete. Peak
values in this figure indicate the spalling point.

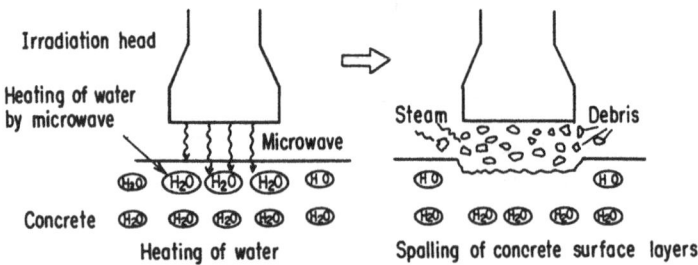

Fig.3 Concrete surface removal mechanism by microwave irradiation

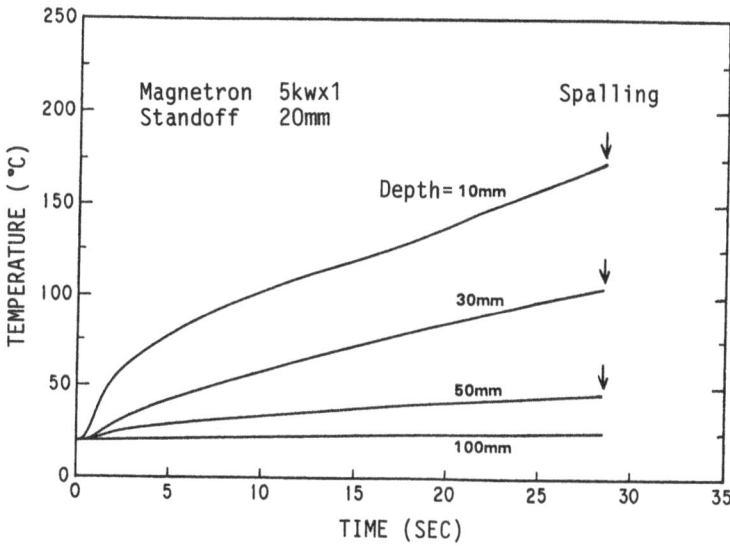

Fig.4 Transition of temperature inside irradiated concrete

2.2 Device

A photograph of the prototype microwave concrete surface remover is shown in Fig. 5. It consists of three 5kw magnetron units arranged in parallel generating 2450MHz microwaves (① in Fig. 5), wave guides (②) and irradiation heads (③). Microwave propagates through the wave guide to the head and onto the concrete surface. The head is shifted at a constant speed for continuous surface removal and can be oriented to walls, ceilings and corners as well as floors. The standoff (distance between the head and concrete surface) is adjustable by raising or lowering the whole system. Concrete debris is collected by the vacuum cleaner attached to the microwave irradiation head.

2.3 Performance test

Performance tests were conducted to obtain operational data for the microwave concrete surface remover. The results were:

(1) Shape of irradiation head
Figure 6 shows cross sectional view of the concrete surface spalled by three types of irradiation heads, such as a spreadtype, a straight type and a tapered type. The spread type removes the concrete surface smoothly compared with the other heads. Therefore, the spread type was chosen as the irradiation head.

283

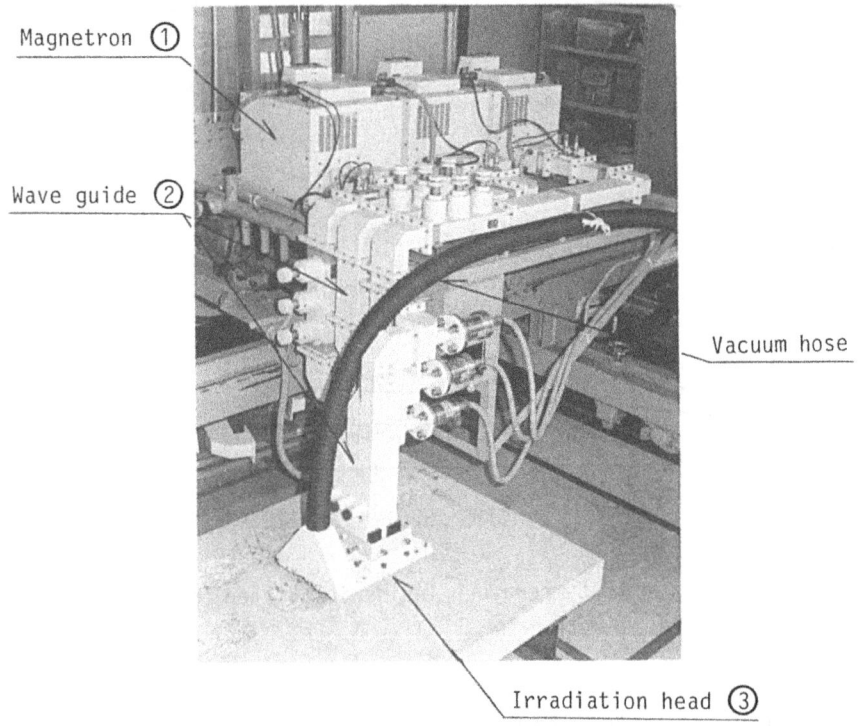

Magnetron ①

Wave guide ②

Vacuum hose

Irradiation head ③

Fig.5 Microwave concrete surface remover

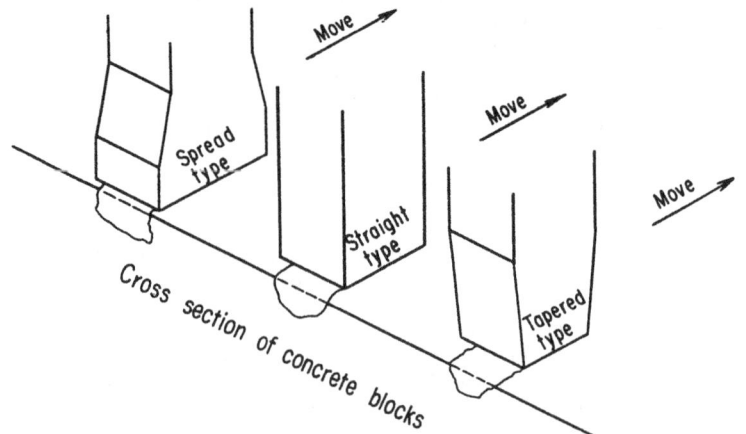

Fig.6 Shapes of cross sections of concrete surface spalled by three
types of microwave irradiation heads

284

(2) Effect of parallel irradiation
The removed material volume was compared for one, two and three mag-
netrons. Three units in parallel removed up to 14 times the volume
of one unit, and 6 times the volume of two units. This demonstrates
the effectiveness of parallel irradiation.
(3) Standoff
The removal depth decreases as the standoff increases as shown in
Fig. 7. A standoff of 20mm was selected to avoid concrete debris
spalled from the irradiated surface during the operation, while
keeping a sufficient removal depth.
(4) Head moving rate
The effect of the head moving rate on removal depth is shown in
Fig. 8. The removal depth varies inversely as the head moving rate.
A concrete surface partly remains at more than 4mm/sec. Then the
head moving rate was determined to be 3mm/sec in this test.
Furthermore, overlap of successive head moving passes is necessary
because a concrete surface between two passes remains without overlap
as shown in Fig. 9.

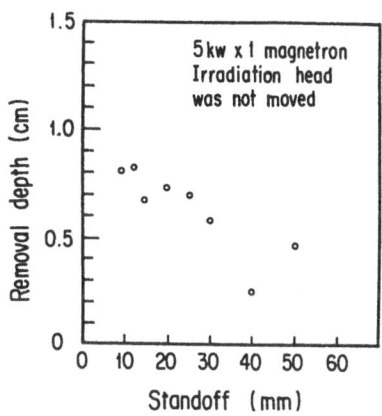

Fig.7 Effect of standoff on
removal depth

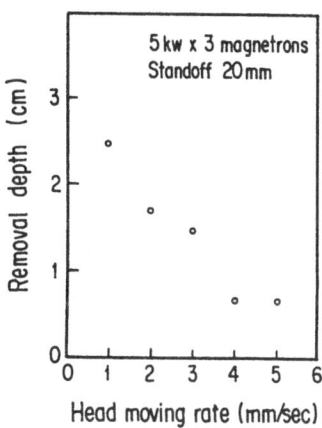

Fig.8 Effect of head moving
rate on removal depth

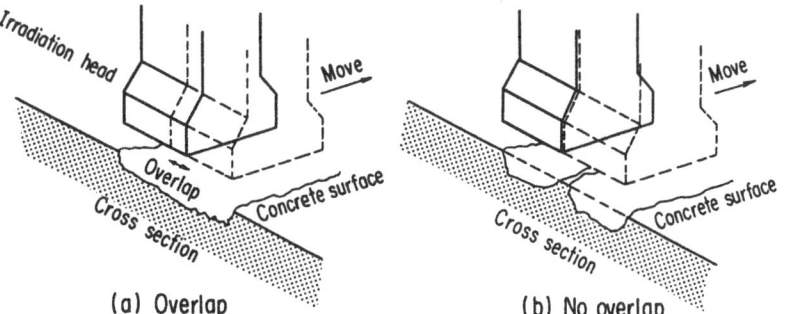

(a) Overlap   (b) No overlap

Fig.9 Concrete surface removal with overlap or no overlap

285

(5) Dust concentration and size of concrete debris
Minimizing dust concentration is important to reduce worker intake.
Dust concentration was 2mg/m³ with the vacuum system, and 200mg/m³
without the system at a height of 0.8m and a distance of 2m from the
irradiated surface. Figure 10 shows a weight distribution of
concrete debris by size. About 70% of concrete debris was larger
than about 5mm in diameter.

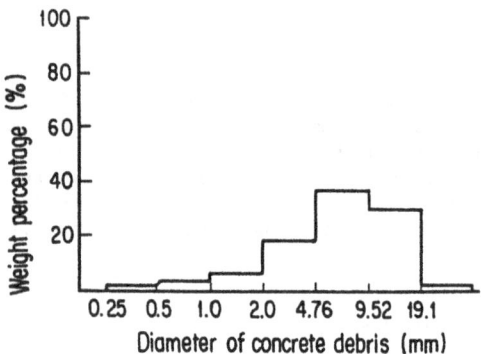

Fig.10 Distribution of concrete debris generated by microwave
concrete surface remover

(6) Collecting of concrete debris by the vacuum system
At first, 50% of the concrete debris was collected by the vacuum
system. But, through the improvement of the shape of the attachment,
this was increased to 70%. An increase of the vacuum system capacity
is necessary to provide more improvement.
    (7) Influence of water content in concrete and cracks on concrete
        surface
Two concrete blocks with different water content were tested to
compare the removal depths. Table 1 shows the composition of the
blocks. The removal depth was 22mm for A, and 16mm for B. The water
content was 4.3% for A, and 3.5% for B. The removal depth increased
in proportion to water content. Cracks also affected removal depth,
occasionally having an infavorable effect on spalling because of the
escape of steam.

Table 1 Composition of concrete for specimens

Mix proportion	A	B
Water	170 kg/m³	139 kg/m³
Portland cement	325 kg/m³	290 kg/m³
Sand	650 kg/m³	784 kg/m³
Aggregate	1200 kg/m³	1205 kg/m³
Water-cement ratio	0.52	0.48
Slump *	18.0 cm	7.5 cm

* Slump is an index of consistency of concrete

(8) Effect of coating on concrete surface
The effect of paint coating was investigated using epoxy and urethane resins. The removal depth was reduced to 80% of that of concrete without coating. Toxic gases were in undetectable level.
(9) Development of the flexible wave guide and the sliding wave guide
In order to put this system to practical use, the improvement of the system function is much required. As the first step for this purpose, a flexible and a sliding wave guides were made. Both wave guides have been newly designed, instead of adopting the existing fixed type square shaped wave guide. The flexible wave guide is shaped into a cylindrical form and can be bent at angles up to 90°. The sliding wave guide is also cylindrical and is able to expand and contract within 10cm. Though the attenuation rate of microwave in these wave guides is a little more than that in the existing guides, it can be very easy for those wave guides to change or adjust the irradiating position. Figure 11 shows the installed flexible wave guide and sliding wave guide.

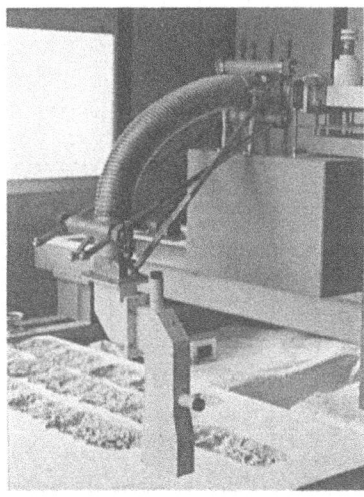

Flexible wave guide   Sliding wave guide

Fig.11 Flexible wave guide and sliding wave guide

287

(10) Performance in comparison with conventional technique
The performance of the microwave concrete surface remover was compared with those of various conventional mechanical equipments. Their
features and results are summarized in Table 2. At present, the
microwave concrete surface remover can remove only an insufficient
quantity of concrete in corners. It will be necessary to develop a
special irradiation head to improve a removal performance in corners.
However, the removal depth and volume by the microwave concrete surface remover are larger than those by other equipments. Moreover,
the microwave concrete surface remover is more effective for wall
decontamination than a scabbler and a plainer because any pressure to
push the equipment to the object is not necessary. A hand breaker
and a steel grit blast are usable for walls and corners, but their
work efficiency is very low. Furthermore a steel grit blast generates a large quantity of secondary waste. As the size of the present
prototype microwave concrete surface remover is larger than other
equipments, any modification is needed to make it compact for the
decontamination in a confined place of nuclear facilities.

Table 2 Comparison of concrete removal characteristics among
various equipments

Equipment	Concrete removal mechanism	Removal depth by single-pass (mm)	Removal efficiency (x 10^{-3} m³/hr)
Microwave decontaminator	by steam pressure caused in concrete	10 – 20	40
Scabbler	by impact of steel head	3 – 7	40
Plainer	by impact of steel blade	2 – 4	30
Hand breaker	by impact of steel chisel	5 – 10	20
Steel grit blast	by impact of steel grit	1 – 5	1.0

3. Conclusion

It was found that more than 80% of the penetration of concrete contamination were within 2cm from the surface in the JPDR. As a
microwave concrete surface remover can remove concrete surface within
1-2cm by a single pass operation and 2-4cm by a double pass operation, it is applicable to the concrete surface decontamination of the
JPDR. The microwave concrete surface remover is being improved
further to be compact and to be equipped with flexible and sliding
wave guides for easy operation. The microwave concrete surface
remover and conventional mechanical equipments will be used for the
JPDR concrete surface decontamination in 1990. Operational data with
various equipments including the conventional remover not only will
demonstrate the usefulness of the microwave concrete surface remover
for the concrete surface decontamination, but also will be valuable
for decommissioning of nuclear facilities in the future.

References

YASUNAKA, H. et al. "Microwave Decontaminator for Concrete Surface Decontamination in JPDR," Proceedings of 1987 International Decommissioning Symposium, Pittsburgh, Pa. Vol.2, pp.IV-109-116, 1987.

TACHIKAWA, E. and YASUNAKA, H. "Development of Reactor Decommissioning Technology in JAERI," Nuclear Egnineering, Vol.32, pp.65-72, 1986.

YOKOTA, M. and YASUNAKA, H. "Concrete Surface Removal by Microwave Irradiation," Cement & Concrete, No.487, pp.87-89, 1987.

SUWA, T. and YASUNAKA, H. "Present State of Reactor Decontamination Technology," Corrosion Egnineering 32, (12), 721, 1983.

IWASAKI, Y. et al. "Research and Development of Concrete Surface Decontamination," Proceedings, 1986 Annual Meeting of the Atomic Energy Society of Japan, Vol.1, D41, 1986.

STUDY OF A METHOD FOR CRUSHING CONCRETE WITH MICROWAVE ENERGY

MASAYOSHI KAKIZAKI, MINORU HARADA   Kajima Corporation, Kajima
Institute of Construction
Technology, Japan
ISOKAZU NISHIKAWA   Kajima Corporation, Japan

ABSTRACT

When dismantling a nuclear power plant, reinforced concrete members
contaminated with activated materials must be safely removed,
demolished, and carried out.

In this study of methods to rationally remove the surfaces of
concrete members contaminated with radioactivity, concrete members
were irradiated with electromagnetic waves and were examined for the
speed of destruction, depth and efficiency of the concrete and the
effectiveness of the method of demolishment.

An electromagnetic wave radiator which consumed 30kW at 915 MHz was
used for this experiment.

Result of the experiment

a. The concrete started to disintegrate 2 $\sim$ 10 minutes after the
electromagnetic wave radiation was initiated and the internal temperature (2 $\sim$ 3 cm from surface) was raised above 150°C. The destruction
depth at this time was 2 $\sim$ 6 cm.

b. The best results were obtained when the distance between the
electromagnetic horn and the concrete was less than 20 cm and the
moving speed of the horn was 10 $\sim$ 20 cm/min.

Key words: Microwave energy, Thermal expansion, Concrete crushing

1. INTRODUCTION

Basic experiments were previously performed in demolishing concrete
with microwave energy using a device which has a maximum microwave
output of 5 kW[1] . These basic experiments proved that crushing
concrete with microwaves was effective for dismantling concrete structures. However, at that time, the destructive force was limited by
the 5 kW microwave output.

The microwave output for crushing the concrete was now increased to
30 kW and the destructive effect observed.

The following reports the results of this experiment. The experiment covered the following items.

a. Measurement of the specific inductive capacity and dielectric
power factor of concrete.

b. Temperature distribution in concrete irradiated with microwave
energy.

c. Measurement of the effect of the water content in concrete, microwave output, irradiation distance and radiation period on the distruction of the concrete.

2. THE PRINCIPLE OF DEMOLISHING CONCRETE

The principle of demolishing concrete with this method is as follows:
Strong microwave energy is radiated from an electromagnetic horn to the concrete, and the concrete is internally heated by the dielectric loss (the phonomenon in which microwave energy is converted into thermal energy in the concrete).
As a result, water contained in the concrete boils and expands on a small scale to shatter the concrete. In addition, the concrete is ruptured or cracked by the difference in thermal expansion between the surface and inside of the concrete and between parts ehated by, and parts not affected by, microwave energy.

3. EXPERIMENTAL EQUIPMENT

The specifications for the microwave generator used in this experiment are shown in Table 1.
The microwave generator consists of a power supply unit and a microwave radiation unit. An external view of the microwave radiation unit is shown in Fig. 1. The power supply unit is connected to the microwave radiation unit by a cable.
The electromagnetic horn used to effectively irradiate concrete to be demolished with microwave energy has a shape (Fig. 2) determined by basic experimentation.

Table 1 Microwave Generator Specifications

Max. output	30kW
Oscillating frequency	915MHz
Source voltage	220V
Power consumption	45kWh
Oscillating methods	Magnetron
Q'ty of magnetron cooling water	9 1/min
Dimension and weight of power supply	1300x750x1080mm, 300kg
Dimensions and weight of radiator	1500x550x 890mm, 150kg
Manufacturer	Tokyo Electronic Giken (IDX) Corporation

4. THE METHODS AND RESULTS OF THE EXPERIMENT

4.1 Measurement of the specific inductive capacity and dielectric power factor of concrete

If microwave energy irradiates concrete, a parts of the electric power is dissipated in the concrete, which is a dielectric due to ohmic losses, molecular vibration, etc. caused by the induced current. The dissipated electric power is finally converted into heat in the concrete which is a dielectric.

The consumption of power in dielectrics is called dielectric loss. The dielectric loss per unit volume is given by the following formula;[2],[3]

$$P = 0.556 \cdot f \cdot E^2 \cdot \varepsilon r \cdot \tan\delta \times 10^{-16} \quad \ldots\ldots\ldots(1)$$

Where,

- P : Dielectric loss per unit volume (W/cm^3)
- f : Frequency (Hz)
- E : Electric field intensity (V/m)
- εr : Specific inductive capacity
- $\tan\delta$: Dielectric power factor

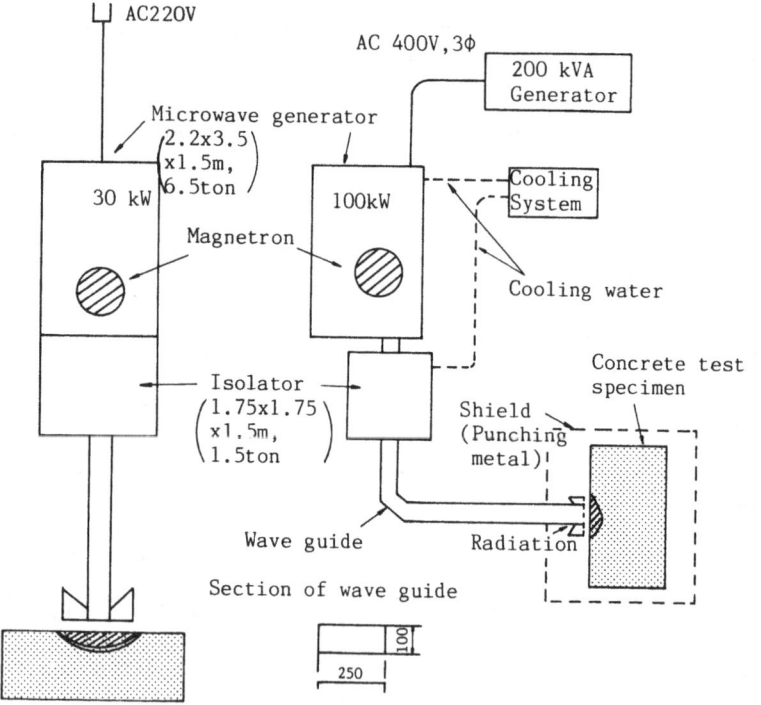

Fig.1   Radiator section of microwave generator

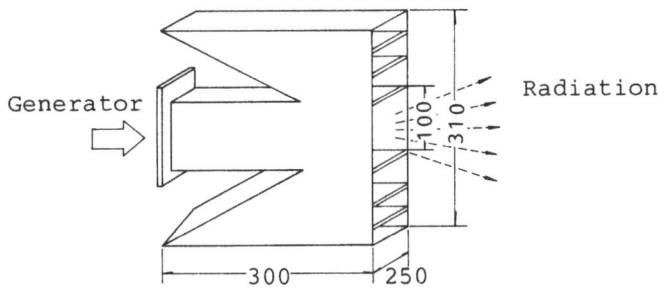

Fig.2   Horn for Microwave Radiation

The dielectric loss per unit volume is increased in proportion to the frequency, the square of the applied electric field intensity, specific inductive capacity, dielectric power factor, etc. Based on this, it is assumed that the destructive force increases with the frequency. However, this is not always correct since the specific inductive capacity and dielectric power factor depend on the frequency.

In this section, a study was made as to how the specific inductive capacity and dielectric power factor of concrete are affected by a change in frequency to determine the frequency which provides the greatest destructive force.

The results of the measurements are shown in Fig. 3 and Fig. 4. These figures provide the following information.

a. The specific inductive capacity is 4.2 ∿ 4.3 and changes little from the 800 ∿ 4,000 MHz frequency range.

b. The dielectric power factor is 0.048 ∿ 0.050 and changes little within the above frequency range.

c. Formula (1) shows that the microwave generated destructive force increases with the frequency of microwave energy.

d. In the case of the device used at this time (Output: 30 kW, Frequency: 915 MHz), about 5 W/cm³ of electric power was converted into heat in the concrete. If the specific heat of the concrete is 0.20 cal/g °C, the temperature of the concrete is increased by 2.57 °C every second. This conforms to the following experimental results.

293

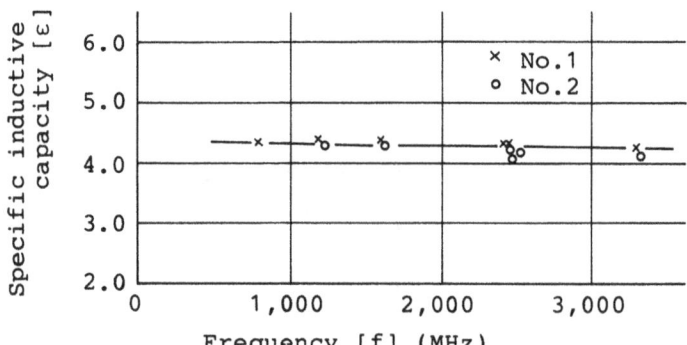

Fig.3　Relationship between the Specific
Inductive Capacity of Concrete
and Frequency

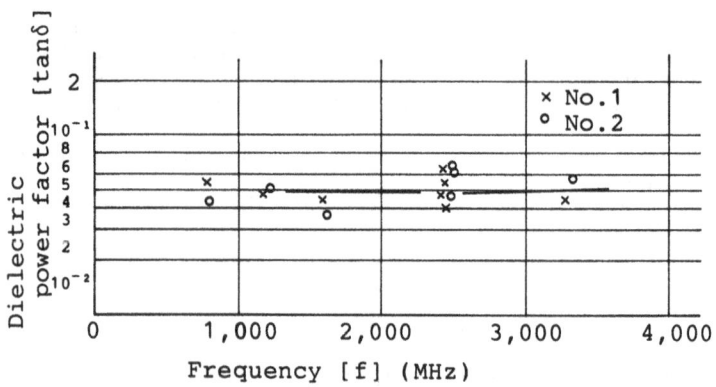

Fig.4　Relationship between the Dielectric
Power Factor of Concrete and
Frequency

4.2 Temperature distribution in concrete

When microwave energy irradiates concrete, the concrete is suddenly partially heated and severe stresses occur. If this stress exceeds the tensile strength of the concrete, it is crushed.

In this section, the temperature in the concrete receiving the microwave energy was measured, and the heating and crushing conditions were examined to determine the most effective irradiation method for destruction.

To measure the temperature in the concrete, thermocouples (Alumel-Chromel) were embedded in a reinforced concrete test specimen measuring 150 mm x 300 mm x 550 mm. Temperature were measured with two automatically balanced recorders.

To measure the temperature distribution, each concrete test specimen was divided into four symmetrical parts about the center of the part receiving microwave irradiation. Temperatures were measured at 17 points set on the partition planes of the specimen. The temperature measuring points are shown in Fig. 5.

An example of the experimental results is shown in Fig. 6. In this example, the output of the microwave radiator is increased to 15 kW ten minutes after irradiation commences, and the concrete is crushed three minutes after the increase in output. Also, the temperature is measured just after the concrete is crushed.

The following items were proved by this experiment:

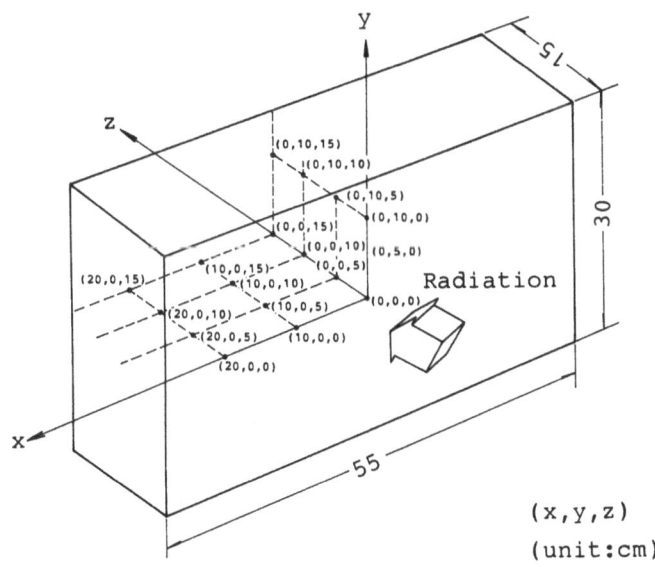

Fig.5　Temperature Measuring Points

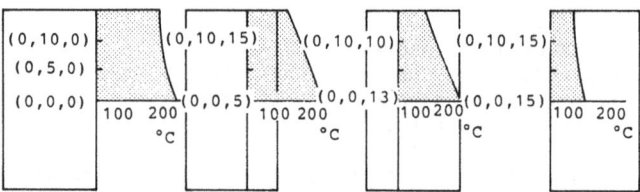

Temperature distribution in y-axis direction

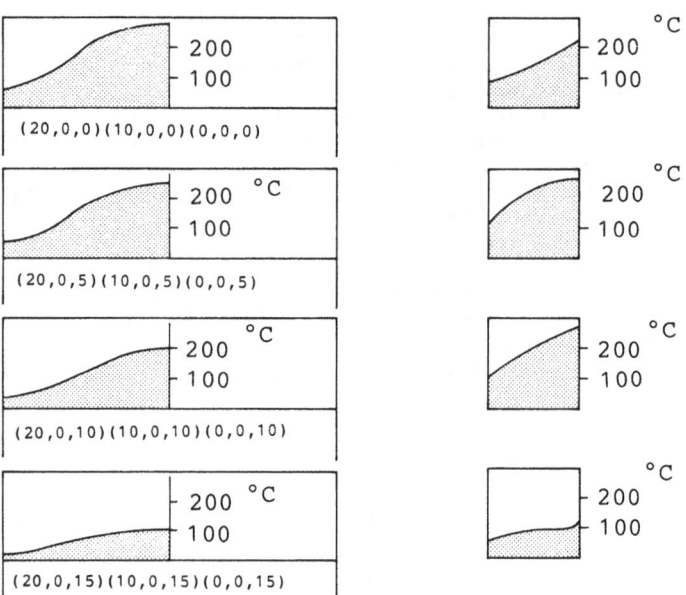

Temperature distribution   Temperature distribution
in x-axis direction   in z-axis direction

Fig.6   Temperature Distribution just after
        Disintegration

a. The temperature of the concrete was the highest at the surface receiving microwave energy, and was lowered as the depth increased. The difference in the temperature between the surface and the inside of the concrete reached a maximum when the concrete was crushed. At this time, the temperature gradient in a right angle direction at the surface was 20°C/cm.

b. The temeprature of the concrete area facing the opening of the electromagnetic horn was evenly high, and the adjacent area was heated little. This occurred because the electric field outside the area facing the opening of the electromagnetic horn was extremely weak.

c. When the output of the microwave radiator was 10 kW, the temperature rise was 20°/min, and when it was 15 kW, the temperature rise was 30°C/min.

d. When the maximum temperature was 270°C, the concrete was crushed.

4.3 Measurement of the effect of water content, microwave output,
      irradiation distance and irradiation period for crushing concrete
It seems that water content of the concrete, microwave output, irradiation distance (distance between the electromagnetic horn and concrete test specimen) and irradiation period affected the scale of
concrete destruction.

In this section, experiments were performed in accordance with the
"Orthogonal Table L₉" of the design for the experiment and the variation was analyzed to see how the above factors affect concrete crushing.

Table 2 shows the result of an analysis of variations in the
factors. Based on this table, the effects of each factor on concrete
crushing is summarized as follows.

a. The microwave output, irradiation distance and time are the
most important factors for increasing the crushing effect. Especially, the volume of concrete crushed is greatly affected by the irradiation period.

b. The crushing depth is affected by the irradiation distance.

c. The diameter of the concrete to be crushed is affected by the
microwave irradiation time in the same way as the volume to be
crushed. Therefore, it is necessary to irradiate to increase the
crushing width area. On the other hand, it is necessary to shorten
the irradiation distance to increase the crushing depth.

d. The temperature condition for the fracturing of concrete was
not affected by the experimental conditions. It seemed that the
concrete was crushed when its internal temperature (2 ∿ 3 cm below
the surface) exceeded 150°C.

e. Concrete deterioration was slight.

Table 2   Analysis of Variations

Examination item	Crushed volume		Crushing depth		Crushing diameter		Crushing temperature		Degree deterioration	
Factor	F	ρ	F	ρ	F	ρ	F	ρ	F	ρ
Water content	3.5	2	4.0**	1	11.0**	6	6.7*	-	3.0	-
Output	15.8**	17	47.0**	20	36.0**	20	1.3	-	3.2	-
Distance	15.8**	17	94.2**	40	32.3**	18	1.6	-	0.1	-
Time	53.9**	61	88.7**	37	92.5**	53	1.8	-	3.6	-
Error	-	4	-	2	-	3		-		-

F : F-value   ** : Significant when ration of risk is 1%
ρ : Contribution ratio %   * : Significant when ratio of risk is 5%

Fig. 7 ∿ 9 show how crushing conditions, including the crushed
volume, crushing depth, etc., are affected by the microwave output,
irradiation distance and irradiation period. Based on these tables,
the following was proved:

a. Concrete is seldom crushed when the applied microwave output
is below 10 kW.

b. Since the volume of concrete crushed by 15 kW microwave irradiation is only slightly different from that crushed by 20 kW microwave
irradiation, it is not always necessary to increase the microwave
output.

c. The test specimens were crushed, by the various factors explained above, about two minutes after microwave irradiation started.

d. The longer microwave irradiation occurs, the more the concrete is crushed.

e. Concrete is not crushed if it is 20 cm or more away from the electromagnetic horn.

f. With a shorter microwave irradiation distance, crushing effect increases.

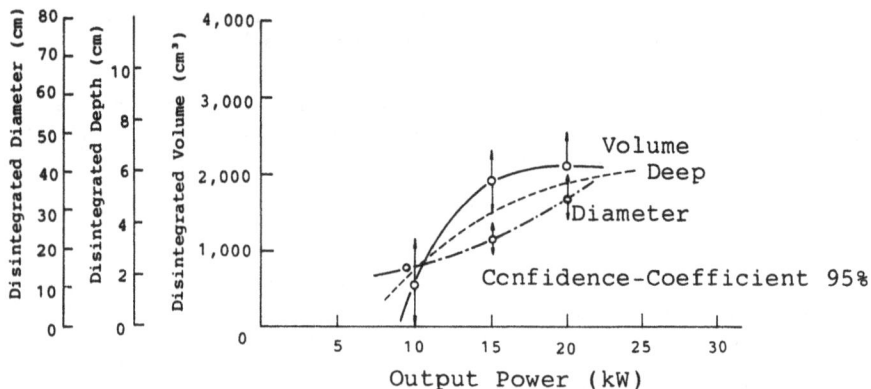

Fig.7　Relationship between Disintegrated Diameter, Depth, Volume and Output Power

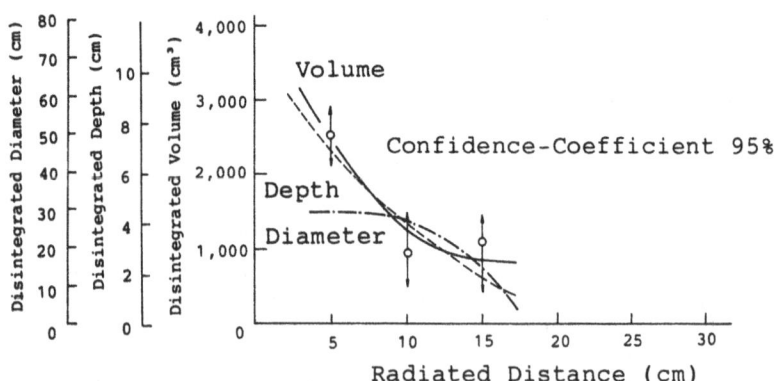

Fig.8　Relationship between the Disintegrated Diameter, Depth, Volume and Radiation Distance

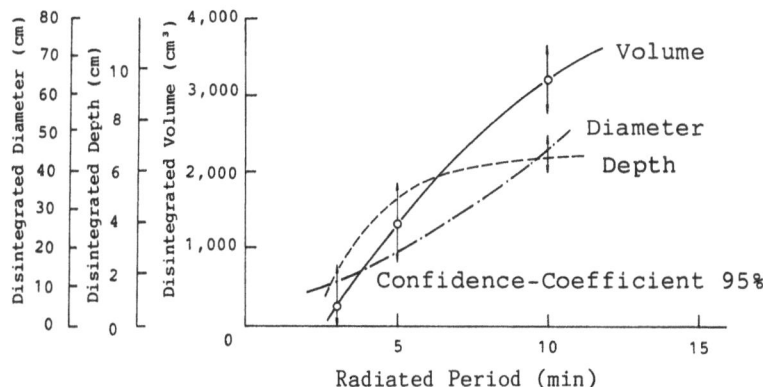

Fig. 9   Relationship between the Disintegrated
Diameter, Depth, Volume and Radiated Period

5. CONCLUSION

A method for demolishing buildings without generating noise and vibration requires stringent demands and the method explained above could
be one example which meets these requirements.

The applicability of the method explained above requires further
examination. For example, trial devices used for these experiments
ahve mechanical and electrical problems.

To resolve these problems, experiments are now being performed
in crushing concrete with a device which can provide 100 kW microwave
output.

BIBIOGRAPHY

Kakizaki, Harada and Nishikawa; Demolishing method for concrete
    structures (Part 4) Method for crushing concrete with microwaves,
    Digests of Architectural Institute of Japan, Kanto Branch Office,
    42 (1971) pp. 221 - 224.
Misawa and Takahashi; Possibility of crushing rocks with microwaves,
    scripts of lectures of Autumn Hokkaido Mining Meeting in 1971,
    A3 - 8.
The Institute of Electrical Engineers of Japan, Electrical Engineering
    Handbook.
H. Püschner; Heating with microwaves, Philips Technical Library,
    (1966), pp. 157 - 163
Justin Scott; Is today's standard for microwave radiationsa fe for
    humans?, Microwaves, (1971, January), pp. 9 - 14.

ON CONCRETE DEMOLITION BY ELECTRIC DISCHARGES

N.T. Zinovyev, B.V. Siomkin and Zh.G. Tanbayev
High-Voltage Institute, Tomsk, USSR

Abstract
The paper deals with a possibility to build installations for demolition of different concretes by electric
pulse currents up to sizes to make them fit for utilization in new concrete structures. Results of investigations of specific energy consumption and granulometric
composition of concrete sample demolition products having
dimentions of 100 x 100 x 100 mm3 and a compression
strength of 18 MPa in a broad variation spectrum of a
pulser discharge circuit parameters are cited. An asses-
ment of peak pressures in electric discharge channel is
made, a way to improve reliability and service life of
generation equipment is indicated.
Key words: Pulse electric discharge, Demolition, Concrete, Energy consumption.

1. Introduction

Intensive developments in industrial and civil engineering with application of elements of ferro-concrete production makes it urgent to utilize non-standard and worn-
out concrete articles. There is also a great amount of
work in closing up cracks, making openings for communica-
tions, preparation of surfaces for facing and plastering
procedures.

One of the promising methods for demolition of such
concrete products is demolition by electric pulse discharges initiated directly inside the articles to be demolished. Because of large scale of building activity
this technology needs naturally reliable and long-living
installations, simple in operation. Solid dielectrics,
including concretes, are known to possess high electric
strength, so in order to demolish them one has to apply
voltages up to some hundreds kV. Judging by data on elec
tropulse drilling of oil wells, crushing ores and non-ore
materials, demolition of ferro-concretes seems to be rea-
lizable only when using highly intensive methods of ener-

gy release in the concrete brekdown channel, that is, discharge currents of some tens kA, momentary powers up to $10^9 - 10^{10}$ W. Main units of the equipment such as capacitances, commutators, electrode systems under so intensive actions are, as a rule, short-living, unreliable in operation and imply high maintenance costs. In this connection it seems to be actual to study granulometrical composition of demolition products, specific energy consumption in a wide spectrum of energy release conditions in order to be able to assess possibilities of technology under low-intensive conditions which alleviate operation of the equipment.

2. Experimental methods

2.1 Subjects
The present work concerns results of an investigation of specific energy consumption and granulometric composition of demolition products of 100 x 100 x 100 mm^3 concrete samples having a compression strength of 18 MPa for eleven conditions of energy release in conformity with Table I.

Table I. Discharge circuit parameters

Condition	U_o (kV)	L (μH)	T (μs)	W_o (J)	$P_k \cdot 10^{-8}$ (Pa)	W (J)	W_1 (J)	η (%)
1	150	13.6	5.6	656	9.5	460.3	359.5	70.2
2	192	13.6	5.6	1080	10.9	743.9	528.0	68.9
3	240	13.6	5.6	1680	12.0	963.5	543.5	57.4
4	300	13.6	5.6	2625	13.5	1476.4	848.8	56.2
5	360	13.6	5.6	3780	14.7	2250.7	1362.6	59.5
6	408	13.6	5.6	4855	15.7	2521.9	1394.1	51.9
7	444	13.6	5.6	5750	16.3	2701.8	1352.5	47.0
8	360	63	12	3780	8.3	2283.4	1187.7	59.9
9	360	210	22	3780	2.9	2337.0	822.7	61.8
10	360	840	44	3780	1.6	2218.9	743.8	58.7
11	360	2650	78	3780	0.95	1709.7	672.5	45.2

The tested samples were made of concrete mixture with a ratio sand/cement being 1:1, that of water/cement = 0.3, the 5-40 mm gravel fractions content was 53 per cent.

2.2 Conditions
An Arkadiev-Marx generator with a maximal pulse amplitude of 444 kV and a stored energy of up to 5.75 kJ was

used as a high-voltage signal source. Variation of energy release dynamics was accomplished by changing inductance L of the discharge circuit and the generator's voltage amplitude U_O, see Table I.

Sample breakdown by single voltage pulses was realized in a point - surface electrode system. Energy W_r released in the discharge channel during the time interval t of the discharge current flow as well as during the first half-cycle W1 was defined by oscillography of current and by voltage drop across the sample. In each combination of experimental conditions 8-10 samples were subjected to breakdown. A picture of demolished samples under energy release conditions is shown in Fig.1. A sieve analysis was made after demolition, the results of which were used to find average granulometric characteristics. An analysis showed that dependences of granulometric composition after a single act demolition of concrete samples by electric discharges could be well described by Rosin - Rammler equation.

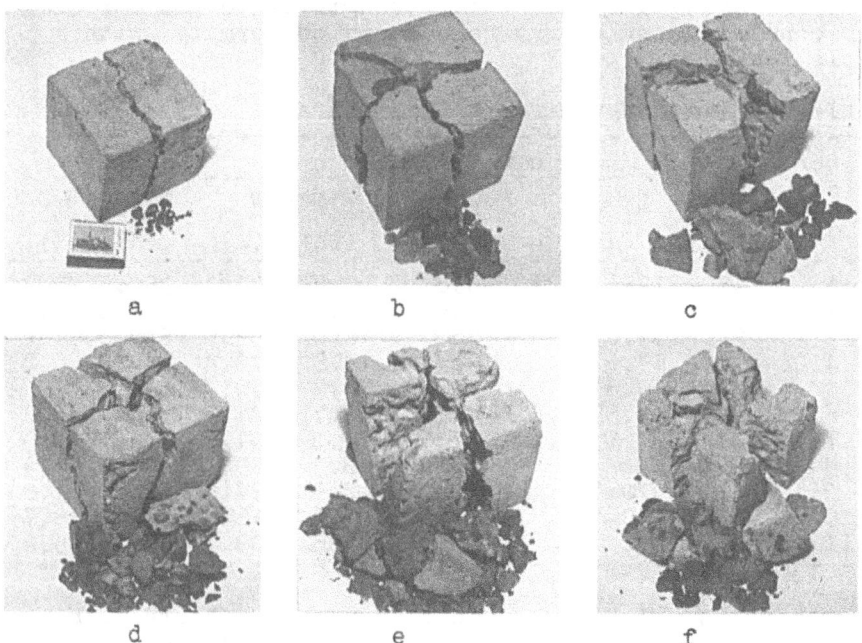

Fig. 1. Demolished samples under energy release conditions corresponding to numbers in Table 1: a-1; b-2; c-3; d-4; e-5; f-6.

The investigation consisted of a set of experiments (conditions 1-7) according to the influence of energy W_O sto-

red in the capacitor and energy W_Σ, released in the breakdown channel on the specific energy consumption ω for demolition of samples and producing definite chip fractions
as well as of a set of experiments (conditions 5,8-11) on
influence of the energy release duration on ω . The value
ω in each type of conditions was defined as an arithme -
tic mean in a set of experiments.

3. Results

In the first set of experiments the energy was varied by
changing voltage U_0, leaving the discharge capacitance
$C=0.0583$ μF and inductance $L=13.6$ μH unchanged. Voltage
rise from 150 to 444 kV was accompanied by an increase of
energy release duration from 8.4 to 25.2 μs, i.e., a shift
of energy release condition from aperiodicity boundary into regions of oscillatory modes was observed. An evaluation of this shift after the similitude criterion Π showed
a change of Π from 0.308 to 0.104.

Table 2. Granulometric composition of demolition products

Condi-tion	Composition after fractions (mm), %					
	0-5	5-10	10-20	20-40	40-70	70-100
1	0.13	0.14	0.35	0-53	0.0	98.85
2	0.24	0.3	1.04	2.0	12.15	84.27
3	0.36	0.41	1.0	2.99	24.75	69.89
4	0.59	0.75	2.68	5.28	28.44	62.26
5	1.13	1.17	4.12	7.28	32.76	53.54
6	0.79	1.59	6.12	11.12	80.38	0.0
7	1.12	1.35	5.17	11.69	80.68	0.0
8	1.05	1.18	3.25	6.16	22.0	66.36
9	0.71	0.81	2.86	7.08	21.45	67.09
10	1.0	0.74	2.57	6.52	19.58	69.59
11	0.85	0.73	2.49	4.1	21.13	70.7

Fig.2 represents dependences $\omega = f$ (W_Σ) for fractions
5-10,10-20,20-40,40-70,5-70 mm, accordingly. In this case we took into consideration Soviet Standard 8268-74 demanding to use in concrete production a filler of the said
 fractions. Minimal specific energy consumption over all
 fractions corresponded to condition 6 in which there were no chips over 70 mm, see Table 2, and ω for fraction
5-70 mm was 1.08 J/g. This condition was characterized
by a discharge current of 17.8 kA and momentary power of
1.04 $\cdot 10^9$ W. The peak pressure P_k in the spark channel
was found to be $1.5 \cdot 10^9$ Pa. It should be noted that spe-

cific consumption in the range of variations of W from 800 to 2500 J was changing only insignificantly, and for condition 3 when ω (fraction 5-70) was 1.34 J/g, the discharge current as compared to the optimal condition (condition 6) reduced by 36.2 per cent, momentary powers reduced by 64.5 per cent, peak pressures by 32.8 per cent, the efficiency η of the discharge circuit increased by 9.6 per cent.

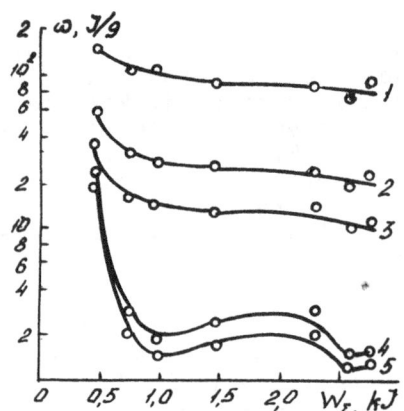

Fig.2. Dependency of specific energy release for yield of fractions on energy release in the discharge channel. Fractions, mm:1=5-10; 2=10-20; 3=20-40; 4=40-70; 5=5-70

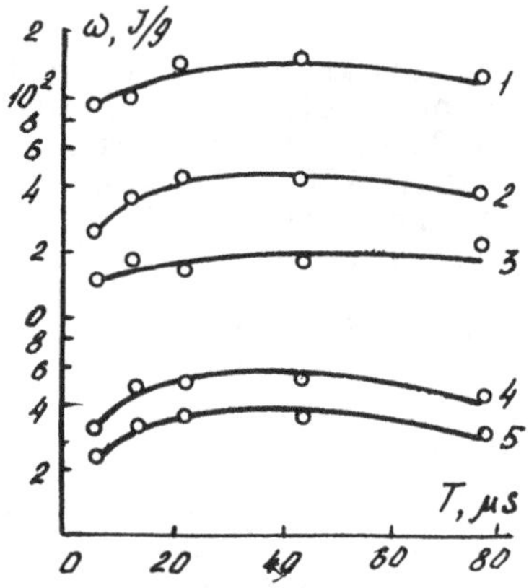

Fig. 3. Dependency of specific energy consumption for different fractions yield on the discharge current fluctuations. Designation of fractions see in Fig. 2

The same efficiency under condition 3 can be achieved by increasing the pulse repetition frequency by a factor of three. Most simple assessments show that service life of capacitors under this condition was raised by about one order of magnitude.

A further easing of the equipment operation conditions in generators with one discharge circuit is associated with limitation of discharge currents by raising the discharge circuit impedance. In the second set of experiments variation of the energy release conditions was carried out by switching on air coil chokes of different inductance L with constant voltage U_0 = 360 kV and a capacitance C = 0.0583 μF. A change of inductance from 13.6 to 2650 μH was accompanied by increase of energy release duration from 22.4 to 546 μs, by decrease of peak pressure from $1.4 \cdot 10^9$ to $9.5 \cdot 10^7$ Pa and by reduction of the discharge current amplitude from 15.22 to 1.45 kA, by reduction of momentary powers from $8.5 \cdot 10^8$ to about $3.2 \cdot 10^7$ W.

Fig. 3 shows dependences of ω on period T of the discharge current fluctuations chosen for numerical characteristics of the energy release condition. Taking into account the fact that changes of the discharge circuit have at the same time substantially changed a number of parameters of the energy contribution dynamics as well as interaction of generated voltage waves with the borders of the samples, it is of course difficult to give an interpretation of regularities represented in Fig. 3. However, of interest is the phenomenon of a substantially weak dependence of the specific energy release in a wide scale of energy release conditions. Usage of a most sparing mode of equipment operation had led to a rise of the specific energy release being less than 25 per cent as compared to condition 5. It is logical to suppose that under such conditions, when the wave length of the generated mechanical stresses substantially exceeds the sample dimentions, its dispersion occurs before formation of machanical disturbances in the breakdown channel. In order to verify this hypothesis, a number of experiments on concrete demolition under condition 11, when the sample was placed in a polyethylene casing to prevent scattering of chips, was carried out. In this case a reduc - tion of content of fractions 70-100 mm to 50 per cent and an increase of fractions 40-70 to 37 per cent was observed. The content of small fractions didn't undergo any substantial changes. The experiments have confirmed a certain role played by disturbances generated in the breakdown channel at late stages of the discharge current oscillations. In other words, application of small intensity modes of energy release with a long duration of the discharge current flow for demolition of large ferroconcrete articles may result in an additional effect as compared to that observed in our experiments because separation of chips from the block at early stages will be limited by the mass of the article and by the reinforced casing.

Thus, it has been shown that when demolishing concretes by pulsed electrical discharges, there exist opportunities for building reliable and long serving installations for dispersion of such concrete and ferro-concrete articles. In this case increase of reliability and service life are achieved due to limitation of the energy release intensity in the breakdown channel and the energy per pulse, and this increase is accompanied by only a small rise of specific energy consumption as compared to electroexplosion under optimal conditions.

References

O.V.Batanov, N.T.Zinovyev. Electric Pulse Demolition
    of Non-standard Ferro-concretes. Proc. of All-Union
    Scientific Conference "Electropulse Technology and
    Electromagnetic Processes in Loaded Bodies", Tomsk.
    November 16-18, 1980, p. 92-93.
N.G.Antonov, V.D.Kazantsev, et al. Outlooks for Electro-
    pulsed Technology in Connection with Usage of Dis-
    charged Gaps of Decimeter Band, Ibidem, p. 5-6.
I.I.Kaljatsky, V.I.Kurets, et al. Principles of Electro-
    pulse Disintegration and Outlooks for its Application
    in Industry. In: Ore Dressing, 1980, No. 2, p. 5-10.
N.T.Zinovyev, B.V.Siomkin. An analysis of energy charac-
    teristics of spark channel in solid dielectrics. In:
    Electron Material Processing, 1979, No. 2, p. 41-44.
Kuchinsky G.S. High-Voltage Pulse Condensers. Leningrad:
    Energia, 1973, 175 pages.

CUTTING CONCRETE BY CO_2 LASER

M. MORI

K. SUGITA

T. FUJIOKA

Technical Development Department
Taisei Corporation
Technical Research Institute
Taisei Corporation
Laser Laboratory
Industrial Research Institute

Abstract
The objective of the experiments is to study the feasibility of using
laser to cut concrete and to find a method to efficiently cut concrete
by understanding the factors which affect the cutting performance.
Authors hope that this technology can be applied in the future to
dismantle concrete contaminated by radioactive radiation at nuclear
power stations. In the experiments, cutting tests were carried out by
changing the intensity of the laser beam, cutting speed, position of
focal spot, and the type and pressure of the gases used to assist
cutting, and the effects of these factors on the cutting performance
were clarified. Furthermore, a reinforced concrete with deformed 10 mm
bar was successfully cut when the beam intensity was 15 kW and the
cutting speed, 2.5 cm/min. From the results of the study, the authors
concluded that although there are many problems that need to be
resolved before practical application, this technology can be applied
for cutting concrete in the near future.
Key words: Laser, CO_2 laser, Cutting, Concrete, Reinforced concrete,
Nuclear power station, Dismantling

1. Introduction

Cutting concrete by laser would seem to be suitable for dismantling
concrete structures that have been exposed to radioactive waves, such
as the biological shield walls of nuclear power stations, because of
its various features, e.g. (1) reinforcing bars and concrete may be cut
simultaneously; (2) water is not required, thus prevents spread of
contamination; (3) it produces only a small amount of fumes and melted
debris at the time of cutting, which can be easily disposed of; and (4)
the torch that guides the laser beam is small, light and easy to
control, etc.

The objective of this research was to determine the feasibility for
practical application by laser cutting specimens of reinforced
concrete.

When laser cutting concrete, the factors that are considered to
affect the depth and width of cutting are: output of laser, cutting
speed, focal length of converging lens (mirror), position of focus,
laser beam mode (intensity distribution), whether the pulse oscillation exists or not, direction of assisting gas nozzle, type and

pressure of assisting gas, etc.

2. Procedure used in the experiments

The effect of increasing laser output, making the focal length of converging lens (mirror) longer, direction of assisting O_2 gas, and pressure, etc., on the laser machinability was investigated. A comparison of cutting was also made with an unreinforced concrete specimen.

2.1 Laser system
Fig. 1 shows the CO_2 laser system used for experiments and Table 1 shows the approximate specifications of the system. The laser system consists of an oscillator power source, a laser oscillator, an operating table and an experiment processing system. Fig. 2 shows the details of the processing head.

The authors have developed the auxiliary system for a high output CO_2 laser that can be used for cutting and have been studying the safety and ease of use thereof with the aim of putting it to practical use. The features of the system developed include: (1) a converging mirror with a long focal length (1 m) which was incorporated into the system so that the laser beam can reach deep into the reinforced concrete without losing energy through diffusion; (2) a protective cylinder to enclose the laser beam so as to keep the fumes produced during cutting from interrupting the laser beam and lowering its output; (3) a method of preventing fumes from entering the protection cylinder by blowing air in the direction of irradiation of the laser beam from inside the cylinder; and (4) an assisting gas nozzle which was incorporated in the system to blow oxygen gas at the pressure of 2-8 kgf/cm² for cutting reinforcing bars, etc.

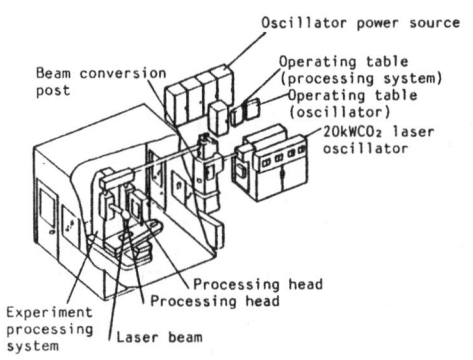

Fig. 1   Large output CO_2 processing system

2.2 Specimen
Fig. 2 shows the shape and dimension of the specimen. The specimen used was a wedge-shaped section of varying thickness so that the depth of cutting could be readily known. There were four types of specimens used: reinforced concrete, steel fiber-reinforced concrete (1-D10 @40), normal weight concrete, and lightweight concrete. Table 2 shows the mix proportion of concrete used and Table 3 shows the results of material tests. The specimens, 2 days after the placement of concrete, were removed from their mold and cured indoors until the day of testing (28 days of age).

Table 1   Approximate specifications of large output CO_2 laser system

Item	Specifications
Wave length	10.6 μm
Resonator type	Unstable type resonator
Mode	Ring mode
Maximum output	26.5 kW
Beam diameter	70 mm
Laser gas composition	CO_2-CO-N_2-He=2-1-6-32
Converging mirror diameter	150 mmø
Focal length of converging mirror	1000 mm, 600 mm

2.3 Method of experiments

Fig. 2 shows the conditions of the cutting experiments, in which the laser beam was fixed and the table on which the specimen was placed was moved at the specified speed.   The laser beam was converged by means of a concave mirror.   Photo 1 shows the cutting of reinforced concrete.

2.4 Experiment items

Items subject to experiments were as follows: (1) Laser machinability of reinforced concrete (relationship among output, cutting speed and cutting depth), (2) effect of focal length of mirror of cutting depth, (3) effect of direction and pressure of assisting gas on cutting depth.

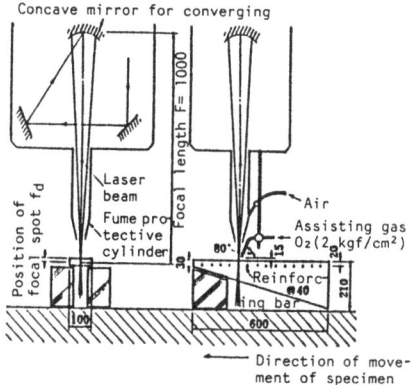

Fig. 2   Condition of cutting experiment by a large output laser

Table 2   Mix proportion of concrete

Type of specimen	Max. size of coarse aggregate (mm)	Compressive strength (kgf/cm²)	Slump (cm)	Water-cement ratio (%)	Sand-gate aggregate ratio (%)	Unit quantity (kgf/m³)				
						Water	Cement	Fine aggregate	Coarse aggregate	Admixture
Reinforced concrete normal weight concrete	25	240	18	50	44.5	177	354	774	986	3.540
Steel fiber-reinforced concrete	25	240	18	50	55	186	346 (36)	900	778	5.730
Lightweight concrete	20	210	18	60	53	190	317	907	502	0.793

Note: 1. Parenthesis shows value of pozzo-mix (silica fume).
       2. Steel fiber is 1.5% (120 kgf/m³) by volume percentage.

The following show the cutting conditions under which experiments were performed:

Laser output (kW):   5, 10, 15
Cutting speed (cm/min):   2.5, 5, 10, 20, 30
Focal length of converging mirror (mm):   1,000, 600
Position of focal spot (mm):   10 (below the surface)
Direction of assisting gas (°):   30, 45, 60, 80
Pressure of assisting gas (kgf/cm²):   0, 2, 5

Table 3   Results of material test

Specimen	Compressive strength (kgf/cm²)	Specific gravity
Reinforced concrete	282	2.29
Steel fiber-reinforced concrete	325	2.36
Normal weight concrete	282	2.29
Lightweight concrete	370	1.94

3. Results of experiments and discussion

The results of experiments and discussion are summarized by test items as follows:

(1) Laser machinability of reinforced concrete - Fig. 3 shows the relationship between cutting speed and cutting depth of reinforced concrete and unreinforced concrete with output as a parameter. The larger the laser output and the slower the cutting speed, the larger the cutting depth. Photo 2 shows the reinforced concrete after cutting.

No remarkable difference can be observed within the scope of this experiment between the laser machinability of reinforced concrete and that of unreinforced concrete mainly because the reinforcing bars (D10) of these specimens were laid out near the surface (20 mm from the surface). However, the laser machinability would differ between two types of concrete where the reinforcing bars were laid out deeper

Photo 1   Cutting reinforced concrete

310

Photo 2   Reinforced concrete cut by laser

or in the case of multi-
stage bar arrangement.

Fig. 4 shows the reinforced
concrete and steel fiber-
reinforced concrete that could
not be cut in Fig. 3. The
concrete was able to be cut to
the depth shown in the figure,
however, D10 reinforcing bars
could not be cut completely.
Marks (●,★) in the figure show
the domain in which normal
weight concrete is cut but not
reinforced concrete. This is
considered due to the
relationship between energy
density and cutting speed. As
far as can be seen from this
figure, unless the concrete
can be cut more than 50 mm
deep, the reinforcing bars
cannot be cut. The same results
were obtained for steel fiber-
reinforced concrete.

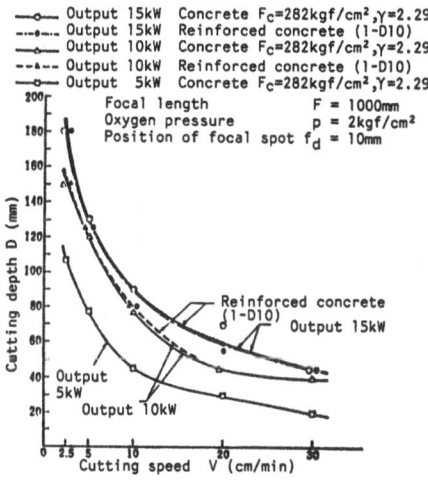

Fig. 3   Relationship between cutting
speed and cutting depth

Fig. 5 shows the domain in which reinforced concrete can be cut,
with the relationship between laser output and cutting speed obtained
by clarifying Fig. 4. The figure shows that cutting cannot be completed unless the cutting speed is slowed down where the laser output
is smaller.

(2) Effect of laser output on cutting depth of concrete - Fig. 6
shows the relationship between laser output and cutting depth of
concrete with cutting speed as a parameter. The data used for lower

311

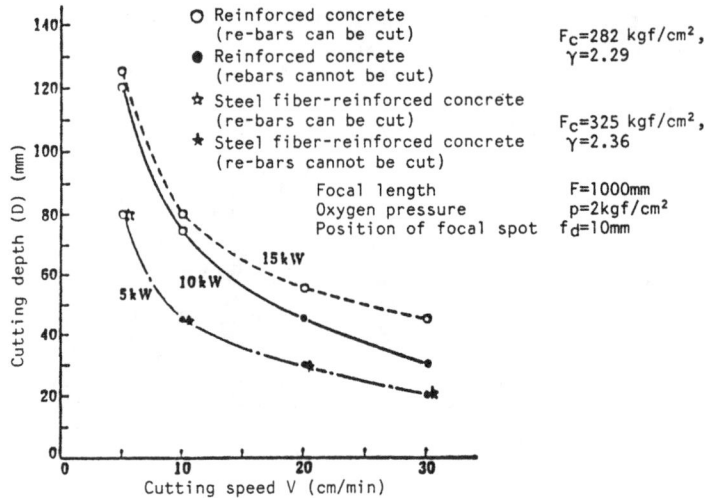

Fig. 4 Relationship between cutting speed and
cutting depth for reinforced concrete

output, such as 1 kW, 3 kW and
5 kW was taken from the paper
which has already been reported
by the author, et al. (See
references.)

According to the results, up
to 10 kW, the cutting ability
increased approximately in
proportion to the increase in
laser output. However, for the
laser output above, no increase
in proportion was observed;
evidently the laser
machinability of concrete,
etc., does not simply depend
solely on output. In other
words, how to discharge the
dross that interrupts the laser
beam is an important
consideration.

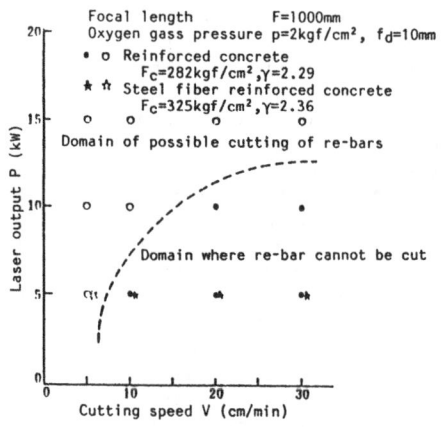

Fig. 5 Domain of possible cutting
speed and cutting depth

(3) Effect of focal length
of converging mirror on laser machinability - Fig. 7 shows the effect
of the focal length (F) of the converging mirror on laser
machinability. According to the figure for F=1,000 mm and F=600 mm,
the values of laser machinability were about the same. For F=600 mm,
as the energy density at the focal spot was about 2.8 times that for
F=1,000 mm ($do=4\lambda F/\pi D$, do: minimum diameter of beam at converging
point, D: beam diameter, λ: wavelength, F: focal length of mirror), it
was estimated that cutting depth would be larger for F=600 mm in the

higher cutting speed domain, but the result showed no difference. This is due to (1) the cutting speed of about 30 cm/min could well be in the low speed domain; (2) using a mirror with a short focal length shortens the domain that allows cutting in front and behind the focal point, etc.

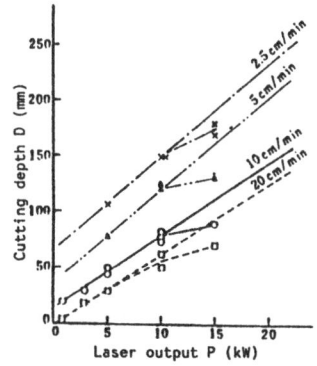

Fig. 6   Relationship between laser output and cutting depth

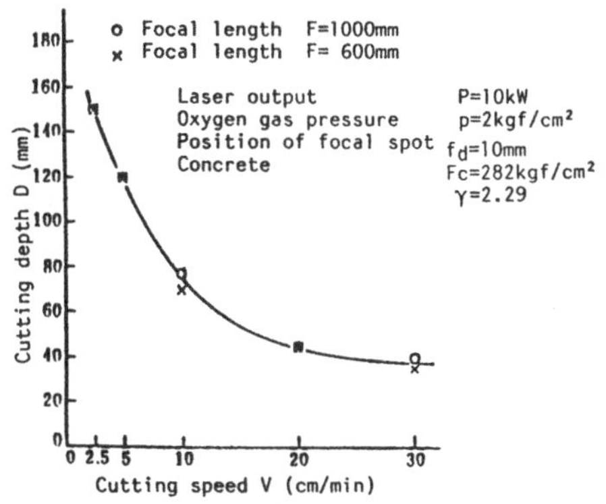

Fig. 7   Effect of focal length on laser machinability

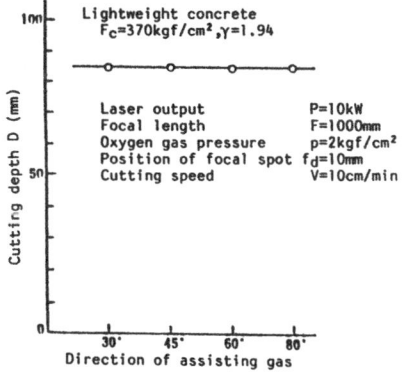

Fig. 8   Relationship between direction of assisting gas and cutting depth

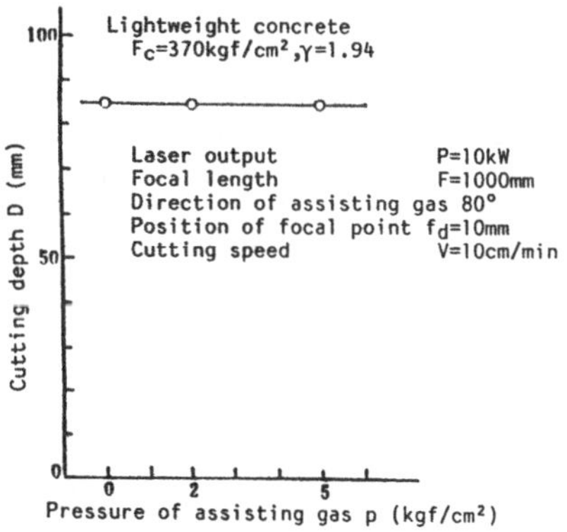

Fig. 9   Relationship between assisting gas pressure and cutting depth

(4) Effect of direction and the pressure of assisting gas on laser machinability - Fig. 8 shows the relationship between the direction of assisting gas and cutting depth, and Fig. 9 shows that between the pressure of assisting gas and cutting depth. As how to discharge dross is important, cutting was performed with the direction and the pressure of assisting gas as parameters. The result, however, indicated approximately uniform cutting depth regardless of the direction and pressure

of assisting gas. Therefore, it can be seen that the direction $(30^\circ-80^\circ)$ and the pressure (0-8 kgf/cm², 0-5 kgf/cm²: the range of this experiment, 5-8 kgf/cm²:, which is the same as reference) of assisting gas do not affect cutting depth. The cutting width, however, tended to widen on both the surface and bottom of the specimens as the gas pressure became higher. This factor is scheduled to be re-examined on a later day by varying the shape of the nozzle.

4. Conclusion

The following summarizes the results of these experiments:

(1) Even a concrete reinforced with D10 bars can be cut up to the thickness of 180 mm with a laser output of 15 kW and a cutting speed of 2.5 cm/min.
(2) In cutting reinforced concrete, there will be cases where the concrete can be cut but not the reinforcing bars, due to the relationship between the energy density of the laser beam and cutting speed.
(3) Up to a laser output of 10 kW, cutting ability increases in proportion to the increase in output. However, higher laser output will not remarkably increase cutting ability.
(4) No difference in laser machinability can be recognized between the focal length of mirror F=1,000 mm and F=600 mm in the low speed domain.
(5) The direction and the pressure of assisting gas do not affect laser machinability as far as these experiments show.

From the experiments performed thus far, the feasibility of cutting reinforced concrete by laser was confirmed. In the future, we will endeavor to improve cutting conditions so as to enhance laser machinability, as well as perform cutting of specimens which are as close to actual structures as possible.

5. Acknowledgment

Lastly, we wish to take this opportunity to sincerely thank Messrs. Norikazu Tabata, Haruhiko Nagai and Masao Hishii of Mitsubishi Electric Co., Ltd. for providing us with facilities for these experiments and for giving us guidance throughout.

References

Kazunao Sugita and Masahito Mori (1986) "Fundamental Research on Cutting Concrete by Laser," Proceedings of the Kanto Division Meeting, Architectural Institute of Japan, pp.285-288.
Masahito Mori and Kazunao Sugita (1986) "Fundamental Research on Cutting Concrete by Laser," Summaries of Technical Papers of Annual Meeting, Architectural Institute of Japan, pp.509-510.

WALL DAMAGE, VIBRATION AND OUT-DOOR SOUND CAUSED BY EXPLOSION INSIDE A CONCRETE STRUCTURE

E. KURODA and M. HAYASHI   Sirakawa plant, Nippon Koki Co. Ltd.

N. KOBAYASHI   Chuo University

T. SAITO Corporated Juridical Person Japan Explosive Safety Association

T. YOSHIDA   The University of Tokyo

Abstract

To obtain the fundamental data necessary to design a practical small magazine in the urban area, we constructed a simple frame magazine at first and then full-sized magazine, and made some experiments in them. After that experiments, we made another experiments in the full-sized magazine to design the explosion sound proof structure and the explosion proof structure. Sound, wall vibration and ground vibration measured in those experiments are explained and discussed.

Key Word : Explosion Sound, Safe Magazine, Wall Acceleration, Ground Vibration, Concrete Structure, Wall Damage

1 . Introduction

We have been investigating a safe storing method of explosive cartridge in a magazine in which the cartridges are put saparately in small quantity in the capsale type containers. In order to design a safe magazine for storing the cartridges of commercial explosives in the urban area, we constructed a simple frame magazine at first and then a full-sized magazine, and made some experiments in them. We exploded commercial explosives in Japan in a separating container buried in sand, on the top of the sand and in mid air inside the magazine, and measured various effects on the structure elements, degree of destruction, out-door explosion sounds, ground vibrations, wall vibrations etc.

As the continuation of these experiments, we made some explosion experiments inside the magazine to obtain the fundamental data to design the explosion sound proof structure and the explosion proof structure. We have changed the explosive charge weight untill the cracks on the wall were observed. After many cracks on the wall were found, we finally broke up this structure by blastings. In the course of these experiments, we continued the measurements of sounds and wall accelerations.

Sound, ground vibration and wall acceleration characteristics seen from the small explosion inside such as an imperfectly or nearly closed-up structure and from the blastings using a small quantity of explosives have been not yet known. So we wish to report the outline of the measured results.

2. Experimental

2.1 Explosives
Commercial explosives; No.2 Enoki dynamite or Energel MA-7 slurry
                       explosive cartridge 100g, 30mm φ
Initiator;   No.6 instantaneous electric detonator.

2.2 Structure
The simple frame magazine is composed of a ferro reinforced concrete U-shaped channel, with the width of 2.18m, the length of 2m and the height of 1.5m, set upside down, and two sand walls. The wall thickness of the structure is 9cm at lower part, and 15cm at the upper part and ceiling. Both open sides of this structure are closed with the sand walls of 2.3m in width, 1.7m in height and 15cm in thickness, which are made in such a way that two veneer boards are fixed parallel and filled with sand between them.

The full-sized magazine is composed of a ferro reinforced concrete of 2m in width, 2m in length and 2.1m in height, and the walls of which are 10cm thick. One side wall is a double row steel bar frame and the other walls are a single row steel bar frame. Steel bars are D10(dia. 10mm), and D13(dia. 13mm) in the corner, of SD30 described in JIS G 3112(steel bars for the concrete reinforcement). Spacing between the bars is 20cm. A steel door, 0.8m in width, 1.8m in height and 4.5mm in thickness, is installed on a side wall and a hole for the ventillation of 20cm in width and 25cm in height with a steel cover is installed on the neibouring side wall.

The sand wall was used as the wall of the magazine in the case of the simple frame magazine and as the sound shelter for the sounds from the door in the case of the full-sized magazine.

2.3 Box and separating container
We used two types of box for containers, a upward opened type and sideward opened type. The upward opened type box is 1m in length, 1m in width and 50cm in height, composed of steel frames and veneer boards, and is filled with sand and set on the concrete floor. The sideward opened type box is 40cm in width, 55cm in depth and 40cm in height, and is composed of veneer boards, and is filled with sand and set on the wooden stand of 80cm in height.

The separating container is buried in sand in such a way that container is perpendicular to the open face, which is arranged vertically for the upward type and horizontally for the sideward type box. The separating container is composed of the VP-30 PVC tube (31mm i.d., 38mm o.d. and 35cm long) and the VP-40 PVC tube, and it is tamped by sand from one end to 15cm lengths.

2.4 Experimental condition
                Explosive condition
A : Separating container without sand tamping in upward type box.
B : Separating container with sand tamping in upward type box.
C : Separating container with sand tamping in sideward type box.
D : On sand in upward type box.
E : Suspended in mid-air.
                In the case of the full-sized magazine
1   : sand wall shelter is not added with ventilator opened.
1'   : sand wall shelter is not added with ventilator closed.

11 : sand wall shelter is added with ventilator opened.
11' : sand wall shelter is added with ventilator closed.
                In the case of the simple frame magazine
1   : with crevice.
11 : without crevice.
In the experiments of the explosion sound proof or the explosion proof
structure, structure is all under 1' condition.

2.5 Measurement
Sound was measured at 10, 30, and 50m distances from each of the four
wall surface of both type structures on flat or linear, and A scales.
Impulse precision sound level meter RION NA-61 was used on flat and A
scales. Low frequency sound level meter RION NA-17 and low frequency
pressure transducer BK-2631 were used on the linear responce(0.1 to over
1000Hz). BK-2631 was set at 1m distance in front of the side wall with
the double row bar frame.

    Ground vibrations were measured at the foundation of the wall and at
30m distances by three orthogonal velocity gages, Geo Space HS-1. Acceleration piezo gages were mounted at the center and the corner of the
two type side walls.

3. Result and Discussion
Out-door sound caused by explosion inside the structure is the impulsive
sound of very short interval. Sound levels by the explosion inside a
structure were much decreased than those in open field. The selected
explosion sound data for a full-sized magazine are shown in Table 1.
Each value, mean value of three or more measurements, is respresented
in the unit of impulse level(dB imp). Lower level sounds are more
observed in the case of the simple frame magazine than in the case of
the full-sized magazine, which seems to owe to the smaller area of crevice per side and heavier wall of the former. The selected typical predominant frequency data, which were obtained in each direction and explosion source conditions, are summarized in Table 2 for a full-sized
magazine.

318

Table.1 Explosion sound (dBimp) data for a full-aized magazine

Explosion	Ventilator			Door		Wall		
	10m	30m	30m	10m	10m	10m	30m	30m
Source	Flat	A	Flat	A	Flat	Linear	A	Flat
A I	1 2 0	1 0 4	1 1 4	1 1 0	1 0 6	1 0 9	9 1	9 1
B I	1 1 4	9 6	1 0 7	1 0 5	1 1 0	1 0 3	8 6	8 8
B Ⅱ	1 1 3	9 8	1 0 6	1 0 7	1 1 0	1 0 5	8 2	9 3
B Ⅱ ′	1 0 3	9 2	9 6	9 9	1 1 0	9 8	8 2	9 0
C I	1 1 4	9 8	1 0 7	1 0 6	1 1 5	1 0 8	8 5	9 6
C Ⅱ	1 1 6	9 6	1 1 0	1 0 2	1 0 9	1 1 0	8 6	1 0 5
D I	over	1 1 3	1 2 6	－	1 2 3	1 2 2	1 0 0	1 1 2
D Ⅱ ′	1 0 9	9 1	1 0 3	1 0 0	1 0 9	1 0 8	9 2	1 0 7

Table.2 Typical predominant frequency data (Hz)

Explosion Source	Ventilator	Door		Wall
A I	85-87,142-143,178-181,204-205,286-288	92-93,142-144,179-180,202-205		74-75,92-93,142-143, 178-181,204-205
A Ⅱ		79-81,92-93,144-145		
B I	80-82,117-118,140-143,177-183	B I	77-81,86-88,116-118,141-143,203-204	80-81,83-85,91-95
B Ⅱ		B I ′		
B I ′	92-93	B Ⅱ	87-93,140-143,178-186,204-205	
B Ⅱ ′		B Ⅱ ′		
C I	79-81,90-95,113-121,162-173,198-204	79-80		78-81,92-96
C Ⅱ		78-81		
D I	82-83,148-152,182-190,204-208,328-335	D I	80-83,92-96	80-81,83-85,92-96
D Ⅱ		D I ′		
D I ′	92-96,180-184,266-272	D Ⅱ	92-96,180-184,266-272	
D Ⅱ ′		D Ⅱ ′		
E Ⅱ	80-82,119-120,124-126,154-156,187-192	118-121,152-159,188-196		82-85,93-97

Table.3 Peak particle velocity data of ground vibration (cm/sec)

Explosive conditions	Wall foundation			2 0 m		
	X	Y	Z	X	Y	Z
A	0.42~0.53	0.25~0.28	0.43~0.45	0.047~0.049	0.056~0.057	0.072~0.074
B I	0.47~0.56	0.30~0.40	0.44~0.49	0.053~0.060	0.045~0.100	0.091~0.100
B Ⅱ	0.49~0.56	0.28~0.36	0.41~0.44	0.052~0.058	0.060~0.099	0.074~0.093
C	0.078~0.139	0.083~0.143	0.13~0.18	0.027~0.038	0.018~0.055	0.050~0.081
D	0.18~0.40	0.23~0.34	0.34~0.41	0.012~0.015	0.014~0.020	0.020~0.027
E	0.40~0.50	0.30~0.48	0.36~0.50	0.009~0.015	0.007~0.011	0.014~0.025

In the case of the former magazine, the frequency ranging below 10Hz, which may be caused by the wall displacement, are much observed. On the other hand, in the case of the latter, the frequency ranging 80 to 96Hz, which may be caused by the wall vibration, are much observed. The sound magnitude and predonimant frequency ranges obtained in both cases were much varied considerably depending on the direction and the explosive conditions.

The most remarkable, intense sounds were caused by explosion gas venting and observed in the directions of the opening and the crevice of the structure. In the direction of the door, sounds were caused by the steel door vibration and explosion gas venting through the crevice and more intense than those in the direction of the wall with no installation but less intense than those in the direction of ventilator opened. These all out-door sounds had some special, discrete frequency ranges, and varied considerably depending upon the explosive conditions as shown in Table 2.

Peak particle velocity values of ground vibration are given in Table

Table.4 Predominant frequency of ground vibration (Hz)

Explosive conditions	Wall foundation			20m		
	X	Y	Z	X	Y	Z
A	11.0 88.5, 28.5 18.5	10.0	10.0	9.0	11.0	10.5~11.0
B I	11.0 102.5, 83 19.5, 30	10.5	10.0	9.5	11.0	11.5
B II	10~10.5 28.5, 19.5 85, 38.5	10.0	10.0	9.0~9.5 15, 19, 23	11.0 9.5, 14.5 19.5, 24	11.5 15.5, 20 23.5
C	11.0 79~85	11.0 31.5, 82	10~13 81	9.5~11	11.0~13.5	15~15.5
D	82~84 92~95 125	83.5~84 127~128 93~96	92~94 82~85.5 128~129.5	9~9.5 23~24	15~19.5	12~16 39~40
E	82~82.5 93~97 121~124	96~96.5 84~85.5 120.5	82.5~84 120.5~122 135	9.0~10.0 15, 82	10.5~11.5	15.0~15.5

Table.5 Peak acceleration data

Explosive conditions	Charge weight(g)	Double row steel bar frame		Single row steel bar frame	
		+Peak	−Peak	+Peak	−Peak
In Sand	30	5.9~6.3	4.4~6.4	7.3~9.3	6.8~7.6
	100	29~37	22~31	30~35	27~37
	200	26~86	29~61	40~78	44~65
	300	60~77	52~75	67~84	64~67
On Sand	30	19~34	24~28	12~26	20~32
	100	65~70	62~76	56~69	69~81
	200	101~118	66~115	73~119	92~208
In mid-air	30	30~50	25~36	26~37	31~41
	100	117~126	88~95	57~63	51~71
	200	129~191	138~170	86~179	118~135
	300	190~230	181~210	198~215	173~195

3 and their predominant frequency data are summarized in Table 4. Ground vibrations at wall foundation are greater in explosion in air and its predominant frequency is coincident with the wall vibration. This shows that ground vibration at wall foundation is excited by wall vibration. Ground vibration at 20m distances is less below 0.1 kine and low frequency below 20Hz. High frequency ground vibration is seldom observed. It is noted that ground vibration caused by falling impact of sand is compared with those caused by explosion of explosives.

In the experiments of the explosion sound proof structure and the explosive proof structure, we exploded the explosives(30~300g) in the order starting from lower charge weight and also in the order of explosion in sand, on sand, and in mid air in each charge weight. Peak acceleration(G unit) measured at mid-wall was summarized in Table 5. The relation of explosive conditions in the order of their acceleration amplitudes and its predominant frequency is shown in Fig. 1. It is noted that the predominant frequencies of mid-wall acceleration are not constant but vary with the explosive conditions. The relations of mid-wall acceleration(+ peak) for a double row steel bar concrete and sound pressure(+ peak, dBLimp) observed at 1m distance are shown in Fig. 2.

The regular interrelation for the both amplitudes deviate in the explosion in mid air of 200 and 300g charge weight, and also in sand. Deviation in the explosion in air shows that explosion gave some damage

321

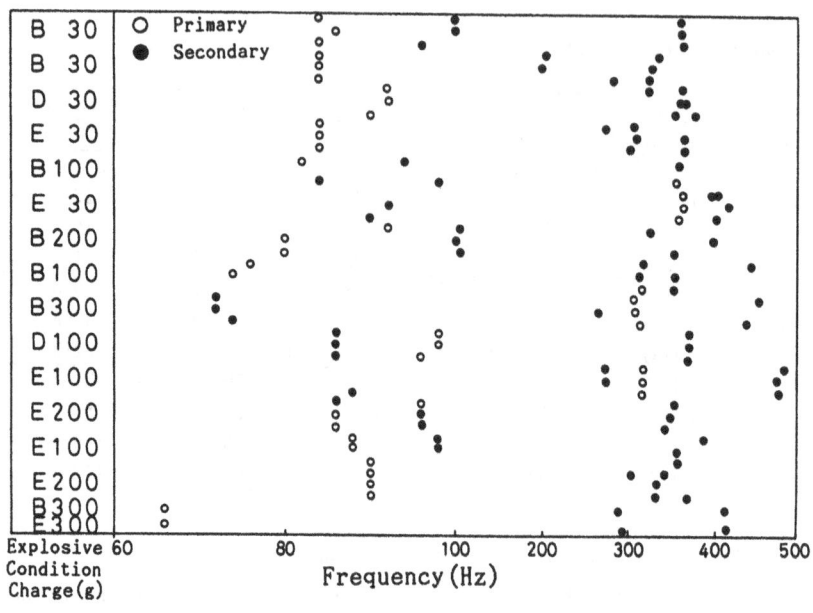

Explosive
Condition
Charge(g)

Fig. 1

Predominant frequency of

mid-wall acceleration

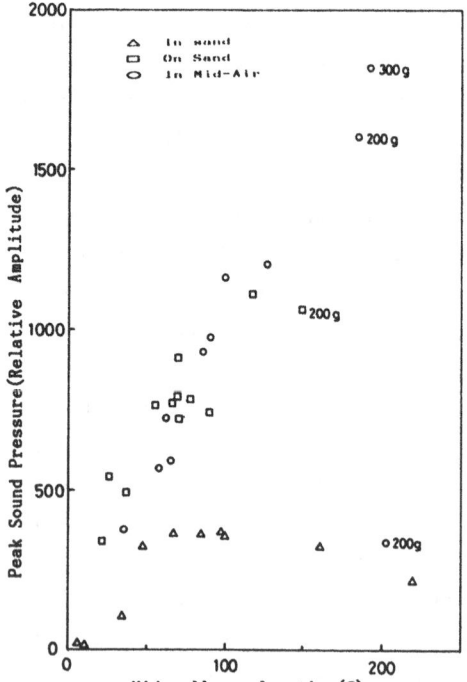

Fig. 2

Relation of peak sond pressure

and mid-wall acceleration

effects on wall. However visible cracks on the wall surface were not observed in those experiments. Mid-wall acceleration in these cases were over 170G. Deviation in the explosion in sand shows that wall vibration in the explosion in sand are locally excited by the sand impact and are not always related to sound generation. After this series of experiment, the similar shots were repeated. Sound observed in the case of the latter series of shots had more magnitude and lower predominant frequency than those in the case of the former shots. Mid-wall acceleration and its predominant frequency become lower, and its frequency not constant, under the same explosive condition. Finally, explosion in mid air of 200g charge weight caused the visible hair cracks on the wall surface and explosion of 300g charge weight caused larger cracks all over the wall surface. The mid-wall acceleration in these cases was under 160G. It seems that the strength of the wall was weakened largely by the repeated blast loads.

After these experiments, we broke up this structure by the confined blasting(2~13g per one hole) and the external charge blasting(72~500g).

4. Summary

(1). Sound caused by the explosion of 100g cartridge in the separating container buried in sand inside a structure seems to have not so much environmental problems. It seems that people can bear the sound without covering their ears at the distance of 10m, and there would be no problem based on noise at the distance of 30m.

(2). The out-door sounds generated owe much to the acceleration or displacement of concrete wall, the vibration of steel door, and the gas venting from the crevices among the sand wall, the concrete wall, and the steel door, and also that from the opening of the ventilator. their sounds have some special, discrete frequency ranges respectively.

(3). The out-door sounds have sharp directionality. In these sounds, especially those caused by the gas venting are highest and varied much depending on the explosive conditions.

(4). In the explosion inside the perfectly close-in structure, the out-door sounds caused by the wall acceleration were only a little low compared with those caused by the gas venting.

(5). Relation of wall damage and mid-wall acceleration was discussed. The wall initially did not show any visible damage by the acceleration over 170G. But after some explosion tests, the wall had visible and large damages by the acceleration under 160G, and the out-door sounds and the mid-wall accelerations showed different values under the same explosive conditions.

REDUCING THE BLASTING NOISE OF HIGH EXPLOSIVES USED TO DEMOLISH CONSTITUENTS OF REINFORCED CONCRETE STRUCTURES

T. YOSHIDA   Department of Reaction Chemistry, University of Tokyo
Y. WADA
N. KOBAYASHI   Department of Precision Machinary, Chuo University
T. SAITO   All Japan Association for Security of Explosives

Abstract
Reducing the blasting noise associated with using high explosives on reinforced concrete structures were carried out using sand coverage. Cylindrical shaped charges of two types were used for blasting concrete blocks with reinforcing H-shaped steel and for cutting steel plates and rods. The sand coverage was found to be very effective for reducing the blast noise. The relationships among the noise and depth of the sand, weight of the explosives and the distance from the point of explosions were examined. Useful information was obtained regarding the successful crashing and cutting of the constituents.
Key words : Blasting, Demolition, Concrete structures, Reduction of noise, Sand coverage.

1. Introduction

Blasting has been said to be one of the most convenient methods for the demolition of concrete buildings. There has been considerable experiences accumulated on the subjects (Kimura(1987)). But in Japan, the high noise level associated with blasting has prevented its use from practical applications in urban areas. In order to solve this problem, sand coverage was applied to the blasting demolition of the constituents of reinforced concrete buildings in order to reduce blast noise.

As the models of the constituents, steel plate and rods, and reinforced concrete with a H-shaped steel were selected. Four experiments were carried out for examining the effects of sand coverage on the reduction of noise and for obtaining additional information on the properties of the blastings.

2. Experimental methods

2.1 *Materials*
The model constituents used in the experiments were steel rods (20mm and 22mm in diameter), steel plate ($10 \times 100 \times 200$mm), reinforced concrete block ($300 \times 300 \times 300$mm) with a H-shaped steel and non-reinforced concrete plate ($60 \times 300 \times 300$mm). The sketches are shown in Figure 1.

Three shaped charges were used in this study and shown in Figure 2. The high explosives were Pentolite, compositionB, C4 and No.21 explosive, which has a detonation velocity of 8100m/sec.

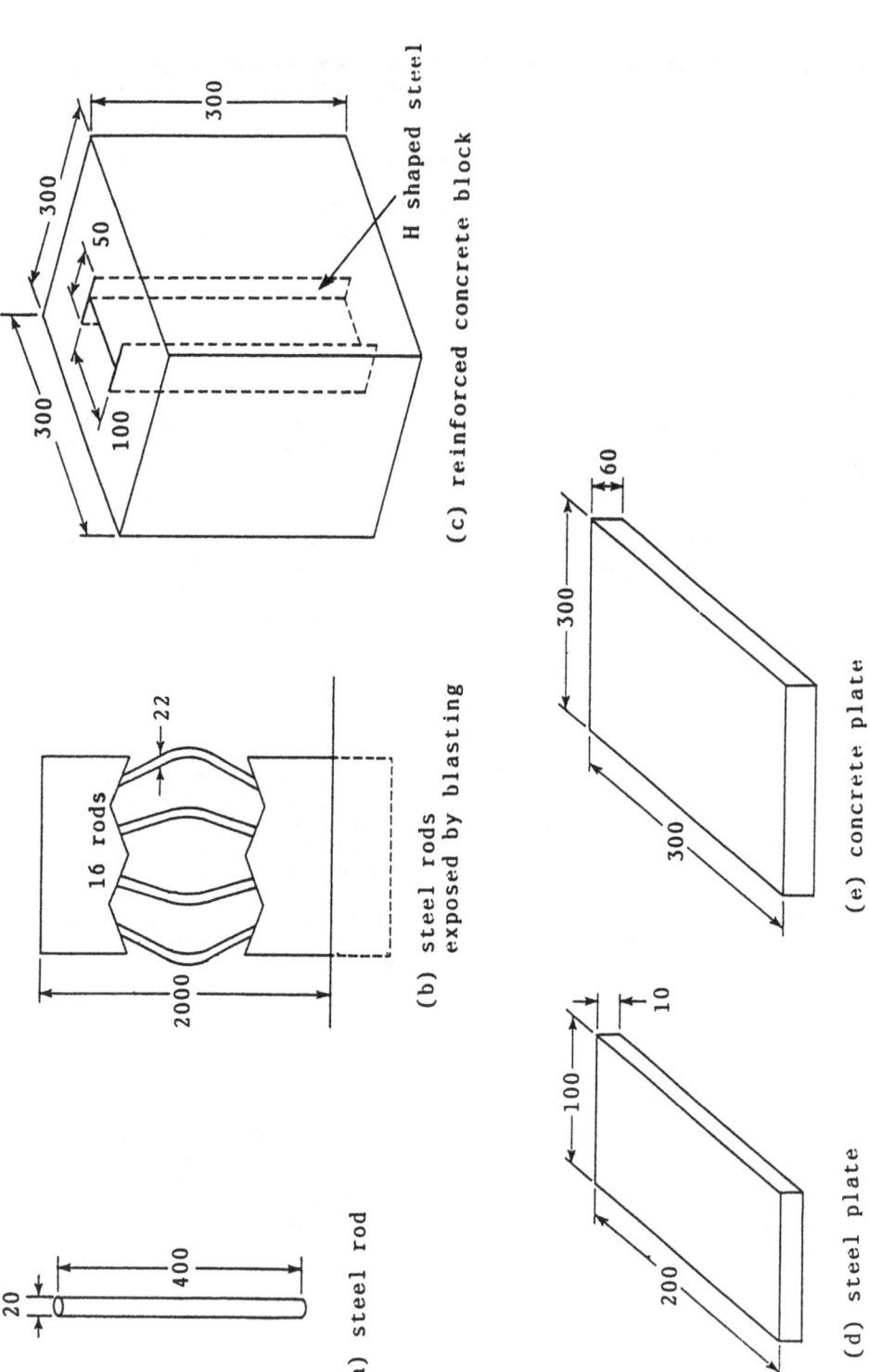

(a) steel rod

(b) steel rods exposed by blasting

(c) reinforced concrete block

(d) steel plate

(e) concrete plate

Fig.1 Sketches of model constituents of reinforced concrete structures

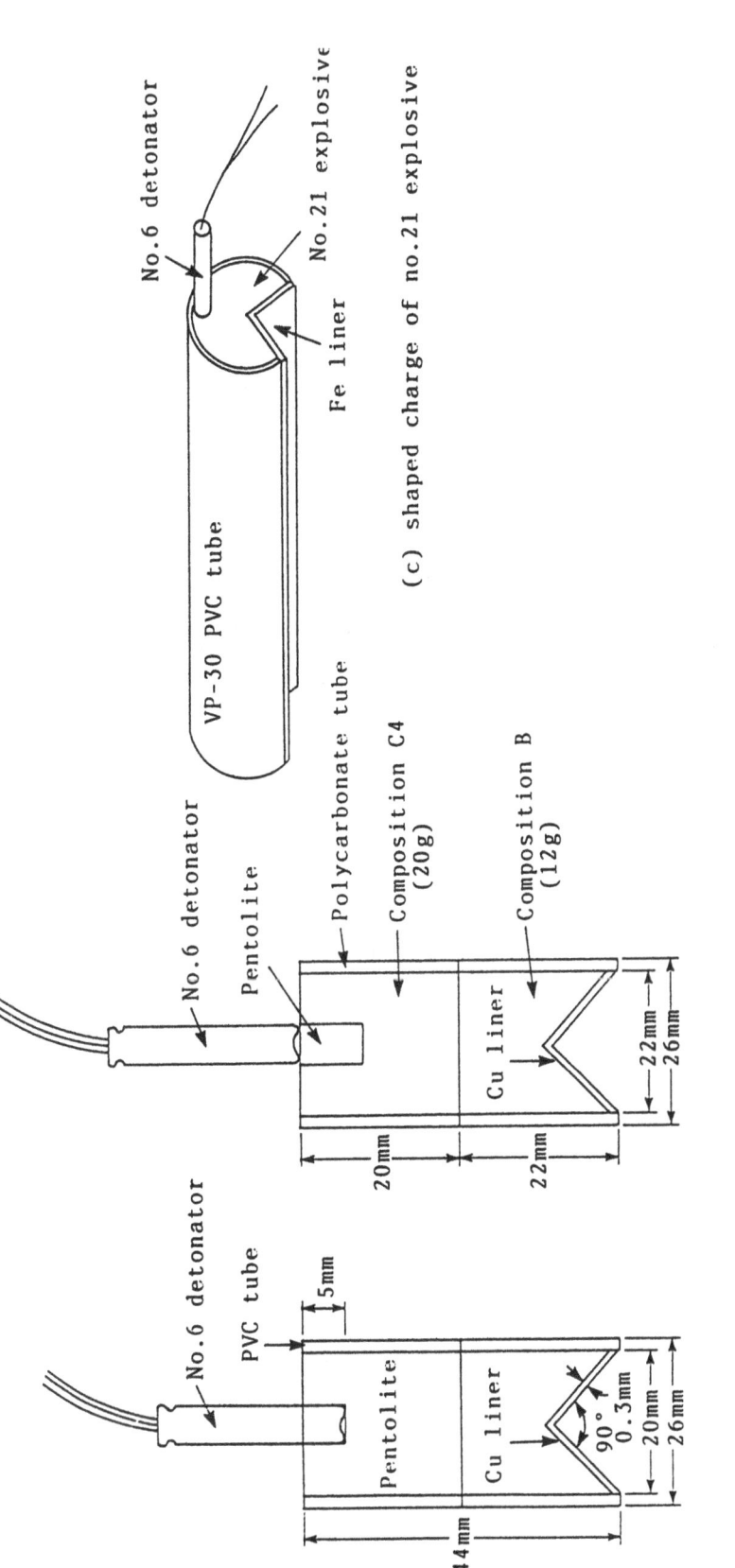

(a) shaped charge of
    Pentolite(20g)

(b) shaped charge of
    Composition B and C4

(c) shaped charge of no.21 explosive

Fig.2 Shaped charges used in the experiments.

2.2 *Measurements of blast noise*

The blast noises were measured using precision noise meters with or without a recorder. Noise measurements are recorded in dB(A).

3. Results and discussion

3.1 *Noise reducing effects of sand coverage*
3.1.1 *Effect of sand thickness(r)*

In the explosive cutting of a steel rod on an open ground, the relationship between the reduction of noise level and the thickness of sand pile was obtained as shown in Figure 3(a) and (b) for 20g explosives at 15m and 30m distance, respectively. In the explosive cutting of steel plates in a half-closed bunker, a similar relationship was obtained as shown in Figure 3(c) for 100g explosive at 15m distance.

These results indicate that blasting noise decreases with sand thickness and the effects of sand thickness are more remarkable in the blasting in the semi-closed space. The blasting in this semi-closed space may be a model of a blasting in buildings. Curves in Figure 3 are represented by the following equations :

(a) S = −32 log r + 131,   l = 15m, W = 20g, open space
(b) S = −32 log r + 126,   l = 30m, W = 20g, open space
(c) S = −44 log r + 145,   l = 15m, W = 100g, semi-closed space

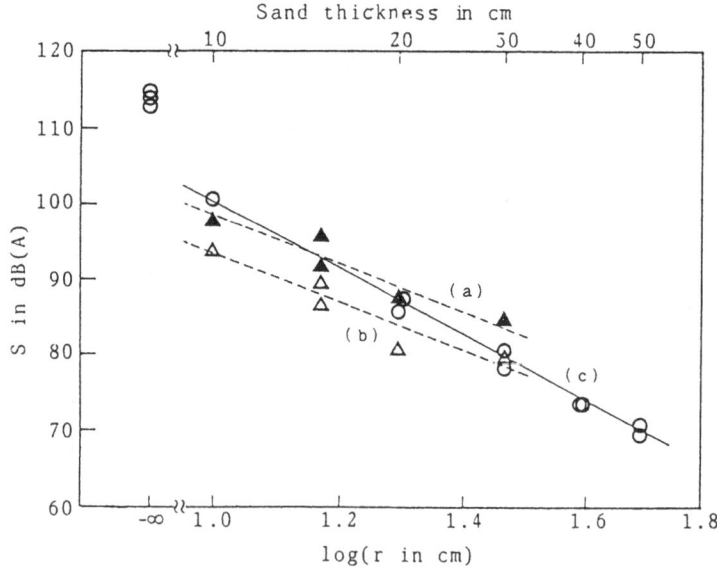

Fig.3   Plot of noise level(S) vs. log(sand thickness in cm) for the blasting with 20g explosives at (a) 15m and (b) 30m on an open ground, and (c) 100g explosives at 15m in a half-closed space.

3.1.2 Effect of distance(l) from the blasting site

Several examples of the effect of distance on the blast noise of sand covered explosions were shown in Figure 4. From these results, it was determined that the noise decreased with distance from blasting site more remarkably in the blasting in semi-closed spaces than in open fields.

The blasting in Tokyo Metropolis is regulated as follows : "The noise level at 30m from the border line of an operation site should not be more than 85dB(A) in the demolition or crashing of buildings or other structures by using explosives."

The explosions from 20g explosives in an open space and 100g in a semi-closed space with the 20cm of sand coverage were found to be in compliance with this regulation. Other blastings were carried out in another open space. Some of the results are shown in Figure 5. Underwater explosion was also effective for reducing blast noise, but the effeciency was less than that of sand coverage. In the open space blasting, 50g explosives and 30cm of sand coverage was also found to be in compliance of the Tokyo Metropolitan regulation. The 540g blasting with 40cm of sand coverage could cut 16 exposed rods of 22mm diameter by blasting a reinforced concrete pillar. The noise levels at 3 points are shown in Figure 5(b). However, the noise level at 30m was 97dB(A) and is not in compliance with the regulation. The problem may be solved by additional countermeasures, such as increasing sand thickness, enclosure by a wall and enlargement of the border line of the operation site.

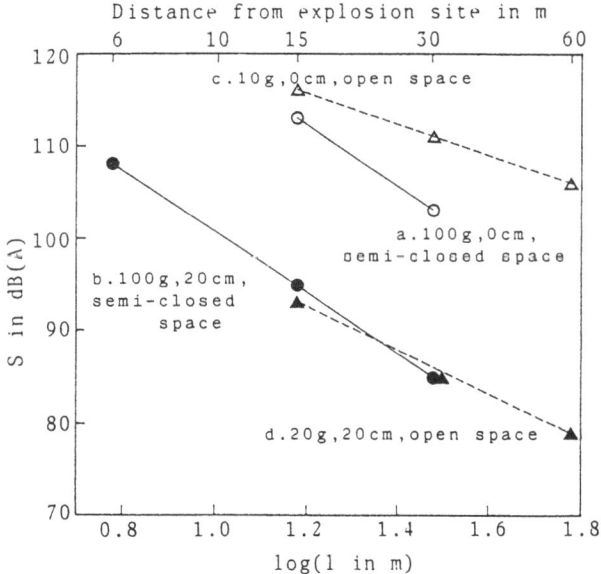

Fig.4 Plot of noise level(S) vs. log(distance
(l) in m from explosion site).

329

3.1.3 *Effect of weight(W) of explosives*

The effect of explosive weight on the blast noise is shown in Figure 6. For small explosive weights, the effect on the noise level was rather small. But, for large amounts of explosives, the effect was much larger.

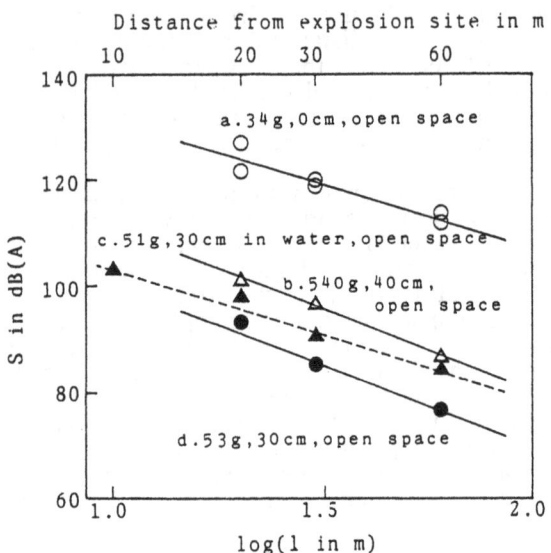

Fig.5 Plot of noise level(S) vs. log(distance (1) from explosion site) for explosions with and without sand coverage, and underwater.

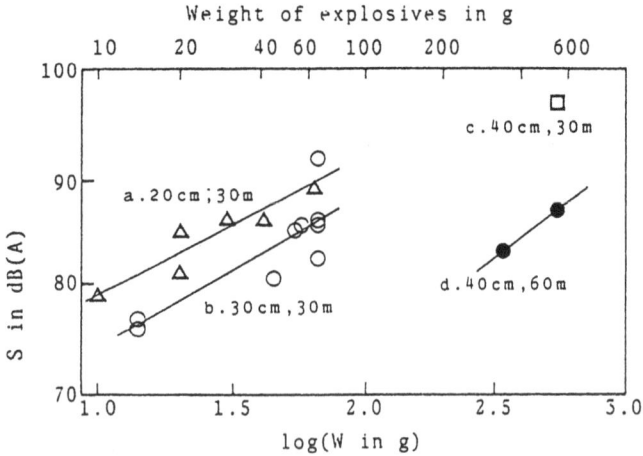

Fig.6 Plot of noise level(S) vs. log(weight in g of explosives) for blasting with sand coverage.

330

3.1.4 *Effect of the shape of sand coverage*

The sand coverage of two shapes shown in Figure 7 were compared in order to determine the noise arresting effect. In one charge(a), the sand was on plywood which was separeted by an air space. In the other charge(b), the sand was placed on the ground directly. The noise level is about 10dB(A) higher for the former than for the latter. The sand coverage was more effective in the blasting at ground level.

3.2 *Some information on the performance of blasting the constituents*
3.2.1 *Cutting steel rods*

By using cast Pentolite(50/50), a 20g explosive was able to cut a steel rod of 20mm diameter. When a cylindrical shaped charge of the same diameter as the rod is used and off-set a few centimeters, the rod is not cut but perforated. For cutting a rod with a shaped charge, it is better to blast with no off-set. The cutting mechanism has not only a penetrating effect of a shaped charge but also a tearing off effect by the explosion of the charge. This was suggested from the structure of cut end of the rod.

A completely exposed rod was able to be cut using a shaped charge. A rod buried in concrete could not be cut by using the same shaped charge, even though a bore hole was made and the charge was attached to the rod directly.

3.2.2 *Cutting steel plate*

In cutting a steel plate by blasting, the material arrangements shown in Figure 8 were used. In the cases of (A) and (C), the plates were not cut with the same shaped charges. This shows that even styrene foam reduces the performance of the shaped charge when a material is present between the shaped charge and the plate. Rigid material behind the plate also prevented the shaped charge perforating the steel plate. The arrangement shown in Figure 8(D) is the best for cutting a steel plate.

3.2.3 *Exposing a H-shaped steel from reinforced concrete*

We could expose a *H*-shaped steel from reinforced concrete by blasting using an external shaped charge. But an internal charge with less explosive was much better for this purpose.

Acknowledgements : The authors are grateful to the people of Nippon Oil & Fats Co.,Ltd., Nippon Kayaku Co.,Ltd., Nippon Koki Co.Ltd., and the students of Tokyo and Chuo Universities for their collaboration.

Reference

Kimura, M. (1987), "Building demolition by blasting", Kogyo Kayaku , 48, 139-150.

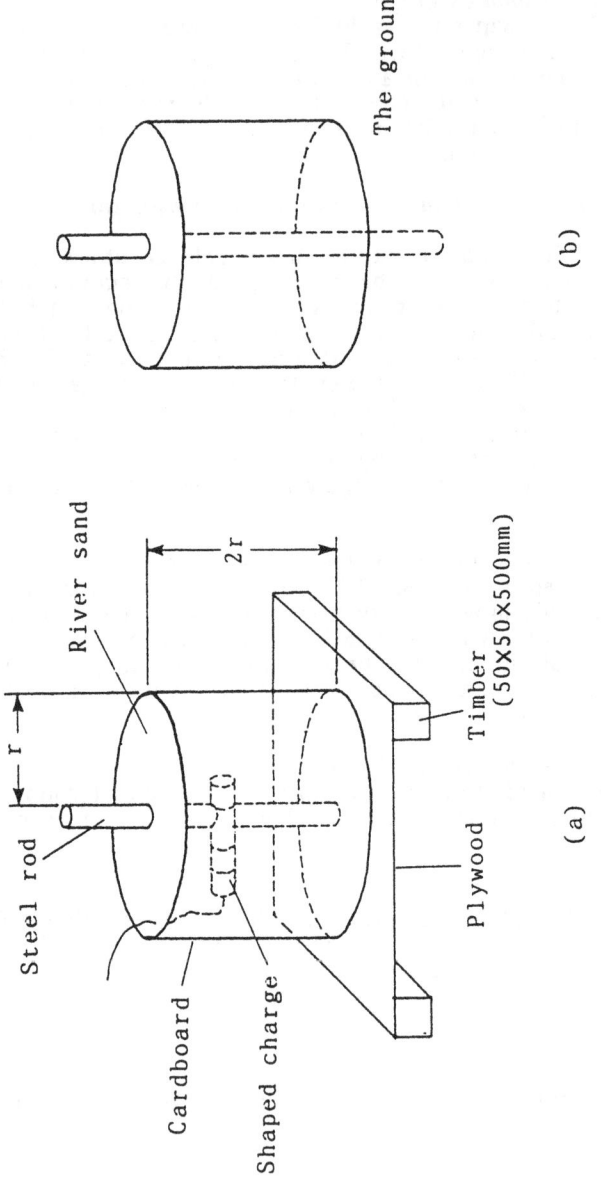

Fig.7 Two types of charges covered by sand.

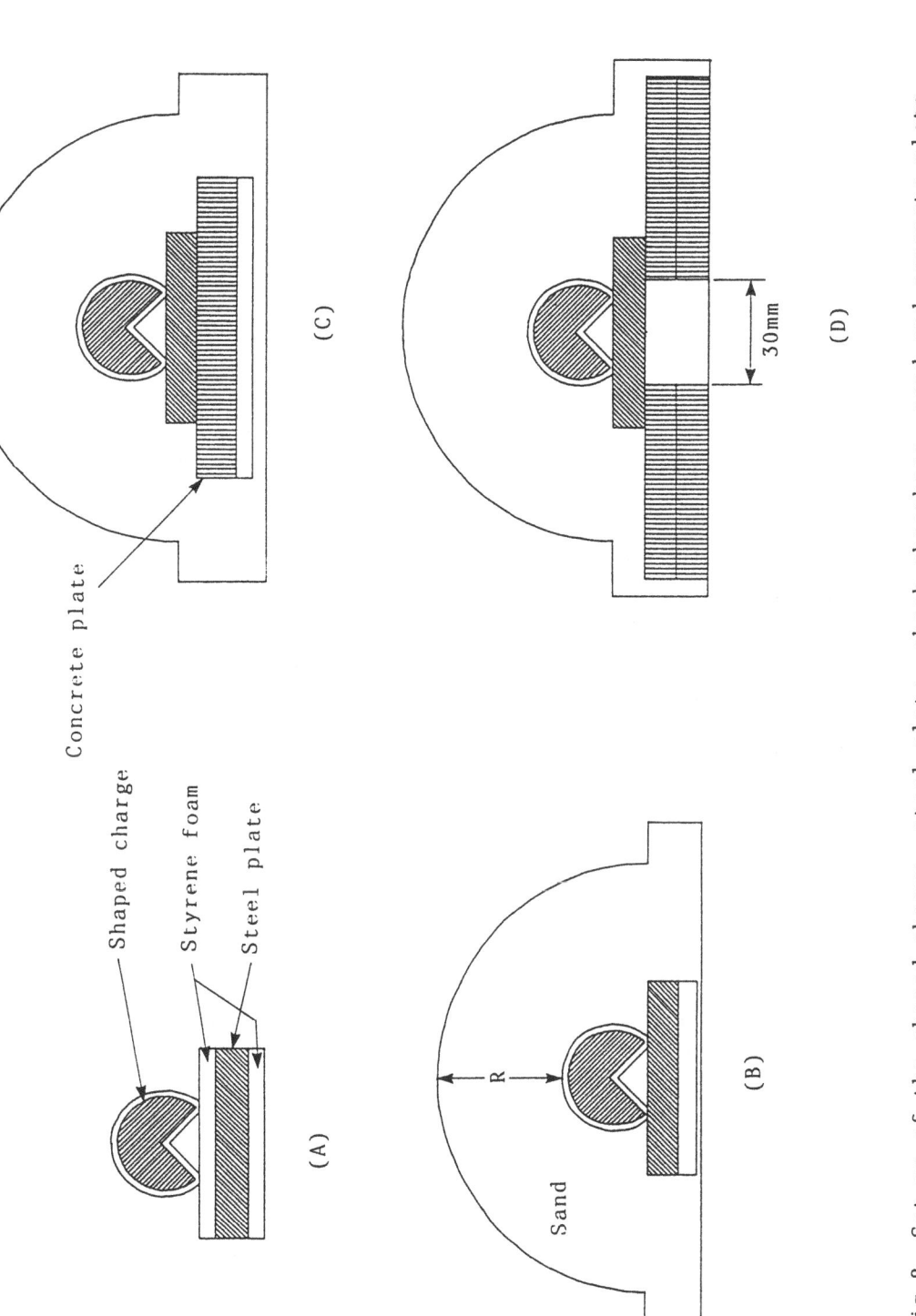

Fig.8 Set-up of the shaped charge, steel plate, shock absorbers, sand and concrete plate.

CHARACTERIZATION AND TREATMENT OF THE CONCRETE DUST
FOR REACTOR BIOLOGICAL SHIELD CONCRETE DISMANTLEMENT

MITSUO YOKOTA, TOSHIMASA NARAZAKI   Japan Atomic Energy Research
                                     Institute
MINORU HARADA, ISAYA YOKOTA   Kajima Corporation, Kajima
                                     Institute of Construction
                                     Technology

ABSTRACT
An abrasive water jet cutting technique, a controlled blasting technique can be applied to dismantling of reactor biological shield
concrete during nuclear reactors decommissioning. A large quantity
of activated dust is generated, when reactor biological shield concrete structures are dismantled by these techniques. To prepare
the necessary treatment method for the dust, we have been researching
fundamentally the characteristics and the controlling of dust. Also
tests to put to practical use were carried out with concrete
model structures under contract with the Science and Technology
Agency. Based on these test results, dust treatment methods such
as dust generation control and ventilation systems were established
for the reactors decommissioning. These treatment methods first
will be applied to the actual dismantlement of reactor biological
shield concrete of JPDR.
Key word: Water jet, Blasting, Reactor, Concrete, Dismantlement,
          Dust control, Ventilation

1. INTRODUCTION

When a reactor concrete structure is dismantled by any method,
including abrasive water jet method and controlled blasting method,
a large quantity of activated dust is produced. Therefore, the treatment for activated dust becomes very important to prevent unnecessary
worker radiation exposure and expansion of contamination. This paper
describes the characteristics and the treatment methods for the dust
produced in cutting, blasting and breaking reinforced concrete by
the abrasive water jet method and the controlled blasting method.

2. SOURCE AND PROPERTIES OF DUST

2.1 Controlled blasting method
When the controlled blasting method is used, dust is produced during
the drilling, blasting, secondary crushing and gas cutting of steel
bars and plates. Since dust is produced on the fractures during
concrete breaking, the quantity of the dust increases as the specific
surface area (surface area per unit weight) of the broken concrete

increases, that is, as the diameter of broken pieces becomes small.
Fig. 1 shows this. This figure was drawn by measuring the quantity
of dust and the diameter of test pieces broken with low-speed blasting
powder.

As the water content in concrete increases, the adhesive force
between the concrete fractures and dust particles and that between
the particles becomes large and less dust is produced. This is shown
in Fig. 2.

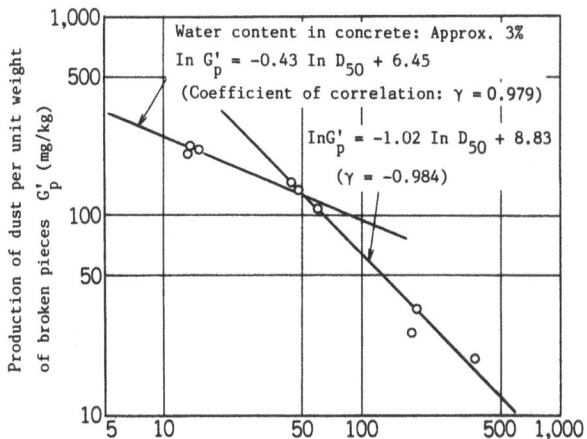

Fig.1 Relationship between mean diameter of pieces
broken by blasting and total dust produced

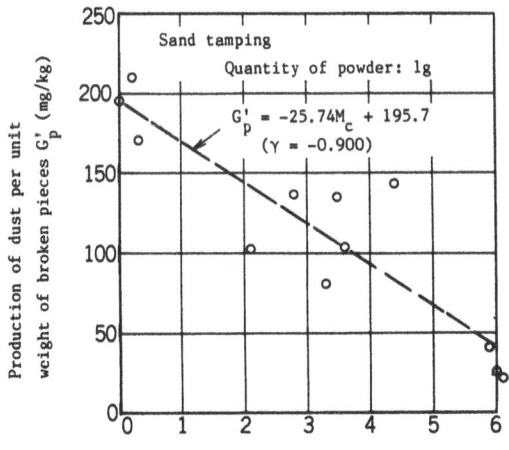

Fig.2 Relationship between water content in concrete
at blasting time and total dust produced

335

Table 1 shows examples of the quantity of dust produced when the controlled blasting method is used. The total quantity of dust produced by blasting is 0.044 g for 1 kg of the broken pieces. The total quantity of dust produced by drilling and secondary breaking are 1.5 g/min and 1.7 g/min respectively, which are at the same level. The quantity of fumes produced by cutting steel bars and steel plates with gas is 0.15 g/min and 0.05 g/min respectively, which are considerably small.

Table 1   Production of dust by the controlled
explosion method

Production of dust / Work name		Production of dust in unit time (g/min)		Production in unit work			Remarks
		10 μm max	Total dust	Unit	10 μm max	Total dust	
Drilling		0.93	1.49	Drilling length (g/m)	35.1	56.2	Drilling machine: Jack hammer
Blasting		–	–	Weight of broken pieces	0.0070	0.044	Explosive: Urbanite
Secondary breakage		0.38	1.72	Same as above (g/Kg)	0.044	0.15	Breaker: Pick hammer
Gas cutting	Steel bar	0.018	0.15	Unit cut area (g/cm^2)	0.0011	0.0092	Combustion gas:
	Steel plate	0.0059	0.049	Same as above (g/cm^2)	0.00021	0.0018	Acetylene Nozzle dia: 0.9 mm

2.2 Abrasive water jet method

When the abrasive water jet method is used, dust and droplets containing dust are produced by slurry being scattered. Most droplets have large diameters, and they settle in a short time. However, droplets smaller than 10 μm in diameter are suspended for a long time, and remain as dust after the water around them had evaporated.

The quantity of dust, droplets and water vapor produced when reinforced concrete is cut with the abrasive water jet method are shown in Table 2. The total quantity of dust produced is 0.38 g/min, which is about 1/4 of that produced by drilling and secondary breaking in the controlled blasting method. Water vapor is produced by a temperature rise of the water jet, and it is condensed into fine water particles (mist) by a temperature drop of the ambient air.

336

Table 2 Production of dust by the water jet method**

Product	Production	
	Per unit cutting time (g/min)	Per unit cutting area (g/m²)
Dust	0.38 (0.10)*	5.2 (1.4)*
Mist	1.7	23.5
Steam	2,000	27,700

  * Value in ( ) is production of dust smaller than
   10 μm in diameter.
 ** Cutting conditions: Water Pressure: 2,000 kgf/cm²
          Water flow: 50 liter/min
          Quantity of abrasive: 5 kg/min

3. TREATMENT METHODS FOR RADIOACTIVE DUST

3.1 Planning procedure for a radioactive dust treatment method

Radioactive dust not only irradiates workers who inhaled the dust but also spreads and contaminates clean places and materials which increases the radioactive waste or has a bad effect on the environment. Therefore, a radioactive dust treatment method must be planned to attain the following basic goals.

 a. Reduction of the exposure level for workers
 b. Prevention of the spread of radioactivity
 c. Reduction of radioactive waste

Fig.3 shows the planning procedure for a radioactive dust treatment method. The work place in a controlled area is classified according to the radioactivity level. The number of times for ventilation N is determined, then the quantity of air to be ventilated Q is obtained. The dust concentration C in the work place is obtained from the quantity of produced dust $G\rho$ by the following formula; $C = G\rho/Q$. When the control concentration is $C\rho$, if $C > C\rho$, that is, if much dust is produced, the source of dust must be properly modified. The local prevention of dust production is performed by wetting the work place or collecting dust with a local dust collector. In many cases, a green house (pollution shield) is installed to enclose the dust source. If C is still larger than $C\rho$ after taking the above countermeasures, the quantity of air to be ventilated must be changed. The air must be ventilated with exhaust blowers, and the disposition of the exhaust air must be designed according to dust property data.

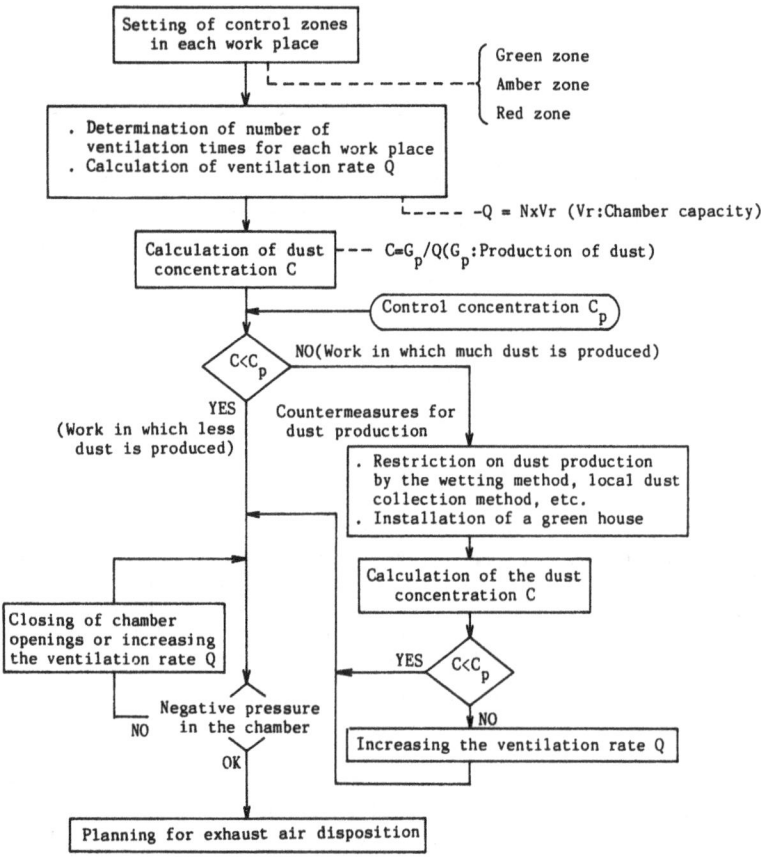

Fig.3 Planning procedure for dust production
countermeasures

3.2 Countermeasures for dust production

3.2.1 Controlled blasting method
As countermeasures for dust production, water supply type drills are used for drilling work, water is splashed over during blasting and breaking work, and hoods and local dust collectors are used during gas cutting work for steel bars and plates.

As an exmaple, Fig.4 shows the effects of a polyethylene bag filled
with water and used as tamping material to restrict the quantity
of dust produced during blasting work. In this example, the quantity
of dust is reduced to 1/2 - 1/5 of that produced by tamping with
sand. The theory of dust restriction in this case seems to be that
gaps among the broken pieces of concrete are forcefully filled with
water because of the blasting pressure and the fractures and dust
particles are wetted momentarily.

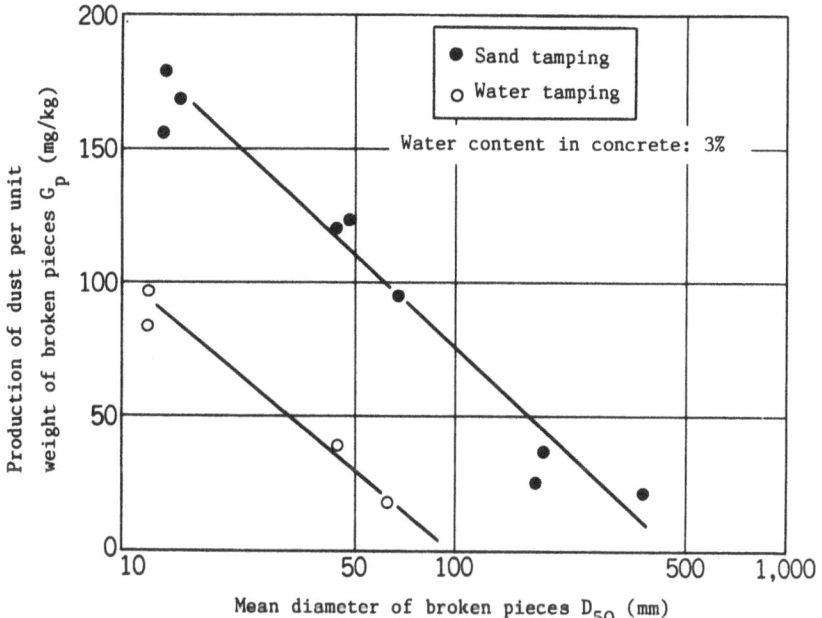

Fig.4 Effects of water tamping on blasting

3.2.2 Abrasive water jet method

When the abrasive water jet method is used, the slurry, droplets and
dust produced by cutting concrete spreads out of the cut groove at
about 60 degrees to the cutting axis up to about 10 m. Therefore,
prevention measures for dust spreading should be taken along this
cut groove.

Fig. 5 shows the cover installed on the water jet nozzle to prevent
dust spreading. This cover has a shape for introducing scattered
slurry smoothly into the slurry chamber, and the collected slurry is
sent by air, etc. to the disposer.

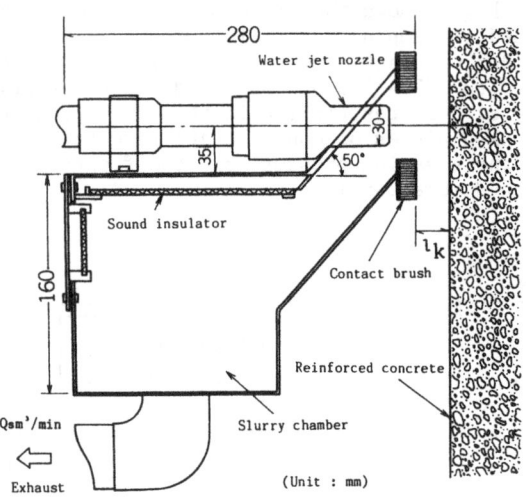

Fig.5 Cover for dust from spreading prevention

(Side view)

Fig.6 shows the collection ratio for slurry collected by this cover while cutting reinforced concrete (under the conditions shown in Table 2). Even if distance l_k between the cover and cutting face is increased to a certain degree, most of the slurry can be collected. If l_k is 0 and the exhaust rate is above 1 m³/min, the collection ratio is increased above 99%.

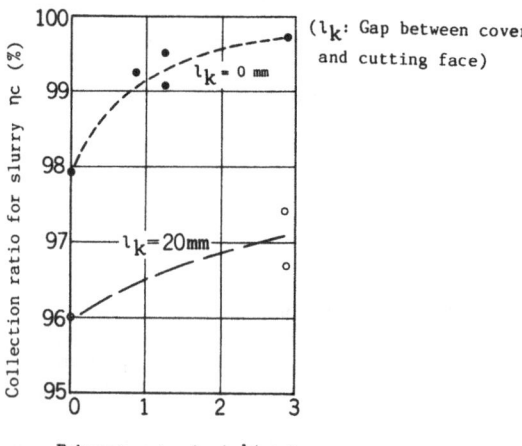

Fig.6 Effects of cover for dust spreading prevention

3.3 Ventilation plan
The purpose of ventilation is to lower the dust concentration in the work place and prevent dust from leaking out of the work place and control area. The general values of the required negative pressure and number of times for ventilation based on the level of radioactive contamination are shown in Table 3.

Table 3   Classification of control area and number of

ventilation control times (example)

	Green zone	Amber zone	Red zone
Negative pressure (mmAq)	-1 ∿ -5	-6 ∿ -10	-15 ∿ -30
Number of ventilation times (times/h)	5 ∿	10 ∿	15 ∿
Control standard area (mrem/h)	2	20	-

Note) Green zone: Zone where men enter constantly and
                  which may be contaminated temporarily
                  by an accident, etc.
      Amber zone: Zone which may be contaminated by
                  dismantling work.
      Red zone: Area in cell and contaminated zone.

The quantity of air to be ventilated is calculated using these values. If the required negative pressure cannot be maintained after the openings and gaps of the control area are closed, the ventilation rate must be increased.

Air is ventilated through intake and exhaust openings. These openings must be installed at the most distant places from each other in the control area so that the air will not take a shortcut. If those places are upper and lower ones, exhaust openings must be installed in the lower ones to prevent particles from setting. Also, a baffle plate (a filter, etc.) or a pressure adjustment valve must be installed on the intake openings to adjust the negative pressure.

As an example, Fig. 7 shows the ventilation method used when reinforced concrete is dismantled by the controlled blasting method. During blasting, combustion gas, as well as the blast, is produced and the negative pressure must be high enough to absorb them. In this example, the required negative pressure is supplemented by an inflated sheet above the steel enclosure.

Fig.8 shows the result of damping the dust concentration after a blast. This figure almost conforms to the expression of the complete dust dispersion in air ($\alpha = 1$).

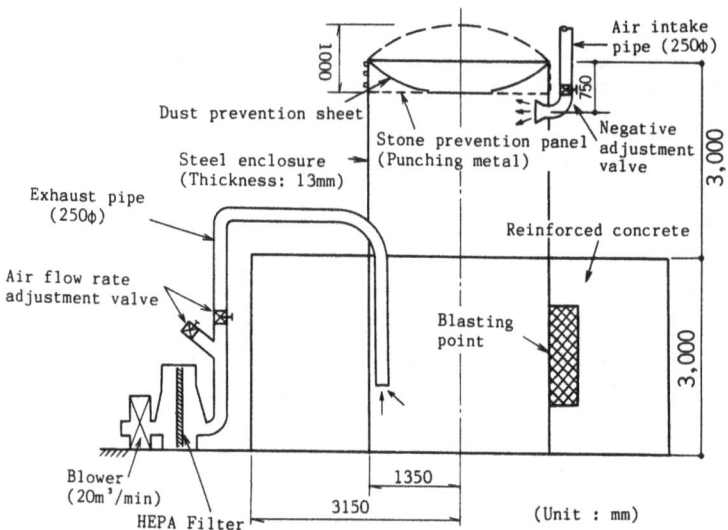

Air intake
pipe (250φ)

Dust prevention sheet

Stone prevention panel
(Punching metal)

Negative
adjustment
valve

Steel enclosure
(Thickness: 13mm)

Exhaust pipe
(250φ)

Reinforced concrete

Air flow rate
adjustment valve

Blasting
point

Blower
(20m³/min)

HEPA Filter

1350

3150

(Unit : mm)

Fig.7   Ventilation for blasting work

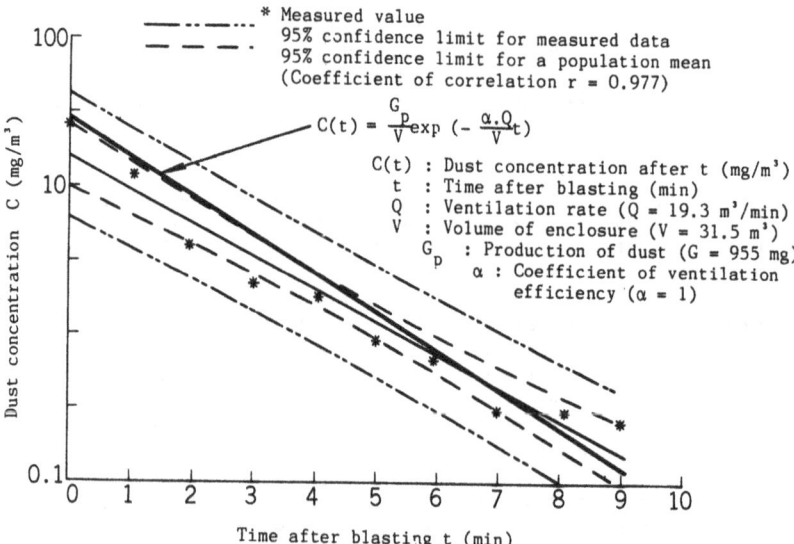

* Measured value
95% confidence limit for measured data
95% confidence limit for a population mean
(Coefficient of correlation r = 0.977)

$$C(t) = \frac{G_p}{V}\exp\left(-\frac{\alpha.Q}{V}t\right)$$

C(t) : Dust concentration after t (mg/m³)
t   : Time after blasting (min)
Q   : Ventilation rate (Q = 19.3 m³/min)
V   : Volume of enclosure (V = 31.5 m³)
G_p : Production of dust (G = 955 mg)
α   : Coefficient of ventilation
      efficiency (α = 1)

Dust concentration C (mg/m³)

Time after blasting t (min)

Fig.8   Damping of concentration after blasting

3.4 Plan for exhaust air disposition

The disposition method for exhaust air is shown in Fig.9. Generally,
for disposing of air containing radioactive dust, a high performance
single-stage or multi-stage filter is installed for the last step.
A HEPA (High Efficiency Particulate Air) filter can collect more than

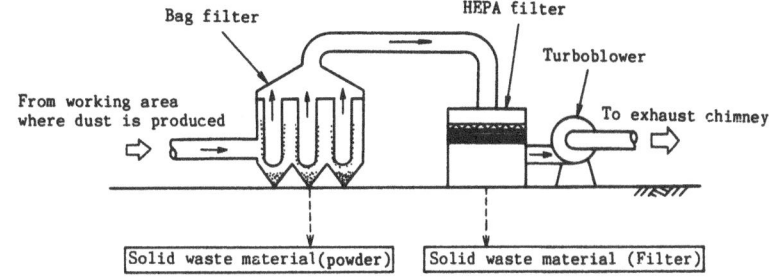

(a) Controlled blasting method

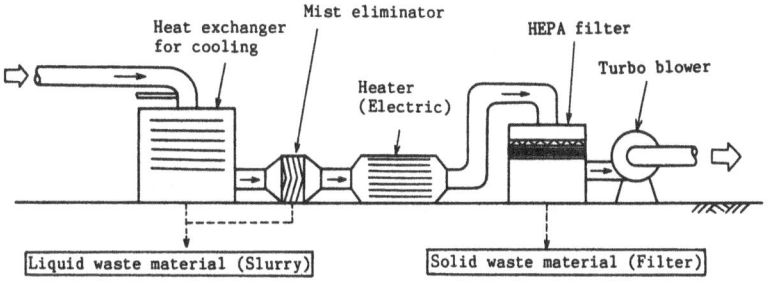

(b) Abrasive water jet method

Fig.9 Method for exhaust air disposition

99.9% of dust that is 0.3 μm in diameter, but its collecting capacity is small (0.5 - 1.0 kg for a 30 m³/min unit). Therefore, it is necessary to install a pre-filter and pre-treat the dust according to its properties and quantity.

For dust pre-treatment, a bag filter is installed when the controlled blasting method is used, while a heat exchanger etc. is installed when the abrasive water jet method is used to protect the HEPA filter from being blocked with steam in large quantities.

4. CONCLUSION

Since the characteristics of dust produced when dismantling reactor concrete structure depends on the method used, it is very important to understand the dust. To prepare the necessary treatment method for dust, cutting and blasting tests using the abrasive water jet method and the controlled blasting method were carried out and the various properties and the controlling methods of generated dust became clear. These treatment methods first will be applied to the actual dismantlement of reactor biological shield concrete of JPDR (Japan Power Demonstration Reactor).

ASBESTOS EXPOSURE TO WORKERS DEMOLISHING ASBESTOS CEMENT CLAD BUILDINGS

S.K. BROWN
Commonwealth Scientific and Industrial Research Organisation
Division of Construction and Engineering, Australia

Abstract
The external surface of asbestos cement (AC) cladding becomes weathered
after many years by the gradual loss of cement such that a loosely bound
surface layer enriched in asbestos is formed. This effect usually
appears pronounced with roof cladding but slight with wall cladding and
may cause asbestos fibre emissions during the demolition of old AC clad
buildings. Such emissions have been measured so that appropriate
precautions to protect demolition workers can be adopted. Asbestos dust
exposures during demolition by removal of whole sheets averaged 0.3 to
0.6 fibres per millilitre (f/mL) for roofs and less than 0.1 f/mL for
walls reflecting the significant difference in extent of weathering
between these elements. Suppression of asbestos emissions from roofs by
wetting or sealing of weathered surfaces was not predictable because
asbestos fibres could be present in dust trapped under sheet laps.

Keywords: Asbestos, Asbestos cement, Weathering, Demolition, Worker
exposure.

1. Introduction

Inhalation of asbestos fibres has been established to lead to specific
physical disorders in exposed workers – notably asbestosis, lung cancer
and mesothelioma. Accordingly, regulations have been enforced in most
countries to reduce and control worker exposure to asbestos dust to
below specific exposure guides (hygiene standards or control limits).
Also work codes have been developed to minimise asbestos emissions and
to specify worker protection where appropriate.
    Work with new asbestos cement (AC) products does not normally lead to
unacceptable exposure to asbestos dust unless the products are cut,
machined or otherwise abraded, and codes of practice exist for such work
(National Health and Medical Research Council (NHMRC) 1982). Handling of
new AC products has been considered to lead to negligible asbestos
emissions because the asbestos is firmly bound into a cement matrix.
However external AC products undergo surface degradation after long
periods of weathering by the loss of cement and the laying bare of a
'fleece' of fibres (Opoczky and Pentek 1975, Brown 1982). The presence

344

of such a layer may be significant for various work operations where weathered AC sheeting is handled, e.g. demolition of AC clad buildings.

The Department of the Environment (1983) noted that demolition and removal of old AC products could fall within UK asbestos regulations and have to be carried out by licensed asbestos removal professionals. In Australia the National Occupational Health and Safety Commission (1987) has specified precautions to be observed when removing AC products from buildings, including protective clothing and respirators for workers, prewetting or sealing of sheets, no breakage of sheets or use of power tools for cutting, and proper site clean-up and disposal of sheets. Such precautions are now typically in use in Australia. However the asbestos concentrations evolved during removal and demolition of old AC products do not appear to have been investigated to determine the adequacy of these or other precautions. This paper describes such an investigation carried out at several work sites; a more detailed report can be found in Brown (1987).

2. Work operations

A series of work trials were carried out at several sites where 30 to 40 year old AC clad buildings were re-roofed or demolished. Roof replacement was carried out by two to six men working on top of the roof who repetitively unfastened and removed small sections (20 to 40 m^2) of old AC roofing and replaced it with steel roofing. Sheets were removed whole and were carried across the roof to be stacked or dropped into a bin at ground level. Trials were conducted for 2 to 6 hours during which 50 to 100 m^2 of roofing was replaced. The buildings that were demolished were several large (90 m x 36 m) warehouses with corrugated AC roofing and flat AC sheet wall cladding. Most of these were demolished by removal of whole sheets from the structure by men confined to elevated platforms beneath the sheeting. In comparison to roof replacement, work conditions during demolition were more confined, involved closer handling of sheeting and were visibly more dusty, particularly as sheets were stacked on the platforms. Repetitive trials over shorter periods (30 to 60 minutes) were generally employed for these cases. In some trials comparative measurements were made after sheeting had been wet with water or sealed on its weathered surface with an acrylic emulsion (28% by weight solids applied at $0.3L/m^2$).

3. Asbestos measurements

Airborne asbestos dust concentrations were measured in each trial by personal sampling within the breathing zones of workers using the Standard Membrane Filter Method described elsewhere (Brown 1987). The techniques to identify the types of asbestos present and their contents in surface layers and trapped lap dusts are also described in the above reference.

4. Results and discussion

4.1 Characteristics of the weathered layer

The extent of surface deterioration of roofing was severe in most cases, each exhibiting a loose surface layer enriched in asbestos with unbound clumps of fibre clearly visible. Identification of the fibres showed the layers generally contained mixtures of chrysotile and amosite but in two cases chrysotile, amosite and crocidolite were all present. In some cases the layer was scraped from the roofing to determine its density and fibre content (Table 1). It is seen that the layer is enriched in asbestos fibre relative to the original product (usually 12 to 15 % w/w asbestos). In comparison, flat wall sheets with similar periods of exposure exhibited little deterioration and no loose surface layer although occasional clumps of fibre usually identified as chrysotile were observed.

Table 1.   Properties of surface layer from weathered roofing.

Roof	Surface layer density (g/m^2)	Fibre content (% w/w)
A	240	32
B	340	25
C	110	22
D	190	24

4.2 Worker exposure to asbestos dust

In Australia exposure to asbestos dust is assessed according to a time-weighted-average (TWA) concentration for an 8 hour working shift and in this study collective measurements will be reduced to a TWA concentration for comparative assessment of work operations. The hygiene standard for exposure to asbestos in Australia is 1.0 f/mL for chrysotile and 0.1 f/mL for amosite and crocidolite or any mixtures containing these types (NHMRC 1983). Work operations with old AC products are likely to involve exposure to mixtures of chrysotile, amosite and crocidolite and so measurements will be assessed relative to the 0.1 f/mL standard.

Asbestos exposures to workers handling AC sheet during roof replacement are summarised in Table 2. At buildings 1 to 3 these measurements were made during the full process of removal and replacement while at building 4 and 5 they were restricted to the period of sheet removal.

Asbestos concentrations while replacing dry roofing at buildings 1 and 3 were similar to the exposure standard of 0.1 f/mL but higher concentrations were measured at building 4 where sampling was carried out only during sheet removal. Sealing the roof surface with an acrylic resin reduced concentrations somewhat at building 3 but appears to have had little effect at building 5. In all these cases except building 2 sheets were stacked during the work operation. Stacking was observed to

Table 2. Asbestos concentrations while replacing AC roofing.

Building number	Work description	Asbestos concentration (f/mL)		
		n	Range	TWA
1	Replacing dry roofing	8	0.03-0.24	0.10
2	Replacing part-painted roofing	2	0.03	0.03
3	Replacing dry roofing	8	0.04-0.27	0.10
3	Replacing acrylic-sealed roofing	8	0.03-0.08	0.05
4	Removing dry roofing	6	0.07-0.32	0.21
4	Removing dry roofing with careful handling and wetting as stacked	8	ND-0.07	0.03
5	Removing acrylic sealed roofing	6	0.04-0.26	0.15

n   = number of measurements
TWA = time-weighted average
ND = non-detectable

create a surge of dust-laden air back across each worker as sheets were dropped. The effect of this on asbestos dust emmission was investigated at building 4 by hosing each sheet after stacking. Asbestos concentrations were reduced to much below the hygiene standard for this case, suggesting that much of the emission from dry sheets occurred during stacking.

Building demolition involved extensive periods of continuous sheet handling at more vigorous rates than in roof replacement and under more confined conditions - workers were usually located beneath sheets and slid them across structural members before stacking on the work platform. Generally, sheeting was removed at a rate of 100 m^2/man-hour in demolition work as compared to 5 to 10 m^2/man-hour for roof replacement. Asbestos exposures to workers removing AC sheeting from several warehouses are summarised in Table 3. Removal of dry roofing led to TWA concentrations of 0.3 to 0.6 f/mL compared to 0.07 f/mL for wall sheets, reflecting the large differences in surface weathering to these elements.

The concentrations during roof removal were reduced markedly for buildings 6 and 7 by wetting or sealing weathered surfaces prior to sheet handling. However these treatments resulted in little if any suppression for roof removal from building 8. It was observed that much of the visible dust which evolved during sheet handling originated from dust accumulated under sheet laps and ridge capping. Samples of these were taken from buildings 7 and 8 and were analysed for asbestos content using infrared spectroscopy (according to absorptions at 3690 m^{-1} and

Table 3. Asbestos concentrations while removing AC roofing and wall cladding for building demolition.

Building number	Work description	n[a]	Range	TWA[b]
		Asbestos concentration (f/mL)		
6	Roof sheets removed from collapsed building[c]			
	- dry	3	0.10-0.47	0.32
	- wet	2	0.05-0.06	0.06
	- acrylic-sealed	3	0.11-0.32	0.16
7	Roof sheets removed at platform			
	- dry	6	0.30-0.53	0.38
	- wet	2	0.10-0.13	0.12
8	Roof sheets removed at platform			
	- dry	10	0.34-1.1	0.60
	- wet	4	0.29-0.68	0.50
	- acrylic-sealed	4	0.41-0.76	0.55
8	Wall sheets removed at platform			
	- dry	4	0.04-0.12	0.07
	- acrylic-sealed	2	ND[d]-0.05	0.02

a   n   = number of measurements
b   TWA = time-weighted average
c   in this case the building had been collapsed with the roof intact to
    approximately 1 m above ground level before sheet removal
d   ND = non-detectable

780 cm^{-1} for chrysotile and amosite, respectively, when calibrated against UICC asbestos samples). Results of these analyses are presented in Table 4 which shows that dusts from building 8 contain markedly higher amounts of asbestos (predominantly as free fibres by visual assessment). This is consistent with the higher asbestos emissions found in this case and the inability of surface treatments to control such emissions. It may be possible that wetting of sheets as they were removed would have suppressed this emission but this was not investigated.

5.   Conclusions

Handling of weathered AC sheeting during roof replacement and demolition work can lead to workers being exposed to unacceptable asbestos dust concentrations. The source of asbestos emission appears to be both weathered surfaces and dust trapped under the sheeting laps and control measures should aim to suppress dust from both these locations. Wetting sheet surfaces as they were stacked was successful in reducing worker

Table 4. Asbestos contents of dusts accumulated under AC roofing.

Building number	Dust location	Asbestos content (% w/w)
7	Ridge dust	0.01
7	Lap dust	0.04
8	Ridge dust	0.4
8	Lap dust A	4.6
8	Lap dust B	1.1

exposure to below hygiene standards in one case. A combination of this practice and respiratory and clothing protection for workers should form part of recommended work procedures during demolition of AC clad buildings. For example, an effective practice for demolishing AC clad building would include the following aspects:

(a)   sheets are removed whole with minimal breakage and without powered cutting tools;

(b)   sheets are carried by two men rather than one, preferably at waist level and without dragging across other sheeting;

(c)   sheets are stacked carefully with thorough wetting of weathered surfaces with a water spray;

(d)   visible residues at the site are cleaned up by wet sweeping or with a vacuum cleaner approved for asbestos;

(e)   workers carrying out these tasks should wear coveralls which are disposed of or laundered after use and either disposable or half-face cartridge respirators approved for use with asbestos.

References

Brown, S.K. (1982) Fibre release from corrugated asbestos cement cladding due to weathering and cleaning. CSIRO Division of Building Research Report, Division of Building Research, Highett, Australia.

Brown, S.K. (1987) Asbestos expsoure during renovation and demolition of asbestos cement clad buildings. Am. Ind. Hyg. Assoc. J., 48(5), 478-486.

Department of the Environment (1983) Asbestos materials in buildings. Her Majesty's Stationery Office, London, pp. 1-33.

National Health and Medical Research Council (1982) Code of practice - working with asbestos cement (fibrocement) building materials. In <u>Report of the Health Hazards of Asbestos</u> (prepared by Asbestos Ad Hoc Subcommittee), Australian Government Publishing Service, Canberra, pp. 51-55.

National Health and Medical Research Council (1983) Approved occupational health guide: threshold limit values (1983-84). Adopted at ninety-sixth session of Council, Commonwealth Department of Health, Canberra, pp. 1-41.

National Occupational Health and Safety Commission (1987) Draft code of practice for the safe removal of asbestos. In <u>Asbestos: Guide to the Control of Asbestos Hazards in Buildings and Structures</u>, Australian Government Publishing Service, Canberra, pp. 33-81.

Opoczky, L. and Pentek, L. (1975) Investigation into the 'corrosion' of asbestos fibres in asbestos cement sheets weathered for long times. In <u>RILEM Symposium on Fiber Reinforced Cement and Concrete</u> (ed. A.N. Lancaster), Construction Press, Great Britain, pp. 269-276.

DEVELOPMENT OF PREVENTION TECHNIQUES AGAINST ASBESTOS HAZARD IN BUILDING DISMANTLING

K. MOTOHASHI, S. YUSA, and Y. TAKAHASHI
Building Materials Department, Building Research
Institute, Ministry of Construction

Abstract
Recently, asbestos hazard problems have been becoming social problems in Japan. Development of the techniques
preventing ambient air from asbestos pollution have been
in strong demand. In this context, it is an urgent subject to establish a standardized specification on asbestos
abatement works for sprayed asbestos layers. Another important subject is to develop the controlling techniques
in use of asbestos cement product which have been utilized
as building materials and are allowed to use at present.
   The purpose of our research is to make the standard
specification of asbestos abatement works for sprayed
asbestos, and to establish the technology which enable us
to utilize asbestos cement products properly.
Key words: Asbestos, Asbestos abatement, Asbestos cement
sheet, Demolition, Encapsulant, Fire safety, Scanning
electron microscope.

1. State-of-the-art of asbestos hazard problem

Asbestos hazard problems have been becoming social
problems in Japan recently. Journalism became aware of
these problems and has started reporting them frequently
since the time EPA RULE-MAKING PROPOSAL TO BAN ASBESTOS
(U.S.A., 1986) was handed.
   One of the important and urgent subjects is abatement
of sprayed asbestos which had been used for imparting fire
safety, soundproof performance, and thermal insulation
performance. According to the Japanese association of asbestos product industries, sprayed materials containing
asbestos had been in use for the duration shown in Table 1.
   Sprayed asbestos and sprayed mineral wool containing
more than 5% of asbestos had been applied before 1975,
when works of sprayed materials containing more than 5% of
asbestos were prohibited by Ministry of Labor. Some of
sprayed mineral wool products had contained less than 5%
of asbestos until 1980. Since 1980, sprayed mineral wool
have contained no asbestos.

351

Table 1   The use duration of sprayed materials containing asbestos fibers.

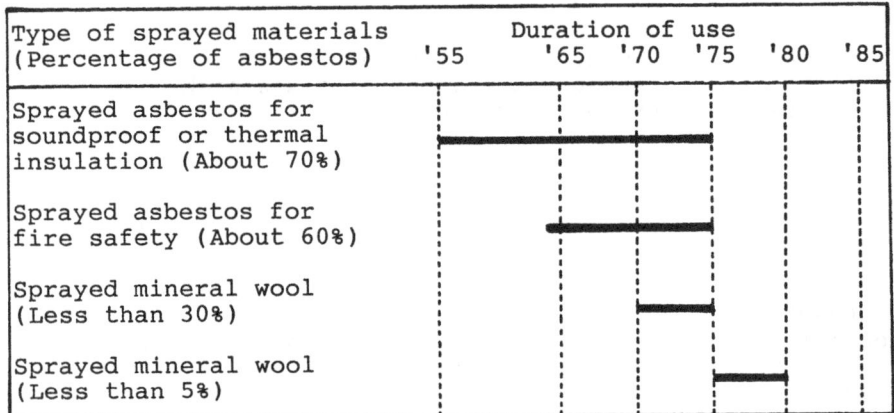

Type of sprayed materials (Percentage of asbestos)	Duration of use
	'55 '65 '70 '75 '80 '85
Sprayed asbestos for soundproof or thermal insulation (About 70%)	
Sprayed asbestos for fire safety (About 60%)	
Sprayed mineral wool (Less than 30%)	
Sprayed mineral wool (Less than 5%)	

Recently, some cases of deterioration of sprayed asbestos layers were reported by journalism. At present, the public organizations are surveying amount of sprayed asbestos in public buildings. According to such survey, it is considered that asbestos fiber concentration in ambient air inside the building where asbestos are used is almost same as that in outdoor ambient air in many cases. These concentration values are 2 to 4 order magnitude lower than the limited value in labor environment (2 fiber/cc for chrysotile and amosite, 0.2 fiber/cc for crocidolite). However, abatement works are necessary for deteriorated asbestos layers.

Asbestos abatement methods can be divided into three types, namely, removal, encapsulation and enclosure. The recommended standard specification is now in preparation for every type of abatement work in order to avoid unqualified works and working conditions. Another important subject is to make a guideline on selection of abatement methods corresponding to building conditions or deterioration states of sprayed asbestos layers. The Building Research Institute is now preparing such a guideline which includes evaluation methods for encapsulants.

2. Evaluation of asbestos abatement methods

In asbestos abatement, qualified works must be performed by following the recommended specification. After abatement works, fire safety and other performance have to meet the required levels.

One of the important subjects in evaluation of asbestos abatement methods is how to evaluate encapsulants. Encapsulated asbestos layers must withstand Japanese

restriction of internal linings to keep fire safety.
Furthermore, the following characteristics are required.

- Resistance to strength
- Resistance to water damage
- Flexibility
- Impact resistance
- Resistance to chemicals, oils, heat, etc.
- Durability

Number of encapsulants are on the market. These encupsulants are both foreign products and Japanese products.
Fig. 1 shows one of the examples of the surface test
results related to the fire protection performance.

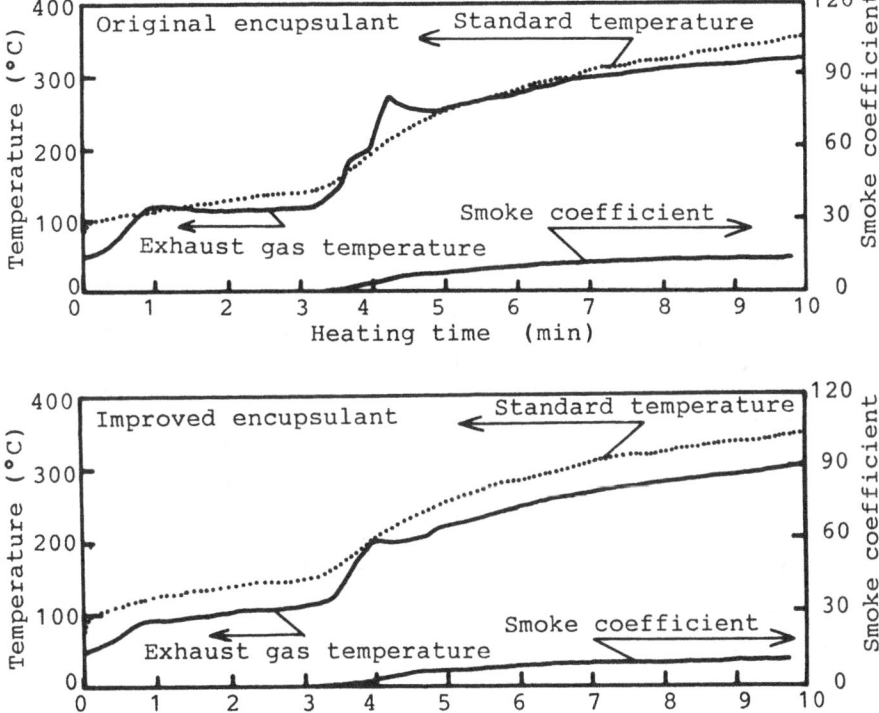

Fig. 1   Example of the surface test for encupsulants.
(Content of aluminum hydroxide of original
encapsulant and improved encapsulant are 30%
and 35%, respectively. Specimens were prepared
by coating encapsulants on asbestos cement
perlite sheet 10mm in thick)

353

Table 2 Classification based on the surface test.

Non-combusible	Semi non-combusible	Fire-retardant
1. No melting through the entire thickness. 2. No cracks through to the back. 3. No sustained flaming lasting for over 30 seconds after each heating is finished. 4. The curves of the exhaust gas temperature, obtained as a result of the heating test, should not exceed the standard temperature curve in the course of 3 minutes from commencement of the test.		
5. tdθ = 1 (°C min)	tdθ < 100 (°C min)	tdθ < 350 (°C min)
6. Ca < 30	Ca < 60	Ca < 120

tdθ : The area enclosed by the exhaust gas temperature
      curve and the standard temperature curve where the
      exhaust gas temperature exceeds the standard tem-
      perature tem curve.
Ca : Smoking coefficient.

The Japanese building law specifies the grade of fire
protection properties of internal linings which may be
used on a particular structure, in accordance with the
use, floor area, and height of the buildings. All
materials except for flooring used as internal linings
must have specific fire protection properties which have
been ascertained by the standardized testing methods. On
the basis of these test results, the materials shall be
classified into three grades, namely, non-combustible,
semi non-combustible, and fire retardant. The surface
test is one of these standardized tests. Consulting with
Table 2, it is considered that original encapsulant cannot
meet the semi non-combustible grade. (The exhaust gas
temperature curve exceeds the standard temperature curve
within 3 minutes.) However, The encapsulant containing
more aluminum hydroxide can meet the semi non-combusible
grade.
The standard test methods for evaluating other required
performance is now under preparation.

3. Controlling techniques for asbestos cement products

Asbestos cement products represent the largest market for
asbestos fibers, and a large amount of asbestos cement
sheet have been used as building materials. At present,
use and production of asbestos cement sheet are not
prohibited. However, possibility of asbestos contamination due to cutting on site or surface deterioration have
been pointed out.

In this context, makers of asbestos cement sheet have been developing substitute materials containing low percentage of asbestos or no asbestos, however, performance of such substitute materials is not so good as the asbestos containing products. Other approaches are to reduce cutting works on construction sites by supplying precutted components on the market and to improve cutting machines.

The association of asbestos cement sheet industries are investigating the fiber concentration values in various conditions. Judging from these data, it is considered that asbestos cement sheet can be used under controlled conditions.

Another important problem is contamination of environmental ambient air due to surface deterioration. Photo. 1 shows scanning electron micrographs of surface of deteriorated corrugated asbestos cement sheet. Photo. 2 shows surface at initial stage. In the deteriorated surface, friable asbestos fibers can be observed.

Influence of friable asbestos fibers due to deterioration is now under investigation, and application of coating materials has been studied as one of the methods to prevent environmental air from such pollution. Photo. 3 shows the surface after application of epoxy primer on the deteriorated surface. Cleaning of the deteriorated surface prior to epoxy primer application must be performed carefully in order to avoid pollution due to removal dust. In this coating process, epoxy primer can attach residual dust as shown in Photo. 3. After primer coating, epoxy coating and acrylic coating are applied. Photo. 4 shows cross section of the coated layers on the deteriorated surface, and these three coat layers can be observed clearly.

4. Problems in building demolition and conclusion

Removal of asbestos is ultimately required in building demolition. At present, Japanese contractors do not possess much experience of removal works in building demolition. However, considering the amount of sprayed asbestos layers, this problem will be an important subject in near future. Therefore, the guideline of removal works will have to be filled up for the various conditions of actual demolition works. Building Research Institute is now carrying out the study on the asbestos abatement works as mentioned above. Specification the standard of removal works in building demolition is in preparation through these research activities.

Photo. 1 Surface of deteriorated asbestos cement sheet.
(Magnification: X30 and X300)

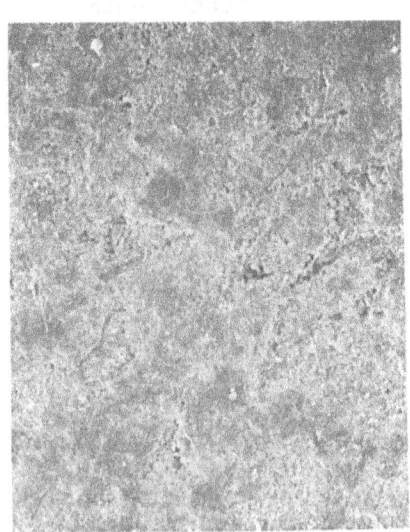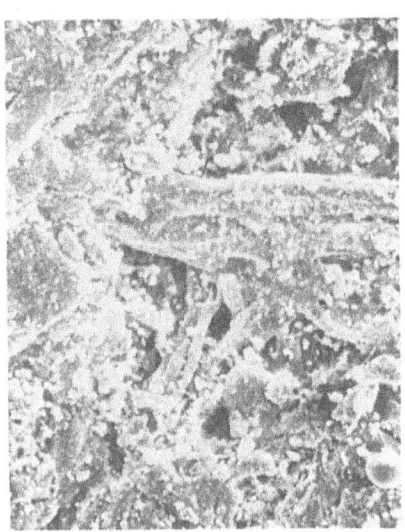

Photo. 2 Surface of asbestos cement sheet at initial
stage. (Magnification: X30 and X300)

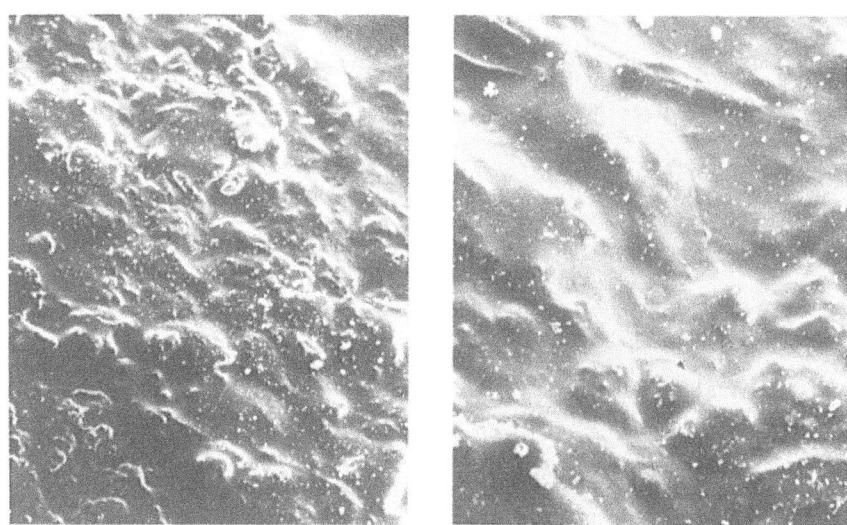

Photo. 3   Surface of deteriorated asbestos cement sheet
            after epoxy primer coating.
            (Magnification: X30 and X300)

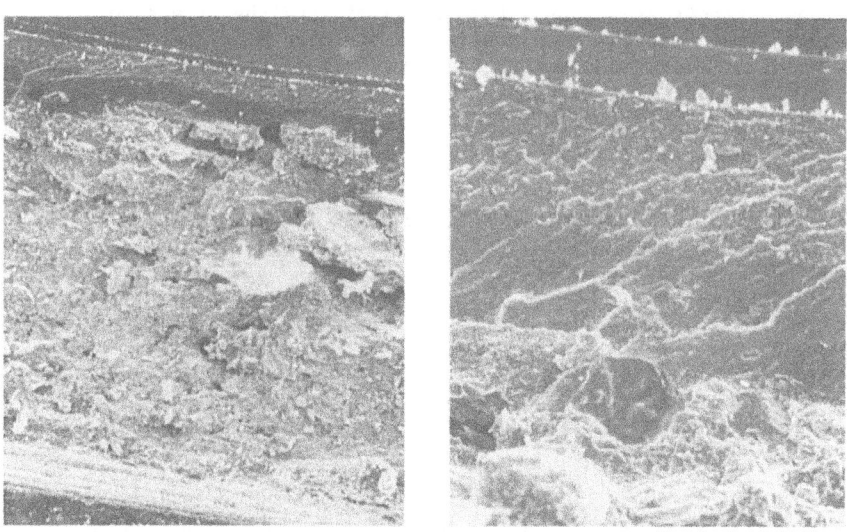

Photo. 4   Cross section of the coating system on
            deteriorated asbestos cement sheet.
            (Magnification: X30 and X300)

CONTROLLED DEMOLITION OF PRESTRESSED CONCRETE STRUCTURES

S.H. BUCHNER and P. LINDSELL
Department of Engineering Science, University of Oxford

Abstract
The safe removal of any structure depends upon detailed knowledge of
the original design principles, but may not be sufficient to predict
the behaviour of a complex prestressed structure undergoing
demolition. Detailed measurements taken on structures over the last
six years have demonstrated that the release of energy in post-
tensioned tendons depends upon a number of parameters.

During demolition, the number of strands or wires in a tendon, the
integrity of the grout and area of shear link reinforcement influence
the damage created in a simple beam. In complex segmental
structures, the debonding length of a tendon, the number of in situ
joints and cables and the residual prestress control the method of
working and the structural response. This paper illustrates these
important aspects and relates them to safety and the effects of
releasing the prestress in a continuous box-girder bridge.
Key words : Demolition, Prestressed concrete structures, Debonding,
Monitoring, Residual prestress.

1. Introduction

There are many forms of prestressed concrete structure, the majority
of which can be removed with little difficulty provided that
guidelines, such as those provided by the Health and Safety Executive
(1984), are followed. However, over the next decade there will be an
increase in the number of prestressed concrete structures containing
sequential prestressing or segmental construction which will have to
be demolished. The demolition or dismantling of these structures is
a complex operation, the safe removal of which relies heavily on full
and expert knowledge of the construction. This knowledge begins with
an initial recognition of the problem and understanding of the
original design principles. All to often demolition contractors are
left entirely to their own devices - mostly without detailed
demolition drawings and sometimes without knowing that the structure
contains prestress. The results of inadequate information could be
disastrous and, as many of the newer structures fall outside the
experience of most demolition men, they must join forces with those
who have a knowledge of the design.

The mathematical prediction of the effects of demolishing any prestressed concrete structure is hampered by variations in the nature and quality of the construction. These problems are, at the moment, insurmountable and assessments of any demolition job are usually left to the contractor after a cursory inspection of the task in hand. Assuming there are no detailed records, a first step in assessing a structure is the recognition of the type of prestressing system used. This information will indicate the form of the prestress and the likely order of magnitude of the forces involved. A multitude of systems have been developed over the years and documents, such as that produced by the Construction Industry Research and Information Association (1985), have provided a certain amount of background information on the different types, although some have inevitably been forgotten with the passage of time. Using knowledge of the prestressing system and any records from the construction, the approximate age of the structure can be assessed and hence the likely code that was used for the design can be arrived at. This information can be important because of the variations in the design criteria which have occurred over the years. Lindsell and Buchner (1988) found that a critical factor in the response of a structure during demolition is the quantity of shear reinforcement surrounding the prestressing cables. Having established the design criteria used, it is possible to make an estimation of the amount of shear reinforcement present without having the construction drawings. Even with all this information relating to the design and construction of the structure it is not possible to guarantee the behaviour during demolition. Hence consideration should be given to the use of structural monitoring for the most complex demolition operations.

2. Pre-Demolition Monitoring

Lindsell and Buchner (1987) have shown that direct and indirect methods of assessing the levels of stress within a structure can be used, before demolition commences, to provide the contractor with information on the forces to be dealt with. This allows for an assessment of the ability of the structure to stand up to the demolition process; low levels of stress within the steel can be as difficult to deal with as high levels. The demolition contractor can then be provided with useful guidance on the best technique for a particular situation.

Direct measurements of the forces can be made by exposing sections of prestressing cables at various positions and attaching small strain gauges directly to the steel. This is a delicate job but once completed the steel can be severed by cutting with an oxy-acetylene torch or a hack-saw, depending on the nature of the job. At the same time an inspection can be made of the condition of the grout.

Buchner et al (1984) have developed an indirect method of assessing the concrete stress levels which does not require direct access to the prestressing cables. This method is based on the removal of small core samples, from the structure, at critical positions. Once the in situ stress assessment has been completed, an

analysis of the levels of prestress remaining within the structure
can be made.

Another important testing proceedure, which can be adopted prior
to demolition, is the use of surface measurements along the line of a
cable duct. This technique can be used in conjunction with the
direct method as it requires a portion, or all, of the steel to be
severed. Once completed, however, the results show how the steel
behaves upon cutting. When a cable is severed there is a release of
force which can break down the bond between the steel and surrounding
grout and result in a loss of tension in the tendon over several
metres from the cut position. The degree to which this 'debonding'
occurs depends not only on the magnitude of the forces in the cable,
but also on the degree of grouting and, where this is good, more
importantly on the amount of shear reinforcement containing the
bursting effect. If there is little shear reinforcement within a
section, severe cracking will occur along the line of the duct.

3. Demolition Monitoring

During demolition a number of checks can be carried out remotely to
ensure that the work is proceeding in a satisfactory manner.
Instrumentation of critical sections can be completed, prior to the
start of demolition, using vibrating wire (VW) strain gauges. These
gauges can be operated from a safe distance and will provide
information on the redistribution of load within the structure or the
likelihood of failure at a given section. This type of
monitoring can also been used to find the effects, caused by the
demolition, on the temporary works. The results from this type
of gauging system can be relayed to the site engineer within minutes
of an operation taking place and hence can provide a reliable method
of controlling the work.

4. Case Study

An example of monitoring during demolition can be reviewed for the
case of the Taf Fawr Bridge which carried the A470 over a tributary
of the River Taff north of Merthyr Tydfil. Built in 1964, the bridge
had a height of approximately 30m above the river and an overall
length of some 144m consisting of three spans of 39m, 66m, and 39m.
Lundgren and Hansen (1967) described the design and construction of
this segmental post-tensioned structure. The deck was assembled as a
balanced cantilever with the weight of the segments on the midspan
section being balanced by those on the side span. Details of the
structure and the segment and joint numbering system are shown in
Fig.1. Each segment comprised four, 3m long, precast I-sections
joined by in situ top and bottom slabs to form a three-cell box
structure. The precast I-sections were temporarily held in place
during erection by a diagonal Macalloy bar passing through the webs
at the joints; which became part of the permanent structure and were
fully grouted. Longitudinal post-tensioning was then applied to the
precast units so that two 19 wire/25mm diameter strands were anchored

at each cantilever end, running through the previously assembled
I-sections. The continuously reinforced slabs were cast and further
post- tensioning was applied.

Tendons were provided in the top of the deck, throughout the
length of the structure, to counteract the effects of moments induced
during erection and under working conditions. Further tendons were
added in the bottom slab, in the centre span and near the abutments,
to resist the sagging moments after the bridge was made continuous.

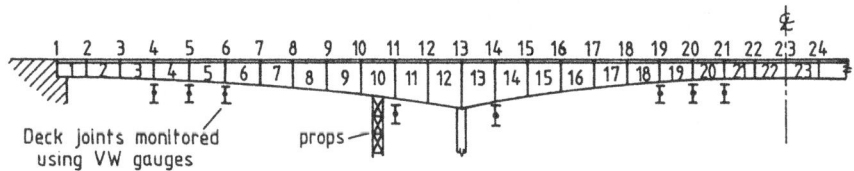

WEST SPAN

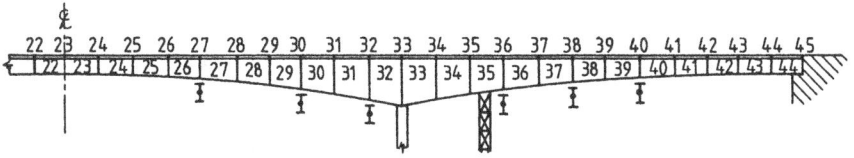

EAST SPAN

Fig. 1   Taf Fawr Bridge - segment and joint numbering system.

In 1985 the decision was made to demolish the superstructure due
to general deterioration of the precast sections. This deterioration
had enabled water and de-icing salts to leak through into the
prestressing ducts causing corrosion of some of the cables. However,
the reinforced concrete piers were still in good condition and it was
considered economically feasible to replace the deck alone. In order
to remove the deck with the least damage to the piers, the
dismantling process demanded a reversal of the construction sequence,
although no attempt was to be made to reduce the prestress in step
with the removal of the dead load. The behaviour during demolition
was not entirely predictable since the extent of debonding on cutting
the tendons was unknown. A number of monitoring systems were
utilised to minimise the uncertainty and to ensure that the
demolition was carried out in a safe and controlled manner. The
monitoring work was extended to include an assessment of the levels
of stress remaining in the superstructure.

4.1 Demolition sequence
Initially the superstructure was divided into two parts by cutting
transversly at midspan. Fig. 2 shows how the central box units were

Fig. 2 Isolation of central box   Fig. 3 Removal of central units
        units.   using lifting frames.

then isolated by removing the outer I-sections. After providing
temporary props under each side span, the connections between the
superstructure and the abutments were severed. The box units were
then lowered into the valley, maintaining stability by ensuring

362

excess weight remained on the side span. Work commenced on the west span side sections and progressed onto the east span. The lowering of the west span central units ran concurrently with the removal of the east side sections. Finally, the east span central units were removed using the lifting frames shown in Fig. 3, which were transferred from the west side.

Initial trials, using a diamond saw to cut the midspan position, proved ineffective as the cantilevers dropped, trapping the blade. Thus, the demolition work proceeded using a combination of jack hammers and a Montabert breaker. The breaker was used to isolate each unit prior to removal by crane or lifting frame.

4.2 Instrumentation

Once demolition began, cutting poorly grouted cables could lead to excessive debonding of the steel occurring over several segments from the cut position. The possible results of such an occurrance were the ejection of an anchorage or problems in the demolition process. Only two cables connected some of the units to the remaining cantilever which had to support both the lifting frame and end unit being demolished. The possible extent of debonding could be assessed by initial trials monitoring the change in concrete surface strains which occurred along the line of a severed cable. However, in order to detect impending failure at the joints during demolition, vibrating wire (VW) strain gauges had to be attached to the inside of the box units across the in situ joints at the ends of both the midspan and side span cantilevers. Any sudden loss of prestress would be immediately apparent at the joints as tensile stresses would develop at the top of the web.

It was anticipated that excessive compressive stresses might develop in the top flange and tensile strains might be induced in the bottom flanges of the units near the piers. This effect would be due to unreleased prestress in the top slab, which could exceed the cantilever moments as self-weight was reduced. Thus VW gauges were placed in these positions and adjacent to the temporary props.

The pier rotations had to be monitored in order to ensure that the concrete hinge at the base did not crack as the weight of the deck was removed. General surveying techniques were used in conjunction with strain gauges at the base of each pier and on the temporary props. In addition, the general stability of the cantilever had to be assessed as each unit was removed.

All structural monitoring of the superstructure was carried out remotely so that the demolition programme was unaffected and the risks to the monitoring team and the Resident Engineer's staff were reduced. Readings from the strain gauges and temperature sensors were taken from the adjacent temporary Bailey bridge. Changes were monitored on removal of each central unit and after movement of the lifting frames.

4.3 Results
4.3.1 Debonding
An initial debonding trial was carried out on the edge sections of unit 14. The cables were exposed at either end of the unit so that the strands could be cut in increments. This also provided an

opportunity to measure directly the level of prestress in these strands. Two similar trials were carried during the demolition of the central box sections 25 and 26. Generally each trial showed that fully grouted tendons would not debond more than a fraction of the length of a unit, ensuring that virtually all the residual prestress force would be present at the next in-situ joint. During progressive cutting of a tendon, the changes in strain on the adjacent concrete surface illustrated the debonding characteristics as shown, by the results from the edge section, in Fig. 4.

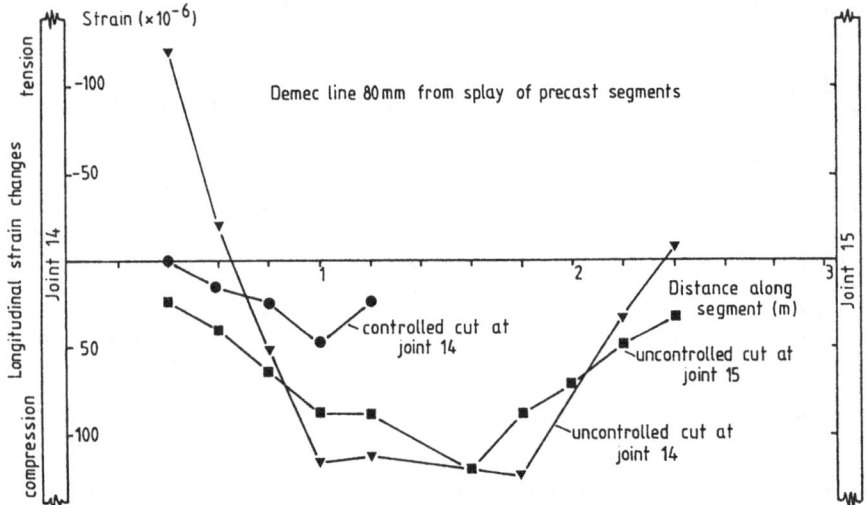

Fig. 4 Typical debonding curve.

The curve for the controlled cut at point 1 relates to the cutting of 3 wires in the nearest 19 wire strand at joint 14. The curves for uncontrolled cutting at points 2 and 3 relate to the severing of all the tendons at joints 15 and 14 respectively. As the cutting at point 3 (joint 14) was carried out last, the corresponding curve shows a symmetrical distribution in the strain changes. High strain gradients near the cut end of the tendon are a consequence of the transfer of force from the steel to the concrete by bond. The transmission length is of the order of a metre. In the absence of bond, a fairly uniform tensile strain change would probably be experienced, relating to the release of the general prestress in the segment.

The results from the central units confirmed the initial findings. It was concluded that for 60% of the tendons, prestress would be re-established well within the length of a unit. Direct measurements taken during the debonding trials had indicated that there was substantial prestress remaining, there was likely to be an increase in the stresses over the piers. As demolition proceeded, this increase could be monitored by the VW strain gauges positioned across the nearest joints.

4.3.2 Behaviour of joints

Generally the strain changes due to the removal of the first box units were relatively small. However, as demolition progressed the strain changes in the joints near the piers increased dramatically. This effect was particularly noticeable for the joints nearest the piers on the centre span and an illustration is shown in Fig. 5.

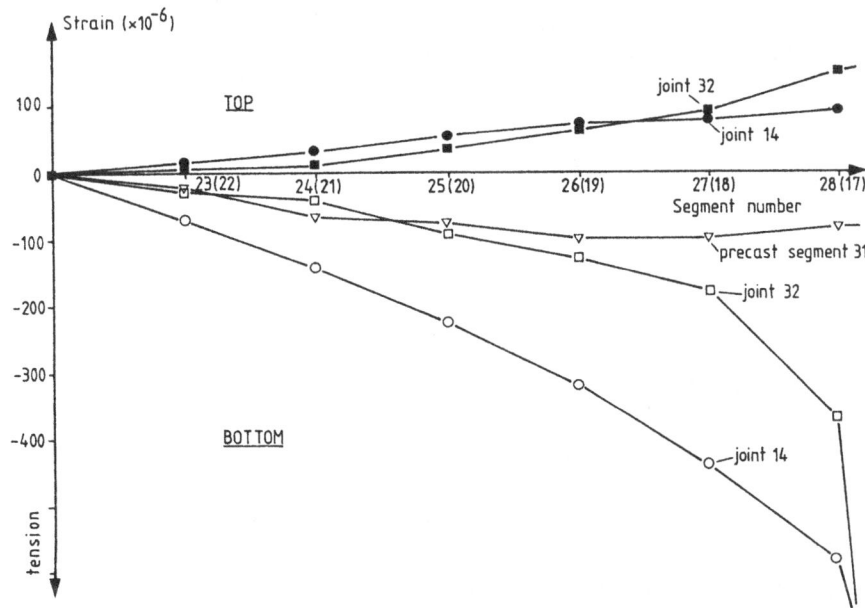

Fig. 5 Comparison of strain changes on south side.

It became apparent that there was a possibility the bottom flange steel would yield well before the demolition sequence was completed and that conventional prestressed concrete theory could not be used to predict the stresses remaining in the cantilever sections.

4.3.3 Stability checks on piers and props

As the piers were such massive structures, temperature variations were likely to swamp the small strain changes caused by removing the central units. Hence the need to monitor the gauges on the steel props where the strain variations were relatively large. The strain changes were converted into equivalent loads on the falsework and the reactions on the piers were estimated. Constant monitoring of the props showed that no unusual effects occurred during the demolition process.

4.4 Conclusions

The demolition of Taf Fawr Bridge provided the opportunity of applying various monitoring techniques. The results contributed not only to the development of these methods but also to the safety of many of the operations. It was shown that when a well grouted tendon was severed, the prestress was probably re-established in the concrete within 1m of the cut and certainly well within the 3m length of a unit. For this reason and because most tendons were found to be well grouted, the units on the central span could be dismantled with an adequate factor of safety.

The total loads on the temporary props, caused by the removal of these units, was closely monitored and corresponded with the reactions anticipated by the contractor. As the demolition progressed and the level of prestress increased in the remaining units near the piers, cracking developed at the joints in the bottom flanges and the strain distribution became non-linear. Warning of the impending yield of the longitudinal reinforcement, in the bottom in-situ concrete at the joint, was provided by the VW gauges, but the residual resistance of the section in flexure and shear could not be predicted.

5. General Conclusions

The effects of cutting post-tensioned tendons has been observed to vary widely from structure to structure; making current mathematical prediction of the response almost impossible. The use of a monitoring system during the demolition of prestressed concrete structures requires expert knowledge of both the design of the structure and of the possible effects of demolition. However, with structural monitoring employed at critical locations, it is possible to achieve a reliable method of control over the demolition operations.

It is recommended that, for complex structures, clients should issue detailed demolition drawings and a specification to potential contractors at the tender stage. At all times full supervision of the work should be carried out by resident engineering staff.

Acknowledgements

The authors would like to thank CIRIA and the Science and Engineering Research Council for their support over the last four years, without which this work would not have been possible. The success of the monitoring work carried out on the Taf Fawr Bridge was due to the close collaboration with Gifford and Partners and Messrs Shephard-Hill, with additional support from the Welsh Office.

The authors are particularly grateful for the valuable assistance given by Gage Technique Ltd., Griffths-McGee (Demolition) Ltd. and Thermic (UK) Ltd.

References

Andrew, A.E. and Turner, F.H. (1985) Post-tensioning systems for concrete in the UK: 1940-1981, CIRIA Report 106 .

Buchner, S.H., Lindsell, P. and Robinson, S. (1984) Prestress losses in full-scale structures. <u>Internal report R/M/CVL 84/2</u> , Dept. of Civil Eng., University of Surrey.

Lindsell, P. and Buchner, S.H. (1987) Prestressed concrete beams: Controlled demolition and prestress loss assessment. <u>CIRIA Technical Note 129</u> .

Health and Safety Executive (1984) Health and safety in demolition work. <u>Guidance Note GS29</u> , Part 3: Techniques. H.M.S.O.

Lindsell, P. and Buchner, S.H. (1988) Dismantling of Continuous Post-tensioned Structures. I.C.E./University of Manchester and U.M.I.S.T. <u>International Conference on Decommissioning</u> .

Lundgren, A. and Hansen, F. (1969) Three-span continuous prestressed concrete bridge constructed of precast units in cantilever construction. American Concrete Institute, First International Symposium on Concrete Bridge Design, <u>ACI Publication SP23</u> , 665-680.

DEMOLITION OF PRECAST PRESTRESSED CONCRETE STRUCTURES

K. TOMITA and K. SUEYOSHI
Building Construction Div., HAZAMA-GUMI, Ltd.
H. YAMADA and K. SUMI
Technical Research Institute, HAZAMA-GUMI, Ltd.

Abstract
We demolished a three-storey lodging facility, which was a precast
prestressed concrete structure, in the grounds of a temple. As the
machine for the demolition work, hydraulic crushers were used because
they seemed to be compatible with our aim to secure the safety of the
work as well as to keep noise and vibration down to a minimum. In
the demolition, we took the precautions to prevent the tendons from
flying out of the members, i.e. crushing the concrete little by
little to release gradually the energy stored in the tendons,
installing the wire nets at the end-anchorages and keeping the
operators away from the anchorages. Further, for the treatment
against the collapse of the building during the demolition, we
supported its girders by telescopic steel props and dismantled it in
the order which seemed to be suitable for this case. As a result of
the above method, we were able to demolish the structure safely with
little noise and little vibration, though the concrete, which had
very high compressive strength, was not crushed so quickly as one
made of ordinary reinforced concrete structure.
key words: Precast prestressed concrete structure, Hydraulic
crusher, Tendon, Grout, High strength concrete, Durability.

1. Introduction

The authors demolished lodging facilities in the grounds of temple,
TAISEKI-JI, in Shizuoka Prefecture. In the demolition work, as a
large number of people visit this temple every day, we had not only
to ensure safety for them, but also to keep noise, vibration and dust
down to a minimum. Further, special attention was paid to the
demolition of the buildings, because they were precast prestressed
concrete structures and the experience in the demolition of this type

of structures is limited.

This paper is concerned with the study of the characteristics of the concrete, prior investigation into the method of the demolition, and progress of the work.

2. Outline of the building

In the ground of TAISEKI-JI, there were five lodging facilities in the total, and our company contracted to demolish all of them. One of them, namely GO-NO-BO, was a cast-in-place prestressed concrete structure, and the other four, namely ICHI-NO-BO, NI-NO-BO, SAN-NO-BO and YON-NO-BO, were precast prestressed concrete structures. In this paper, the demolition work of SAN-NO-BO erected in 1966, which was dismantled first, is summarized.

Fig. 1, 2, and 3 shows the second floor plan, the west elevation and the south elevation, respectively. It was a three-storey construction without a basement floor, whose building area and total floor area were 1285 m^2 and 3595 m^2.

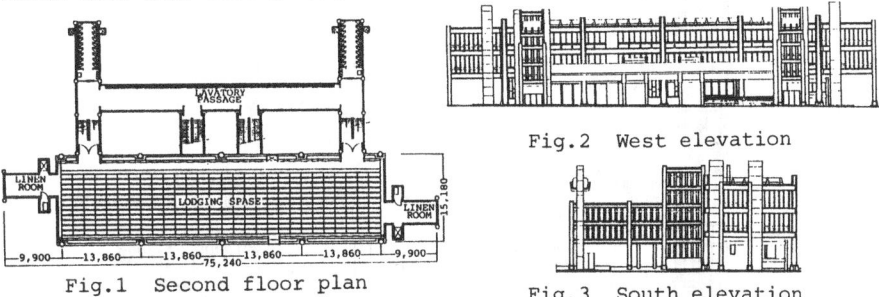

Fig.1   Second floor plan

Fig.2   West elevation

Fig.3   South elevation

Fig. 4 shows the structural drawing of a typical floor. As this figure shows, it had three kinds of slabs, namely A, B and C slab, all of which were erected by lift-slab construction methods. The arrangement of rebars and prestressed tendons is illustrated in

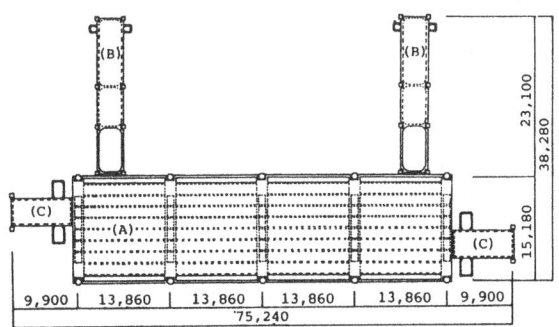

Fig.4   Structual drawing of typical floor

Fig. 5. Though the foundation was an ordinary reinforced concrete structure, the main structural components were precast prestressed concrete members with grouted tendons. The columns were united to the foundation by means of the high strength steel bars (prestressing steel bars) after a prestressing operation. The beams and girders were cast at the same time as the slabs, which were erected by the above method and were united to the columns.

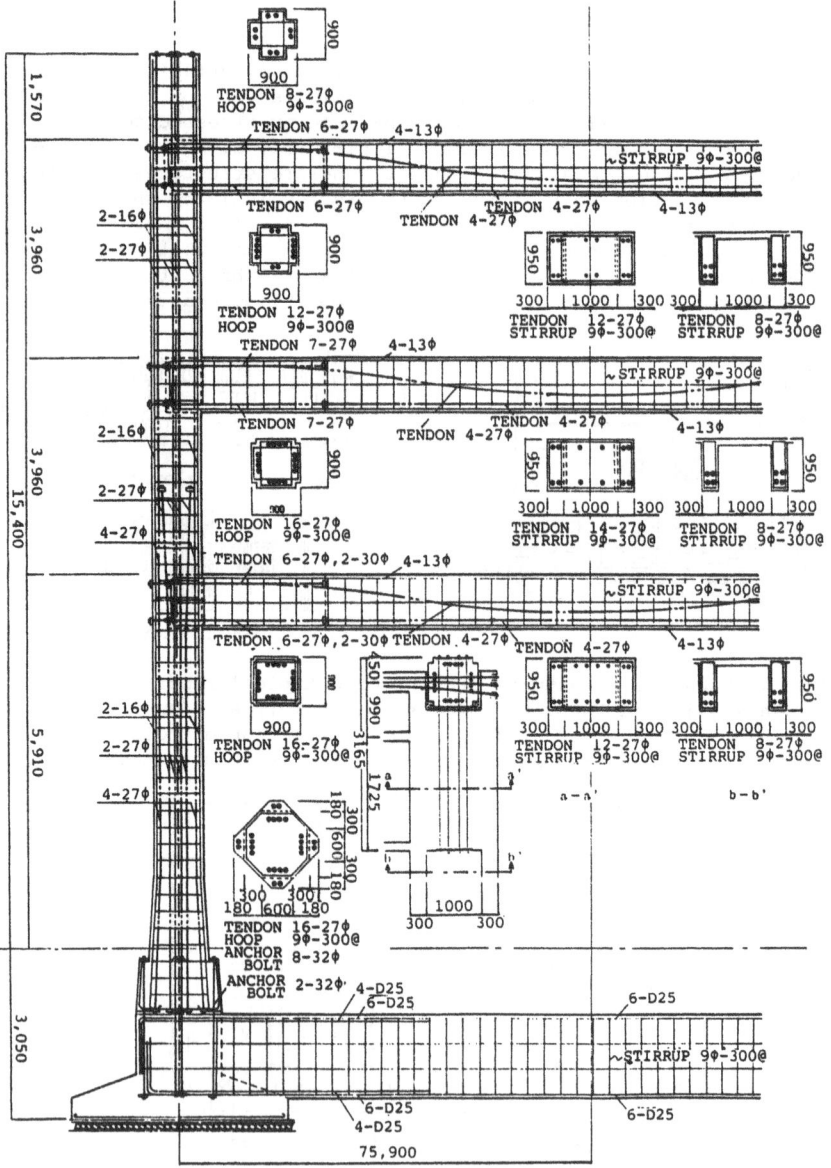

Fig.5 Arrangement of rebars and prestressed tendons

370

3. Study on concrete

3.1 Mix proportion and materials

According to the data at the time of construction, cast-in-place concrete for precast prestressed members was high strength concrete whose specified strength was 400 kgf/cm^2.

Mix proportions and compressive strengths of the cylindrical test specimens are shown in Table 1.

Table 1 Mix proportions and compressive strengths (A siab)

Maximum size of coarse aggregate (mm)	Slump (cm)	Air content (%)	Water-cement ratio (%)	Weight per cu m of concrete (kg)				Compressive strength at 28 days (kgf/cm^2)
				Water	Cement	Fine aggregate	Coarse aggregate	
25	13	1	37.7	181	480	544	1200	501 –523

Materials of concrete were as follows:
(1) cement - high early strength portland cement
(2) coarse aggregate - gravel from the Fuji river (Max. size 25 mm)
(3) fine aggregate - sand from the Fuji river (Max. size 2.5 mm)

Table 2 shows mix proportions and compressive strengths of the cement grout.

Table 2 Mix proportions and compressive strengths (cement grout)

Water-cement ratio (%)	Water (kg)	Cement (kg)	Chemical admixture (g)	Aluminium powder (g)	Compressive strength at 28 days (kgf/cm^2)
42	21	50	125	3.2	306

3.2 Purpose and procedure

The purpose of this study was to obtain basic data to judge the durability of structural concrete from the results of examining the high strength concrete whose age was about 21 years. The items of the examination were the mechanical properties and the extent of carbonation.

The specimens were cored from six columns and one beam by drilling with a 100 mm core diameter. Then the cores were formed to 20 cm in length and both ends were ground.

The examination was made in the following manner.

(1) Mechanical properties

Tests for compressive strength and one for splitting tensile strength were performed. By means of the former test static modulus of elasticity and Poisson's ratio of the specimens were determined.

(2) Carbonation

The depth of carbonation was measured by means that the test specimens were split and 1%-Ethyl Alcohol Phenolphthalein solution was applied to their split surfaces.

In addition, thermogravimetric analysis and differential thermal analysis were performed to examine the amount of CaCO$_3$ in the concrete cores.

3.3 Results

The results are shown below.

(1) Mechanical properties

371

Compressive strength of the test specimens cored from the columns was from 490 to 640 kgf/cm^2 and that of the test specimens cored from the beam was from 628 to 696 kgf/cm^2, with an average of 590 kgf/cm^2. Compressive strength of each of them exceeded the specified concrete strength of 400 kgf/cm^2. The mean tensile strength was 34.4 kgf/cm^2 and the mean Poisson's ratio was 0.188.

Fig. 6 shows the relation between compressive strength and static modulus of elasticity of the test specimens. Static modulus of elasticity was considered an adequate value.

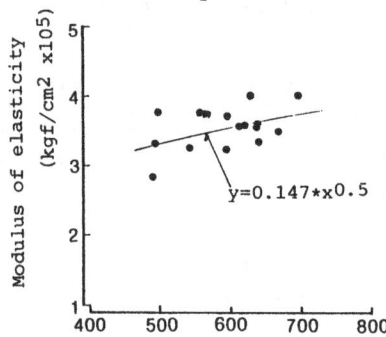

Fig.6 Relation between compressive strength
and static modulus of elasticity

(2) The depth of carbonation

The depth of carbonation of 16 test specimens was in the range of 1-3 mm with an average of 1.7 mm. No special difference in the presence of plaster finish or between indoors and outdoors was detected.

Fig. 7 shows amounts of CaCO$_3$, Ca(OH)$_2$ and T-Ca in the concrete which was the results of thermal analysis. For the amounts of CaCO$_3$

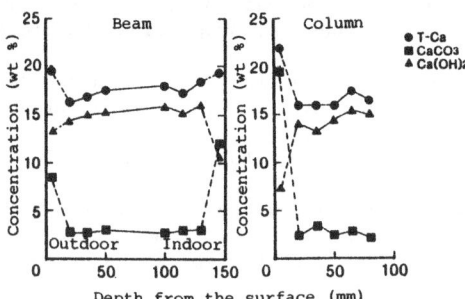

Fig.7 Depth of carbonation

in the concrete close to the surface, the concrete cored from the columns contains larger amounts than the concrete cored from the beam. It is considered that the effect of compressive strength and environmental conditions is apparent.

The results described above confirmed that the carbonation rate of high strength concrete is extremely slow.

4. Planning the method of demolition

For a sequence of demolition, the following two methods are generally considered:
(1) demolishing the building in the order to which it is easy to break, and
(2) demolishing the building in the reverse order to which it was originally erected.
At first, it seemed that the second method was more suitable for the demolition of SAN-NO-BO because it was a precast concrete structure, so we investigated the method of dismantling the slabs according to the construction process, that is, taking down the second floor slab, the third floor slab and the roof floor slab by jacks in that order. But, after deliberation upon this method, it was judged that it had disadvantages since the work of setting the jacks at the tops of columns was very dangerous and the slabs taken down would need some secondary demolition. Consequently, we concluded that the first method should be adopted.
    As machines for the demolition, it was decided to use hydraulic crushers which break concrete by bending, snapping and crushing it, because they seemed to be compatible with our aim to ensure safety of the demolition work as well as to keep noise and vibration down to a minimum.
    The demolition of conventional reinforced concrete structures has been carried out for many years. On the contrary, the experience in the demolition of precast prestressed concrete structures is limited. In demolishing SAN-NO-BO, whose structural form was the latter, the following problems had to be taken into account:
(1) scattering of the tendons and the anchorages
(2) accidental collapse of the building during the demolition work
(3) reduction of the demolition efficiency
In the case of the demolition of SAN-NO-BO, safety rather than quickness and cheapness was required, so we were concerned with the first and second problems.
    Though this building was post tensioned structure with grouted tendons, it was very risky to demolish it with the assumption that grouting was perfectly effective.   Accordingly, we took the precautions to prevent the tendons from flying out of the members, i.e. crushing the concrete little by little to release gradually the energy stored in the tendons, installing the wire nets at the end-anchorages and keeping the operators away from the anchorages.
    Because the elements of this building were jointed to each other by high strength steel bars (prestressing steel bars), it seemed that cutting them carelessly could have caused an accidental collapse. So, we determined to support the girders by telescopic steel props and to demolish it in the following order:
(1) crushing the girders in x direction, the slabs and the beams
(2) crushing the central parts of the girders in y direction
(3) crushing the both ends of the girders in y direction
(4) dismantling the columns
(5) demolishing the basements structures

373

5. Demolition sequence

The working process is shown in Table 3.

Table 3 Working process

Items	Term 1987 Jun. 30	Jul. 10	Jul. 20	Jul. 30	Aug. 10	Aug. 20	Aug. 30	Sep. 10
Temporary works — Temporary enclosure	▬							
Protective scaffold		▬▬▬ ·						
Demolition — Interior finish		▬▬▬ ·						
Passage, Lavatory		▬						
South staircase, Latrine			▬▬ ▬▬					
North staircase, Latrine				▬				
Lodging space (slabs)			▬▬	▬ ▬				
" (beams, columns)				· ▬▬	▬▬▬ ▬▬▬			
Basements						▬▬ ▬▬	▬▬ ·	
Fragmentation				· ▬▬▬	▬▬▬▬	▬▬▬	▬▬ ▬▬ ·	
Transportation				▬	▬▬ ·	▬ ▬▬▬	·	
Leveling of ground								▪

5.1 Temporary work

Fig. 8 indicates the plan of the temporary work. The temporary

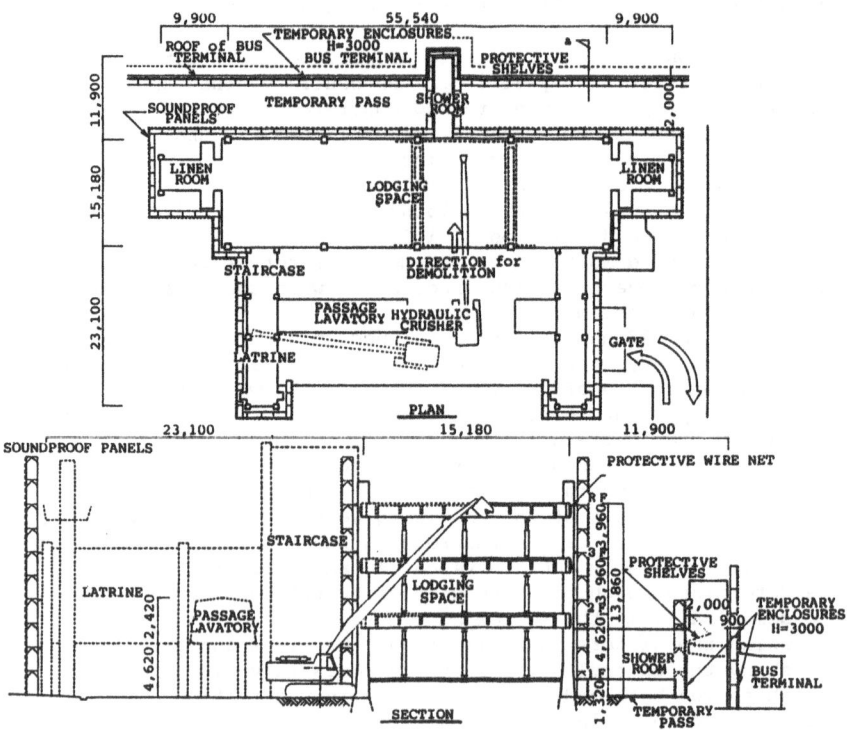

Fig.8 Plan of temporary work

374

enclosure having height of 3 m was installed and the sound-proof panels were fixed to all the scaffolds not only to prevent the tendons and the concrete debris flying out of the structure but also to cut off noise.

5.2 Demolition of the construction
In the demolition of the additional space, i.e. passage, latrine, and linen rooms, we crushed them from the upper parts. Because the members had small cross sectional area and a number of the tendons in them was few, the demolition work could be done easily.

During demolition of the lodging space which was the largest part of the construction, we carried out the work according to the foregoing methods as a general rule(see Fig. 9, 10). The slabs, the

Fig.9 Demolition of slab

Fig.10 Demolition of girder

beams and the girders were dismantled safely in the sense that neither scattering of the tendons nor accidental collapse of the building occurred. But, as to the working efficiency, the concrete was not crushed as quickly as an ordinary reinforced concrete structure, for the reason that the concrete had very high compressive strength and a lot of large diameter tendons were embedded in them. In releasing the tendons, not all of them could be cut by the hydraulic crushers but a few of them were cut by burning with an oxy-acetylene torch.

After the release of the prestress, the columns were dismantled with a back hoe shovel pulling one end of the steel wire rope whose other end was anchored to the column(see Fig. 11). Vibration in

Fig.11 Demolition of column

pulling down the columns was not felt since the concrete debris absorbed the impact. Because the columns had a large cross sectional area, hydraulic crushers were impractical and hydraulic percussion breakers were used(see Fig. 12). The latter method is generally

Fig.12 Demolition of column

accompanied with noise, but it was not felt in the outside of the job site dew to the temporary enclosure as well as the sound-proof panels. Though a few of the tendons came out of the columns during the pulling down, this demolition work was carried out in a satisfactory manners. Also, the basements was broken by the same machines.

6. Conclusion

Considering the whole process of this demolition work, the following conclusions are pointed out:
(1) By the foregoing method, the most dangerous state of affairs, the collapse of the building, did not occur. And, a serious accident was able to be avoided though a few tendons came out of the members, and
(2) While the hydraulic crushers were used for the demolition of the beams, the girders and the slabs, the hydraulic percussion breakers were used for the demolition of the columns and the basement. The latter method is generally accompanied with noise, but it was not felt in the outside of the job site due to the temporary enclosure as well as the sound proof panels. Also, the vibration in the work of pulling down the columns was not felt because the concrete debris acted as a absorbing layer.

References

Niimi, Y., Hisatomi, Y. and Fujimori, T. (1966) Prestressed lift slabs in TAISEKIJI. Reports of The Research Laboratory of The SHIMIZU Construction Co., Ltd., Vol. 8, pp 117-150.
Ejima, T. and Maetani, M. (1966) PC construction work of TAISEKIJI lodging facility. Journal of Prestressed Concrete, Japan, Vol. 8, No. 3, pp 30-36.

THE DEMOLITION OF PRESTRESSED CONCRETE BY EUTECTIC GENERATION THERMAL CUTTING

Y. MALIER
Public Works Research Institute (L.C.P.C.), Ministry of Construction, France

Abstract

Prestressed concrete structures constitute the greatest challenge to demolishers. They are the best test of our capacity to solve difficult and sometimes dangerous problems.

With our thermal cutting process, we demolished a post-tensioned prestressed concrete girder bridge in France in 1969 (Côte d'Azur Motorways). We then applied our technique to different structures (arch bridge, industrial structures, delicate modifications on offshore structures and nuclear power plants, modifications of pre-tensioned structures, etc.).

For all these structures, our general method consisted in gradually "reducing" the prestressing by local thermal cutting of cables until stable elements are obtained which can be readily handled and transported.

In the research laboratory, from 1971 to 1975, we carefully investigated (theoretically and experimentally) the modelling of the behaviour of the cable during cutting as a function of the quality of the injection, of variations in this quality, the curvature of the element, etc. Many measurements carried out in the field then made it possible to validate our models and to provide all the required safety.

We will then describe more particularly the demolition of the roof of the warehouse of the Presse Parisienne, an organization which handles daily some 600 metric tons of newspapers and magazines sent throughout France. With a surface area of 11.000 square metres, this prestressed concrete roof which exhibited cable stress corrosion was completely taken down and loaded using a strictly programmed mobile installation. This was done without disturbing the activities of the 600 (day and night) employees of the warehouse, or its operating conditions (temperature, humidity, noise, fire safety).

In addition to safety, the process made it possible to comply strictly with deadlines (69 days) and with cost estimates.

(The projection of the second part (8 minutes) of a short film will illustrate the subject.)

Key words : Demolition of prestressed concrete

1. INTRODUCTION

The demolition of prestressed concrete constitutes the best challenge for demolition firms. It is always within this area that we find the most delicate and dangerous operations calling for the most precise economic estimates.

The problems are compounded when, in addition, there are strict requirements regarding the conservation of part of the structure (structures modified and not demolished) and compliance with completion dates linked with those of other contractors on the site (reconstruction, installation of technical equipment, and so on).

Our thermal demolition method was used to demolish a prestressed concrete bridge on the French Riviera in 1969. We then applied our techniques to many structures (bridges, industrial structures, offshore installations, nuclear power plants, etc.), most often involving local modifications during construction or more extensive adaptations during the life of the structure.

In our research laboratory, from 1971 to 1975, we were engaged particularly in the modelling of the behaviour of a prestressing cable during cutting as a function of tension, curvature, nature of grouting, type of anchoring, and so forth. Many in-situ measurements conducted on real projects have made it possible to validate our models and to deduce safety procedures specific to the demolition of prestressed concrete.

2. OUR PRESTRESSED CONCRETE DEMOLITION METHOD

After having set up shoring to ensure the static equilibrium of the structure, the general idea is to "reduce" the prestressed concrete into reinforced concrete by cutting cables locally through a hole obtained thermally in a few seconds.

Let us examine briefly (without calculations) the behaviour of cables and anchoring. Different assumptions are to be considered when the cable is being cut :

Assumption 1 : The cable is anchored again perfectly in the duct and a second prestressing system is created (whose length is of course smaller than the previous one but whose ends are in this case free because they are detached from the structure).

Assumption 2 : The cable slides in the duct and is totally stress-relieved ; the element is now ordinary reinforced concrete.

Assumption 3 : The cable is partially anchored, partially free. Its behaviour is capable of evolving further during handling and transport operations.

As anchoring is involved, let us note that during the cutting of a cable the potential energy stored during tensioning is instantaneously released. If the grouting material is then not capable of anchoring the wires (assumption 2), there is the danger of seeing the wires projected forcefully (with only friction to offer resistance) over the jobsite, so that protection is necessary.

Obviously, a calculation similar to the one carried out during construction will make it possible to determine the positions and the order of cuts to be made to obtain the desired "reduction".

3. DESCRIPTION OF AN APPLICATION

3.1 Description of structure

What was involved was the demolition (for reconstruction) of the roof of the sorting and shipment warehouse of Nouvelles Messageries de la Presse Parisienne (NMPP). With a total surface area of 11.000 m², located in Paris, this warehouse is used by 600 people handling 150 forklift trucks and receiving 300 lorries and wagons required for the daily shipment of 500 metric tons of newspapers and magazines (Figs. 1, 2 and 3).

The structure to be demolished, about 260 m long, was made up of 29 concrete sawtooth frames each 40 m wide, 10 m high and 9 m long.

The framework was prestressed by cables tensioned at 800.000 N.

The new roof had to be supported by glued laminated wooden beams with a span of 42 m.

3.2 Prestressing operations

There were many prestressing operations of some difficulty. We shall examine the main ones below.

IMPORTANCE OF MAINTAINING SERVICE IN THE WAREHOUSE IN 24 BAYS OUT OF 29

Only five bays could be disrupted during the project. They were handled in the following order :

- One was stopped for the removal of technical equipment networks.
- One was being demolished (prestressed and reinforced concrete of girders, walls and shells).
- One was used for storing equipment and materials required for demolition and reconstruction.
- One was used for the reconstruction of the framework and roof.
- One was used for the assembly and putting into service of technical equipment networks.

COMPLETION TIMES

This mobile project on five bays had to move at the rate of two bays per week.

Reception and transfer operations could be carried out only on Saturday morning applying a very precise timetable provided in the planning of the NMPP.

This made is necessary for reinforced concrete and prestressed concrete demolition operations to be perfectly planned and coordinated based upon a very strict schedule, which, moreover, had to be absolutely compatible with all the other operations of the project, with the stringent sefety requirements and with the handling equipment utilization programme.

Planning established on the basis of the PERT method, with unit times, had to allow for the completion of each phase of the demolition-reconstruction operation to within 30 minutes.

DEMOLITION OF 200 GIRDERS (OR TILE-RODS) IN PRESTRESSED CONCRETE

This was the most delicate part of the project from the viewpoints of

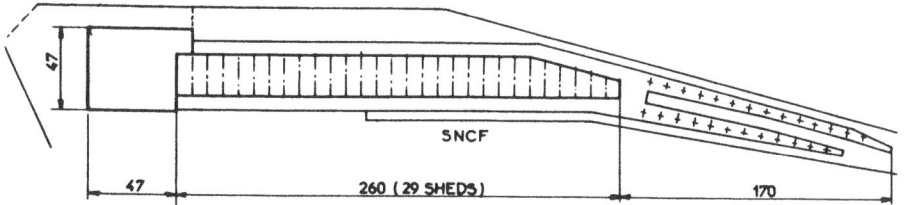

Fig. 1 Plan view of NMPP warehouses

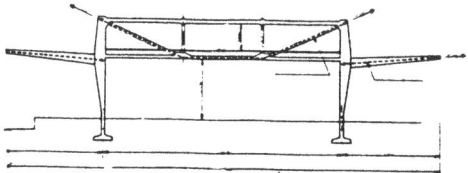

Fig. 2 Cross-section of building

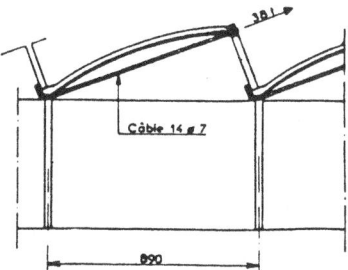

Fig. 3 Longitudinal section of
one of the 29 bays

Figs. 4 and 5 Examples of shields offering protection against the
projection of cables during cutting.

temporary stability of structural members, worker safety and compliance with completion dates.

CONSERVATION OF LOWER STRUCTURE
From the mechanical viewspoint, foundation footings and columns (cut at the head) were kept in order to receive the new roof. Demolition (and reconstruction) operations had to avoid any additional stresses on these elements (problem of windbracing provided by the roof only).

Further, most of the technical equipment (radiation heating among others) had to be kept in perfect condition on the demolition site so as to allow installation after the construction of the new roof.

PROJECT SUPPLIES AND MOVEMENTS
Movements into and out of the project site constituted a significant constaint owing to the small floor area and requirements relative to the highway and railway traffic of the warehouse, leaving only narrow time limits for these operations.

ENVIRONMENTAL REQUIREMENTS
The specifications set very strict limits on the use of noisy mechanical equipment on the project. The proximity of NMPP employees required maximum reduction of all forms of nuisances specific to this type of work.

3.3 Execution
The general idea of the method we proposed was to "reduce" (after having totally braced the entire roof) the prestressed concrete into simply reinforced concrete by the cutting of cables.

To avoid any risks, the three assumptions previously discussed concerning cable behaviour were taken into account simultaneously. Twenty-eight cables had to be cut per shed (or 812 cuts for the entire project).

Further, the ends of the girders were all equipped with shields to receive the cable heads (Figs. 4 and 5).

The inspection of the outlet of wires at the head and the measurement of changes in strains (and stresses) in the prestressed element at the time of the cutting made it possible to determine the validity of the assumptions made and the need to consider the different cases previously mentioned.

Once the prestressing was neutralised, the structure was partially supported by the shoring. This shoring moreover had to be suitably designed to meet conditions relative to loads, clearance (need to envelope the shell), mobility (the assembly, mounted on jacks, had to allow dismantling and transfer from one bay to another in less than 3 hours) and, of course, personnel safety (personnel working constantly at heights of about 10 metres).

The last demolition operations consisted simply of cutting all the reinforced concrete elements (girders, walls, tie-rods, etc.) to size to allow their removal by highway. This involved simple dismantling, carried out quickly and accurately by thermal cutting (Figs. 6, 7, 8 and 9).

3.4 Assessment of project
Thermal cutting operations required 78 workdays (six men), exactly the time predicted and used in project planning.

Figs. 6 and 7   A work station

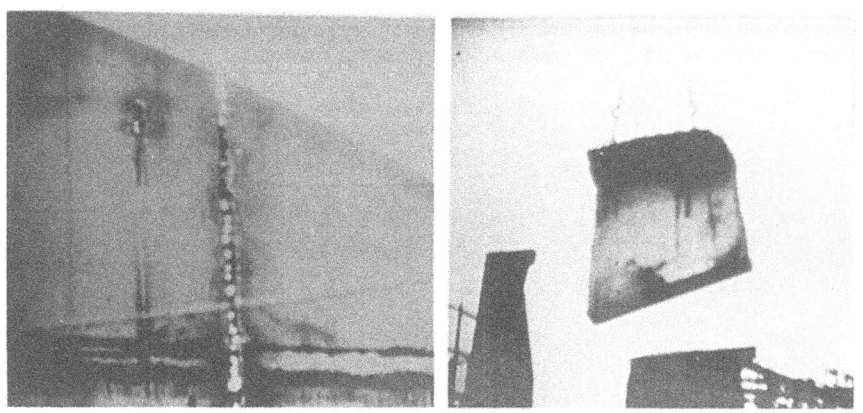

Figs. 8 and 9   Cutting and removal   of a concrete panel

Reconstruction operations were carried out at the same rate, and two new bays were ready each week.

For this work, thermal cutting required 24.000 m of thermal lances. Some 14.000 m³ of oxygen were needed for combustion.

In civil engineering, cost is always a decisive factor which, whatever the difficulty or originality of the technical problem involved, must not be ignored : it is of interest to note that demolition by thermal cutting represented, here, less than 9 % of the total amount of this complex demolition-construction project.

For centuries, one of the primary concerns of man has been to build. On the other hand, he has always left to time, natural catastrophes or the actions of wars or criminals the painful task of destroying. Thus, this natural selection, imposed on old structures, has with time eliminated whatever was unnecessary, unbecoming or grotesque, so as to leave us as a heritage in most cases only the beautiful or sacred.

During recent decades, this truth has no longer held. The advent of concrete has wrought profound changes, making all structures practically indestructible, whether they be beautiful or ugly, original or commonplace.

It must be acknowledged that, in light of these achievements, and of the importance of the quality of life and of the environment, it would be unfitting for us to leave future generations an unworthy heritage because of our technical inability to demolish whatever no longer corresponds to our present needs.

REFERENCES

Malier Y. (1978) Le découpage de la couverture en béton précontraint d'un centre de tri de 11.000 m², Construction Paris 5, 3-7.
Malier Y. (1980) Une méthode de démolition des ouvrages en béton précontraint par post-tension, Symposium F.I.P. Bucarest, proceedings part 2.
Acker P., Brachet M., Malier Y. (1982) Les enseignements à tirer de la démolition d'une construction en béton précontraint, Congrès F.I.P. Stockholm, 346-351 (Ed. Travaux).

DEMOLITION OF MOTORWAY BRIDGES BY BLASTING

Helmut ROLLER
Roller Sprengtechnik GmbH, Wuppertal, West Germany

Abstract
For more than twenty years now demolition of bridges crossing over motorways has been done for the most part by blasting. The main advantages are the restriction of traffic tie-ups to a few hours at a time on a specific date, with blasting times being selected when traffic is light.

Below the demolition blasting of bridges made of reinforced concrete and prestressed concrete as well as the drilling and blasting parameters and safety arrangements used are demonstrated on the basis of a few examples.

Key words: road blocks, drilling and blasting parameters, reinforcement work, safety arrangements.

1. Introduction

The major part of traffic in the Federal Republic of Germany between towns and cities goes over motorways. These motorways are comprised of four or more traffic lanes and are without any junctions or intersections, although numerous bridges cross over them. Setting up road blocks for construction work poses a great number of problems due to the fact that traffic on these roads is heavy. Demolition of bridges crossing over motorways without obstructing the traffic passing through underneath would only otherwise be possible by setting up soffit scaffolding, which is not only costly but is also of a disadvantage due to the lengthy limitation of clearance involved.

During the past few decades motorways have been extended from four to six or more lanes, due to steadily growing road traffic. Each extension requires the construction of new and larger bridges and as a consequence the demolition of old structures. Hundreds of motorway bridges have been demolished in West Germany during the past few decades.

2. General Description of the Procedure

Since about 1964 a process has been employed in which the demolition of road and railway crossings over motorways, for example, has been done in an economical manner, without motorway traffic being unduly obstructed.

A road block is set up during periods of very light traffic, e.g. late Saturday evenings, the remaining (minimal) traffic is diverted to by-roads. Immediately after the road block is set up, the bridge is blasted and the motorway quickly cleared of debris by employing a large amount of machinery.

For demolishing a bridge consisting of approx. 1,000 m^3 reinforced concrete, a general road block lasting for six hours is normally required. The permissible road block period is determined by the Autobahn Administration Authority putting the job out to bid. The penalty for overrunning the contractual period can amount to as much as DM 5000.00 per half hour and portions thereof.

The old structure can be blocked as soon as the new bridge is completely finished or half the width of the road (= two lanes, for allowing traffic in both directions) has been completed. Now there is sufficient time for making preparations for blasting. These preparations are as follows:
- possible removal of the deck slab
- drilling the shot holes
- laying bare the abutments and wing walls by removing earth and backfill
- filling the shot holes with explosive and preparing the electric ignition system
- covering the object to be blasted so as to protect against flying debris, possibly taking measures for protecting the new bridge

These preparations take anywhere from one to four weeks to complete, depending on the size of the object to be blasted. Only sections with considerable wall thicknesses have to be blasted. Generally speaking, this applies to the entire substructure along with the longitudinal and transverse beams of T-beam structures, but not to deck slabs or cantilever plates ranging from 0.20 to 0.30 m thick. These plates are nevertheless demolished when located near supporting beams, so that the demolition machinery can more easily break them up when clearing the site.

Details of the method are demonstrated below on the basis of two examples involving the demolition blasting of motorway bridges of various sizes and different degrees of difficulty.

Example 1

Demolition of a road bridge crossing over motorway A5 Karlsruhe-Basel:

Picture 1: Bridge over A5 to be demolished along with the new one

Picture 2: Wing walls and abutment

This was a double-span bridge with a continuous deck slab (picture 1). The firm bearing was situated over the centre column, the movable bearings were situated over the abutments. The structure, which was built in 1961, consisted of approx. 1,400 m³ reinforced concrete, the visible substructure was encased in red sandstone. The wall thickness of the abutments was 1.20 m to 3.20 m (picture 2), the centre pier measured 1.30 m. The structure had to be demolished down to 1.0 m below the motorway surface. The new bridge had been constructed at a distance of only 10 m to 15 m away.

The drilling of shot holes was begun approx. 1.5 weeks before blasting was to take place. One blasting engineer and one helper drilled the holes covering an overall distance of approx. 600 running meters using an ATLAS Copco ROC 712 hydraulic crawler drill. The deck slab, which had a thickness of 1.06 m, was drilled in a grid measuring 0.90 m x 0.90 m to a depth of 0.80 m with a hole diameter of 34 mm (figure 1). A vertical series of holes with a depth of 7.25 m, a diameter of 43 mm and a spacing of 1.27 m apart were drilled in the centre pier. Due to the fact that wall thickness increased towards the bottom, two series of holes were drilled into each of the abutments and wing walls to a depth of 7.25 m and 4.50 m respectively and with a diameter of 64 mm.

The specific explosive requirement was approx. 0.300 kg of blasting explosive per 1 m³ reinforced concrete. The blasting explosive used was gelatinous ammonium nitrate (Gelamon) filled in cartridges of 25 mm, 30 mm and 50 mm in diameter.

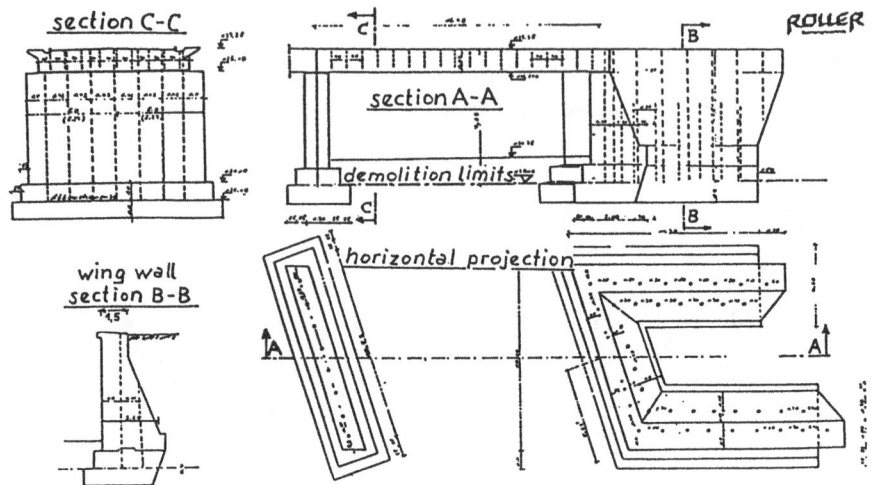

Fig. 1: Drilling and blasting scheme

Primacord with a blasting explosive weight of 12.5 g/m was used for the deeper holes. The blasting explosive was placed at various depths, with wood strips being used as spacers in distributing the explosive over the entire depth of the hole. Electric short-time detonators with intervals of 30 ms in 18 stages where arranged in such a manner that the lowest stages were placed in the side facing away from the new bridge.

Blasting was set for 5 p.m. on a Saturday. Filling the shot holes was started the Friday before, the detonators were placed starting that Saturday morning.

A road block was set up on the motorway about one hour before blasting was to take place. Traffic was diverted to the preplanned detour routes. The pavement of the motorway could then be covered with a cushion of sand 0.5 m thick. The object to be demolished had already been covered with felt matting in order to protect the new bridge as well as the neighbouring buildings located approx. 50.0 m away. Now the last matting was lowered onto the pavement of the motorway.

The surrounding area was cordoned off by police and construction workers in a radius of 300 m just before blasting was to take place. This affected detoured traffic as well, however this obstruction lasted only a few minutes.

Immediately after the dust clouds had dissipated, demolition machinery and other vehicles parked nearby started carting off the debris (pictures 3 & 4).

Picture 3:   Picture 4:
Results of blasting

About 6 hours after blasting took place, the motorway could be opened again for general traffic.

Generally speaking, the concrete structure is loosened by blasting, the reinforcement is at all events damaged by drilling the shot holes. In order to facilitate the clearing of demolition debris it is important that reinforcing steel be laid bare by blasting in such a manner that it can be easily accessed and cut up by the demolition machinery. Cutting the reinforcing steel in structures containing conventional reinforcement by drilling into it poses no problem, however great care must frequently be taken in connection with structures made using prestressed concrete in determining which prestressing tendons, if at all, can be drilled into, without unduly impairing the stability of the structure. On the other hand, it is extremely important in this connection to place the blasting charges in such a manner that the prestressing tendons are laid bare of concrete.

Example 2

The following example of the demolition of a Federal highway bridge crossing over the A1 motorway Dortmund-Cologne (pictures 5 & 6) is a very typical example of difficulties encountered.

Picture 5:   Picture 6:
Bridge crossing over A1, view new bridge
of old crossing from below

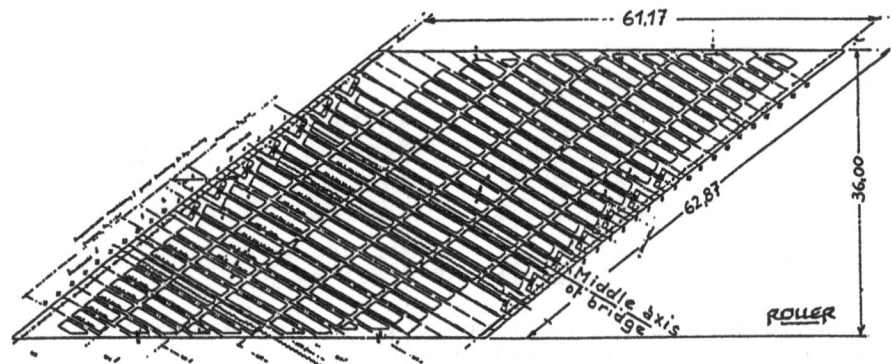

Figure 2: Ground plan of deck slab

In its time this structure was Europe's largest single-span bridge, built in 1958-1959 (figure 2). The superstructure consisted of a hollow body plate which was longitudinally and diagonally prestressed, having a plate thickness of 1.80 m and an area of 2,285.0 m². Longitudinal and diagonal stressing was effected via 1,465 prestressing tendons according to the Leoba system and via 14 prestressing tendons according to the Baur-Leonhardt system with 400 single strands each. The diameter of the box sections amounted to approximately 1.15 m x 0.80 m. The substructures were made of 2,600 m³ reinforced concrete.

The superstructure of the new bridge was a composite steel structure, set on sliding rails placed on temporary abutments located approx. 1.50 m away, next to the old structure. The new structure accommodated traffic during demolition and construction. After the old bridge had been demolished, new abutments were to be constructed in its place onto which the new superstructure was to be slid.

36 hours was planned for blasting and demolishing this bridge, which consisted of approx. 5,000.0 m³ prestressed concrete and reinforced concrete. During this period a railway bridge consisting of approx. 500.0 m³ prestressed and reinforced concrete was to be demolished, however this bridge had to be cleared away first, due to the fact that debris could only be carted off in this direction.

The bearing capacity of the old structure had been severely limited for quite some time, as a consequence it had been additionally supported in some places by steel columns. Although there was no loading by traffic and the pavement had been removed, a structural analysis had to be carried out in order to indicate which prestressing tendons could not be allowed to be damaged in any case in order to ensure

stability up until blasting took place. Measuring the box
sections and beams positioned at an angle of 36.03 degrees
to the axis of the bridge and locating the prestressing
tendons in the beams proved to be time-consuming (fig. 3).

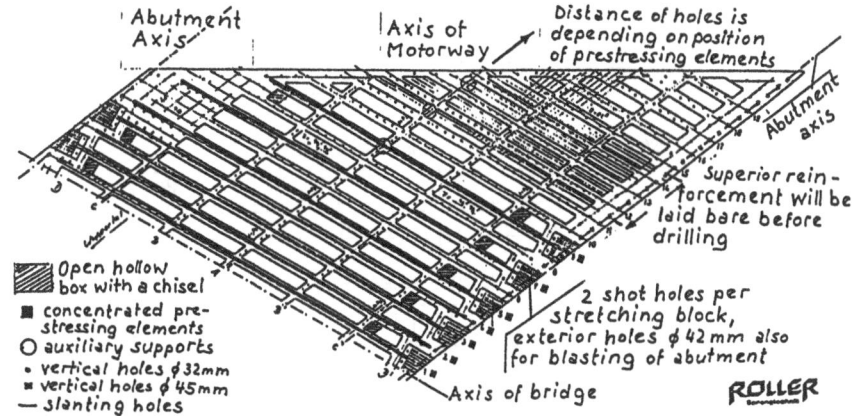

Figure 3:
Drilling and blasting scheme, view of superstructure from
above (sectional view, 1/2 plate).

According to the structural analysis, the prestressing
steel, in particular the prestressing steel located in
the lower beams, could not be allowed to be damaged by
drilling (figure 4).

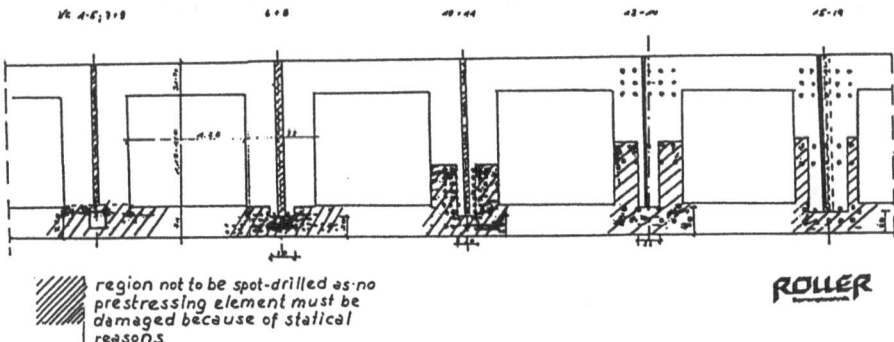

Figure 4: Drilling scheme, sectional view of the
longitudinal beams

The concentrated prestressing tendons had to be preserved
at all costs, thus the vertical holes could only be drilled
up to sheet-steel casing of the tendons. As a consequence,
the tops of the box sections were opened in the areas
adjoining the abutments and the beams below the concen-

trated prestressing tendons were spot-drilled at an oblique angle starting at the box sections using hand-held drills (figure 5).

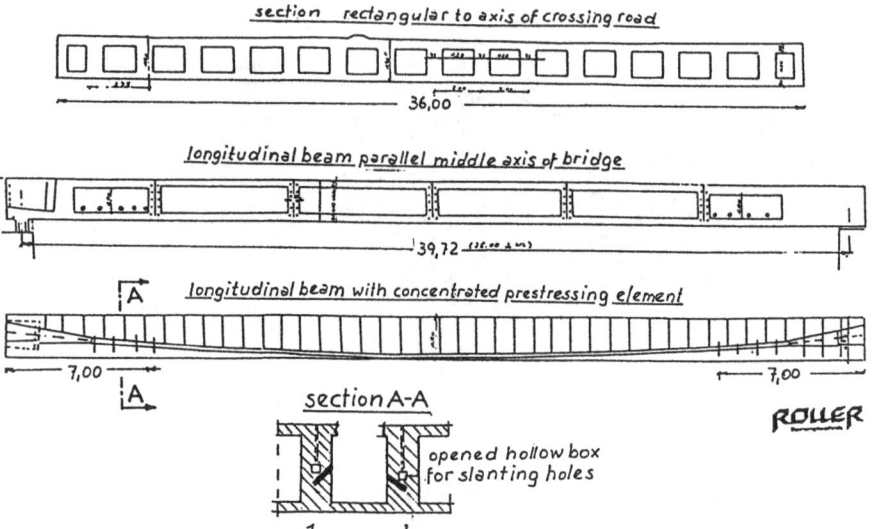

Figure 5: Drilling scheme, sectional views

Using five ATLAS Copco crawler drills, type 301, 302, 600 and 601, a total of 2,600 holes with an overall length of 6,500 drilled meters were made (picture 7). Drilling work took more than 4 weeks to complete.

Picture 7: Drilling of shot holes.

Specific blasting explosive quantities of 0.6 kg to 0.7 kg per 1 m^3 prestressed concrete were calculated for the superstructure, 0.40 kg per 1 m^3 reinforced concrete was calculated for the superstructures. The total amount of blasting explosive required for the bridge amounted to 2,260 kg of ammonium nitrate (Gelamon 30 and 40,

Ammongelit 2).

For safety reasons the filling of the shot holes and most particularly the placement of electric detonators was to be done as late as possible. Sixteen blasting engineers along with eight trained helpers started filling the blasting explosive two days before the planned blasting date, with all charges being fitted with primacords with a blasting explosive weight of 12 g/m. The primacords were also placed along the plate beams and the ends of the primacords projecting from the shot holes were connected to the guide cords (picture 8). These guide cords were covered with sand so as to prevent overturning between guide cords, with the detonation blast noise of the 1,100 m of openly placed cords being abated and the shot holes being covered at the mouth so as to protect against splinters. A total of 3,900 m primacord with a blasting weight of 12 g/m and 300 m primacord with a blasting weight of 40 g/m was used.

Picture 8: Holes drilled in the deck slab fitted with primacords, partly covered with sand

1,050 short-time electric detonators with intervals of 30 ms in stages of 1 to 18 and equipped with polyamide-insulated ignition wires were placed the day before blasting was to take place. In order ensure reliable as well as complete detonation of the primacords, the guide cords on the deck slab were fired using two detonators from either side each and the deep shot holes in the abutments and wings were fired using two detonators from the mouth and innermost of the shot hole.

Blasting was planned for ca. 8:00 a.m. on a Saturday. During the early hours of the morning the electric measurements of the 4 ignition circuits were checked again and sand filled under the bridge as a drop bed. Then the cover matting was lowered from the railings, followed by the police cordoning off the danger area. The residents of neighbouring buildings located 40 m - 70 m away in some cases had been informed in writing of the procedure to take place some days previously.

393

The large road bridge was blasted approx. 2 minutes after the railway bridge was blasted (picture 9). The results of blasting were optimal and there was no damage whatsoever to the new bridge structure or its sliding rails (picture 10).

Picture 9: Explosion   Picture 10: Result of explosion

However, it was not possible to clear away both bridges within 36 hours despite the 7 giant shovel excavators (40 t - 90 t), 5 loaders and 50 workers employed. The roadway could not be reopened for traffic until 48 hours after blasting. A contractual penalty was not imposed as clearing the area in a shorter amount of time was not possible due to local conditions.

Conclusion

In view of the brief obstruction posed to traffic, the demolition of bridges via blasting has been shown to be a reliable as well as economical method. This procedure is also being increasingly employed for demolishing bridges crossing over railway lines.

COMPUTER SIMULATION FOR FELLING PATTERNS OF BUILDING

N. TOSAKA   Department of Mathematical Engineering,
             College of Industrial Technology,
             Nihon University
Y. KASAI   Department of Architectural Engineering,
             College of Industrial Technology,
             Nihon University
T.HONMA   Information Systems Department,
             Technical Research Division,
             FUJITA CORPORATION

Abstract
A new procedure in computer simulation of the felling pattern of reinforced concrete framed structures is proposed
with the direct stiffness matrix method in conjunction with
the variable stiffness technique in structural mechanics.
In order to trace effectively the felling behaviour and
process of the structures, a developed algorithm for
numerical implementation is described schematically. The
method is exemplified with numerical simulations of felling
in several simple framed structures.
Key words: Demolition, Felling pattern, Reinforced concrete
framed structure, Computer simulation, Stiffness matrix
method, Variable stiffness technique.

1. Introduction

Recently, it is become the center of interest in the field
of civil engineering that we must pay attension to the manner of not only construction but also demolition of buildings. There are many papers and reports on the felling
demolition of high stories buildings demolished by the
blasting of explosives in many countries. In the near future, we will start to study on the instantaneous felling
demolition by means of the milli-second blasting of explosives which are charged to the selected parts of structural members.
    When we want to fell down reinforced framed structures,
it is very important to understand in advance the felling
patterns of the structures in the case that the column-feet
and ends of beam are cutted (or nearly cutted) or broken.
The classification of the above felling patterns was given
by Kasai and Shibata (1970) about 17 years ago. In their
paper the patterns were classified into the following five
patterns according to control of the blasting order of
column-feet and beam-ends:1. One side felling, 2. Inside

felling, 3. Out side felling, 4. Zigzag felling, 5. Stationary felling.

The aim of this paper is to show one possibility on computer simulation for analysis of felling of building and to propose the related numerical procedure. In order to trace the felling behaviour and determine the felling pattern, a new computational procedure is developed with the direct stiffness matrix method in conjunction with a variable stiffness technique in structural mechanics. This technique achived a great success in analyses of tension field problems by using the finite element method in papers by Nishimura, Tosaka and Honma (1985,1986,1987).

The numerical procedure of a computer simulation proposed in this paper is based on the following fundamental assumptions: 1. The reinforced concrete building is modelized as a plan framed structure system. 2. Each member of the structure is a linear homogeneous isotropic elastic body. 3. The kinematics and deformation behaviours of the structure are governed with the force-displacement relation under a static loading.

Let us now briefly describe the content of this paper which is divided into five chapters. In Chapter 2, we review the fundamental stiffness equation in matrix method which is useful in analyzing felling problems. Then, a new algorithm for numerical implementation of the resulting global equation of our structural system is described in Chapter 3. Several simple examples are shown to verify validity and adaptability of the proposed procedure in computer simulation in Chapter 4. Finally, we conclude this paper with some coments.

2. Fundamental equation

Recently various kinds of mechanical behaviour of framed structures are analyzed effectively by using the direct stiffness matrix method. It is well-known that this method is one of the most useful tools of structural analysis to emerge from the current development of computer technology.

Following the excellent text bookes by Martin(1966) and Meek(1971), we can set the fundamental stiffness equation which means the force-displacement relation under static loading for some typical element e of our beam-column model in plane framed structures as follows :

$$f_e = k_e \, u_e \tag{1}$$

where u_e denotes the nodal displacement vector in the local element system including the rigid body freedoms, f_e is the nodal force vector and k_e is the element stiffness matrix. The nodal displacement vector and the corresponding vector of force are given for the element with nodal points i and j as follows :

$$\boldsymbol{u}_e = \{\ \bar{u}_i\ ,\ \bar{v}_i\ ,\ \bar{\theta}_i\ ,\ \bar{u}_j\ ,\ \bar{v}_j\ ,\ \bar{\theta}_j\ \}^T \tag{2}$$

$$\boldsymbol{f}_e = \{\ \bar{X}_i\ ,\ \bar{Y}_i\ ,\ \bar{M}_i\ ,\ \bar{X}_j\ ,\ \bar{Y}_j\ ,\ \bar{M}_j\ \}^T \tag{3}$$

where each variables and its positive direction in the above vectors are shown in Fig.1.

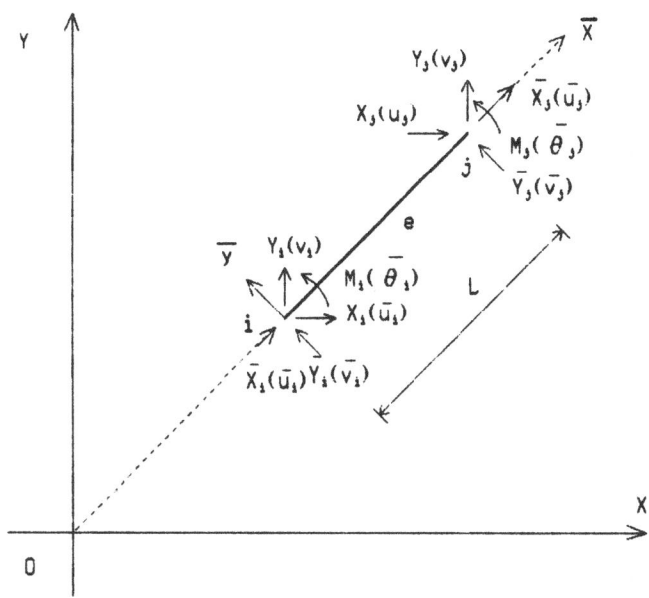

Fig.1   Beam-column element in global system X-Y
showing positive nodal forces and nodal
displacements in local system x-y

In the case of beam subjected to a uniformly distributed load (such as a self weight), we must consider with the effect of such loading system in the fundamental stiffness equation(1).   The stiffness equation(1) in this case is modified with the initial force vector $\boldsymbol{f}_e^0$ as follows :

$$\boldsymbol{f}_e = \boldsymbol{k}_e\,\boldsymbol{u}_e + \boldsymbol{f}_e^0 \tag{4}$$

In order to determine explicitly the element stiffness matrix $\boldsymbol{k}_e$, let us assume the displacement components $\bar{u}$, the axial displacement, and the transverse displacement $\bar{v}$, in the linear form and the cubic form, respectively

$$\bar{u} = a_1 + a_2 \bar{x} \tag{5}$$

$$\bar{v} = b_1 + b_2 \bar{x} + b_3 \bar{x}^2 + b_4 \bar{x}^3 \tag{6}$$

Here, the coefficients a_1 and b_1 mean the rigid body translation, and b_2 is the rigid body rotation. With these expressions, we can obtain easily the following element stiffness matrix :

$$\boldsymbol{k}_e = \begin{bmatrix}
\dfrac{EA}{L} & & & & & \\[2ex]
0 & \dfrac{12EI_{\bar{z}}}{L^3} & & \text{Sym.} & & \\[2ex]
0 & \dfrac{6\,EI_{\bar{z}}}{L^2} & \dfrac{4\,EI_{\bar{z}}}{L} & & & \\[2ex]
-\dfrac{EA}{L} & 0 & 0 & \dfrac{EA}{L} & & \\[2ex]
0 & -\dfrac{12EI_{\bar{z}}}{L^3} & -\dfrac{6\,EI_{\bar{z}}}{L^2} & 0 & \dfrac{12EI_{\bar{z}}}{L^3} & \\[2ex]
0 & \dfrac{6\,EI_{\bar{z}}}{L^2} & \dfrac{2\,EI_{\bar{z}}}{L} & 0 & -\dfrac{6\,EI_{\bar{z}}}{L^2} & \dfrac{4EI_{\bar{z}}}{L}
\end{bmatrix} \tag{7}$$

where E is the Young's modulus, A is the cross-sectional area, L is the length of member and $I_{\bar{z}}$ is the moment of inertia of cross section.

For the overall framed structure, the unconstrained stiffness matrix in global co-ordinate system X-Y is assembled by entirely standard procedures (see, Martin(1966)) and concequently the overall stiffness equation can be written as

$$F = K U \tag{8}$$

where **F** is the global force vector, **U** is the global displacement vector and **K** is the global stiffness matrix.

With use of equation(8) in considering with some given
loading condition and the prescribed boundary conditions,
we can analyze mechanical behaviours of plane framed structures.

3. Computational implementation

For analysis of the fundamental stiffness equation(8)
developed in the preceding section we have set up some procedure suitable for numerical simulation of felling behaviour of framed structures.

Let us explane simply on our procedure of computational
implementation. The fundamental treatment of our procedure
is as follows :

1. The operation starts with the input of initial data on
the initial configuration of rigid frame structures and the
related compulation of basic coefficients in equation (8).

2. We set up the point which are cutted or broken partially. When we wish to fall down the reinforced concrete
buildings, the column-feet or (and) ends of beam in the
framed structure are cutted or broken partially with use of
some kind of means. If we look mechanically these points
as the pin points, we can not determine the inversion of
the global stiffness matrix in equation (8) by the reason
that the corresponding structure reduces to an instable
one. According to this situation, we estimate in this
paper these points as the nodal points on which the only
bending stiffness reduces to gradually zero in the process
of solving equation (8). This procedure is same as the
variable stiffness technique proposed in tension field
problems by Nishimura, Tosaka and Honma (1985, 1986, 1987).

3. We may check the values of slope   of each nodal
point. If this value is equal to and larger than some
prescribe value, we assume the corresponding nodal point as
the cutted point and go on with our computation.

4. In order to trace sufficiently the felling kinematics
of structures, we must consider with a geometrical change
based on the deformed configuration of the structural system in each stage for solving equation (8). Concequently,
we reset the configuration of structure with the deformed
state at the time of input of initial data.

We cannot describe the above-mentioned procedure in this
paper in detail and then we present only schematically the
main flow computer diagram in Fig.2.

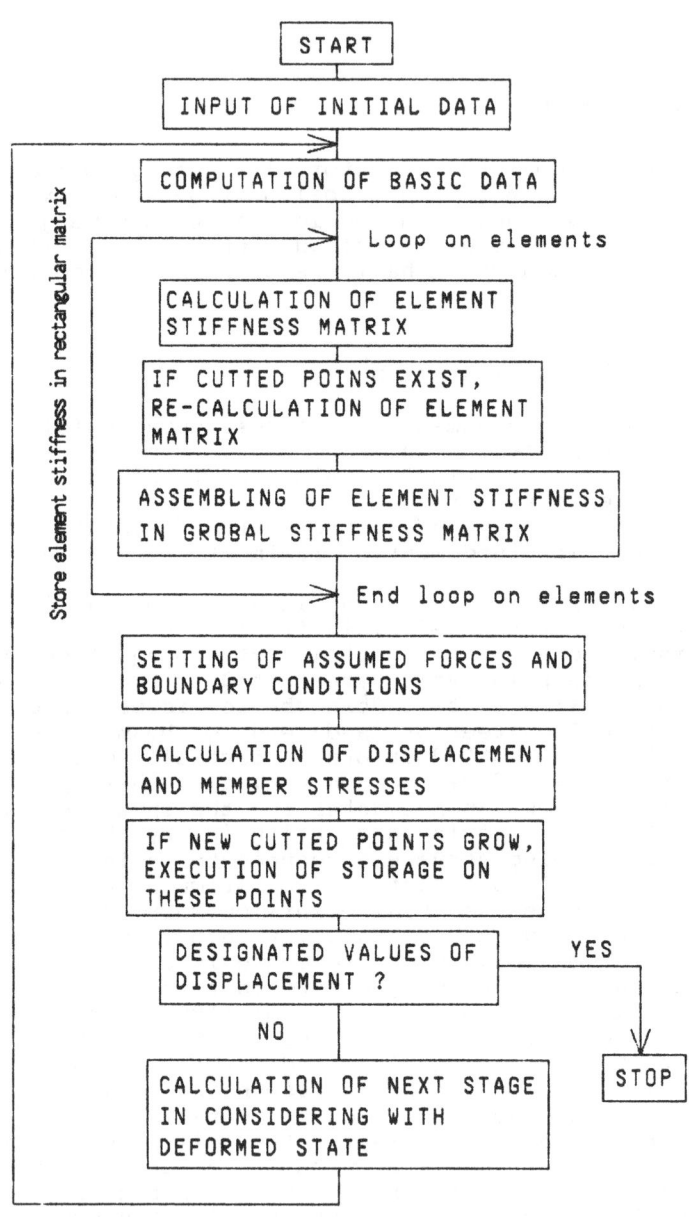

Fig.2　Main flow diagram for felling pattern
　　　　analysis of structures

4. Numerical example

The proposed procedure in computational implementation is applied to examples in order to show some applicability and varidity. For example we select the following simple models of a framed structure.

4.1 One story-one span frame
Let us consider to analyze the typical one side felling process and felling pattern in which a structure falls toward left side. Numerical simulation is started with the assumption that this structure has two nodal points with the V-cutting of column foot indicated in Fig.2. The obtained states for each stage from the initial configuration to the almost final one are also illustrated in Fig.3.

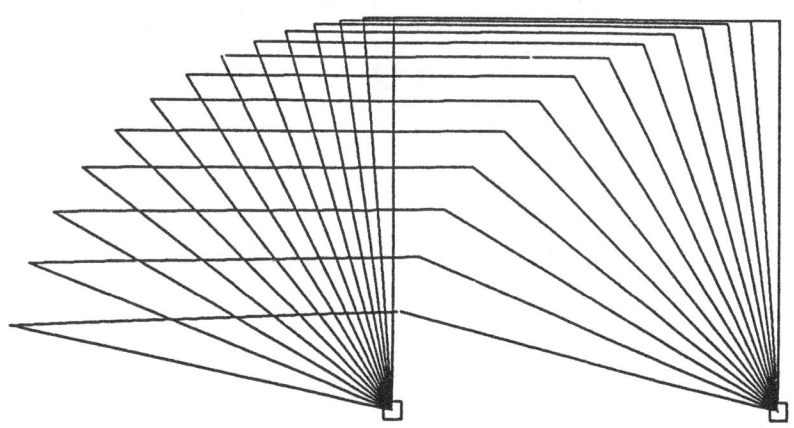

Fig.3. Felling pattern for one-story frame

4.2 Two stories-one span frame

Next, let us show the one side felling pattern in two stories framed structure with two cutted points. The configurations for each stage of computation are described in Fig.4.

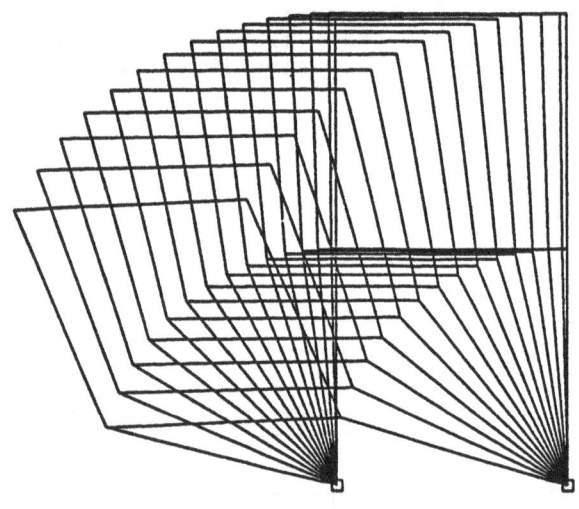

Fig.4.   Felling pattern for two-stories frame

5. Conclusions

Numerical procedure and the related algorithm of computer simulation for felling problems of reinforced concrete framed structures have been developed with the direct stiffness matrix method in conjunction with the variable stiffness technique. This approach is systematically applicable to many kinds of framed structure system. Felling patterns of one story and two stories framed structures are solved numerically by the method proposed in the present study our method could be improved by taking the more consideration and performing many examples in future study.

Acknowledgement

The authors wish to express their appreciation to the collaborators who assisted in the preparation of this paper. In particalar, we wish to appreciate Dr. Kazuhiko Kakuda, Department of Mathematical Engineering, College of Industrial Technology, Nihon University, preparing this paper.

References

Kasai,Y. and Shibata,S. (1970) Felling patterns of the instantaneous demolition of reinforced concrete structures (in Japanese), Summaries of Technical Papers of Annual Meeting of Kanto-Chapter of Architectural Institute of Japan, pp.289-292.

Martin, H.C. (1966) Introduction to Matrix Methods of Structural Analysis, McGraw-Hill.

Meek, J.L. (1971) Matrix Structural Analysis, McGraw-Hill.

Nishimura,T., Tosaka,T. and Honma,T. (1985) Finite element techniques in tension field problems (in Japanese), Jour. of Structural and Construction Eng. (Transaction of AIJ), No.351, pp.76-83.

Nishimura,T., Tosaka,N. and Honma,T. (1986) The planed tension field analysis by finite element techniques (in Japanese), Jour. of Structural and Construction Eng. (Transactions of AIJ), No.368, pp.27-36.

Nishimura,T., Tosaka,N. and Honma,T. (1987) The curved tension field analysis by finite element techniques (in Japanese), Jour. of Structural and Construction Eng. (Transactions of AIJ), No.376, pp.10-18.

DEMOLITION METHOD FOR REINFORCED CONCRETE UNDERGROUND STRUCTURES IN JAPAN (PART 1 GENERAL)

A. Enami   Nihon University
T. Kemi   Toda Construction Co., Ltd.
T. Suzue   Ohbayashi Corporation
J. Mase   Taisei Corporation

Abstract

In recent years, the concentration of the population into specific cities (Tokyo, Osaka, and Nagoya especially) has continued unabated. Consequently, existing reinforced concrete structures are increasingly being demolished for urban redevelopment, effective utilization of space or reconstruction to update old building functions.
In 1971, the Building Contractor's Society established a research committee which has prepared a series of guidelines for dismantling work from planning and execution through to disposal of leftover materials. These guidelines are part of the technical research subject "Research into Noise-free, Vibration-free Destruction of Reinforced Concrete Structures" for the Ministry of Construction.
In particular, the research that is being introduced here on the "Guidelines for Demolishing Reinforced Concrete Underground Structures" has been in progress since 1978 and a guideline was compiled in 1984. Please refer to the "Guidelines for Demolishing Reinforced Concrete Underground Structures" edited by the Building Contractor's Society for details.
Key words: Reinforced concrete structure, Existing underground structure, Existing pile, Demolition.

1. Introduction

Frequently, underground structures such as basement floors, foundation piles, etc. are not the subject of demolition when the above ground portion of the building is demolished, but when a new building is to be constructed.

This may be mainly because demolition of basement floors is in many cases closely related to the retaining wall work for the new building (which prevents the surrounding ground from deformation and collapse) and that the disposal method for the piles of the existing foundation is determined with regard to the design of the foundations of the new structure. In other words, many factors are involved so the demolition of basement floors cannot be determined independently. Features of the demolition method for underground structures are:

(1) In general, columns, beams, footings, etc. of the underground portion are large concrete members with much larger cross-sections than those of the above ground portion.

(2) It involves the treatment (removal or leaving) of members

404

installed deep in the ground, such as existing piles.

(3) The surfaces of many parts are in direct contact with the ground and ground water, such as underground external walls, footings, pressure bearing slabs, etc., that are related to the retaining wall construction method.

This report is Part 1 of a three-part report studying the methods of demolishing reinforced concrete underground structures. In Part 1, (1) the results of investigation of demolition conditions are examined and (2) the demolition methods are classified and selected. Part 2 describes (1) the total plan of a demolition method, (2) the demolition methods for piles, and (3) the demolition methods for basement floors and foundations. Part 3 gives some examples of actual demolition work.

Today various water jet methods are being researched and developed in Japan as methods for demolition of large section members. Thus, it is expected that these methods can be used in actual demolition in the foreseeable future.

2. Examination of demolition conditions and survey results

When planning to demolish existing structures, the purpose of the works and contents of the demolition work and new construction work must be thoroughly grasped. Quality, cost, program and safety control, etc., must be reviewed based on the results of surveys performed in advance.

2.1 Confirmation of client's requirements and the scope of work
Prior to examining the demolition work plan, it is essential that the requirements of the client, the reason for demolition and the scope of work are confirmed.

Particularly, attention must be paid to the case where the new structure is planned to overlap the existing building or come close to the boundary with the road.

In recent years, it has become difficult to obtain approval to construct a substructural retaining wall within a road for the purpose of demolition or new construction work. Even if it is possible to construct the retaining wall, it would be more beneficial to leave the existing underground external wall if the space is limited, considering ground settlement, effects to buried utilities due to excavation and removal at the time of construction of the new retaining wall. Therefore, it may be necessary to consider changing the design of the new building after consulting with the designers.

Although removing existing piles may be different from dismantling massive concrete structures, such as underground structures, it constitutes an important point in the work planning. Therefore, it must be thoroughly examined based on the survey results so as not to cause problems in workability and safety. Consequently, when planning a new building, existing underground structures and piles should be taken into account at the time of preparation of the basic plan.

2.2 Examination of survey results on demolition
In the planning of demolition work, the following problem points must
be analyzed, and the demolition method examined based on the results
of the survey conducted in advance and the notification:

a. Particulars of the building
(1) Construction, size, depth.
(2) Type of foundations, type of piles, numbers, positions.
(3) Cross-section of members, bar arrangements, types and strengths
of the concrete.
(4) Date of execution, method of construction, degree of
deterioration with the passing of time, history.
(5) Comparison between design documents and actual conditions (by
actual measurement, core sample, test-pit, etc.).
(6) Existing dangerous objects, such as utilities and others.

b. Site and surroundings
(1) Site area, configuration, undulations.
(2) Layout of building, relative position to the boundary line,
obstacles in the area of underground external wall.
(3) Properties of ground, existence of groundwater.
(4) Type, configuration, position, etc. of buried utilities within
site.
(5) Type, configuration, position, etc. of buried utilities within
the road.
(6) Positions, types of construction, degrees of deterioration,
etc. of adjacent buildings.
(7) Examples of demolition work in vicinity.

c. Environmental conditions
(1) Adjacent residences, public facilities and special-purpose
buildings.
(2) Construction of adjacent roads, traffic conditions, traffic
controls, etc.
(3) Noise, vibration, etc. in the immediate vicinity.
(4) Related legislations and controls.

d. Relation with new construction work
(1) Layout, plan, cross-section, configuration, dimensions, position.
(2) Type of foundations, dimensions and positions.
(3) Position of underground external wall.
(4) Sizes, positions, cross-sections of piles.
(5) Layout of temporary structures, such as retaining wall, struts,
protective platform.

e. Disposal of waste materials
(1) Type and quantity of waste materials.
(2) Method of collection, hauling, disposal, etc. of waste
materials.

For trouble-free demolition, it is important to analyze and examine
factors turned up by the survey and prepare to offset problems before
they occur.

2.3 How to proceed with demolition plan
In the planning of demolition work, it is important to prepare an
optimum plan to take into account the demolition of the existing
building as well as the construction plan for the new building.
    Based on the survey data, the following items must be gone into
thoroughly:

    a. Identifying conditions for demolition work: See Part 2 Examina-
tion of demolition conditions and survey results.

    b. Identifying the content of new construction work:
    (1) Examining the content of any piling work, etc., for the new
building, that is to be performed before, after or at the same time as
the demolition work.
    (2) Examining an outline plan for temporary structures required for
the new construction work, such as retaining wall, struts, protective
platform for truck, etc.

    c. Determining demolition procedure: The sequence and methods
should be examined and compared between the new construction work and
the demolition work.

    d. Planning temporary facilities: A layout plan of the temporary
facilities for both the demolition work and the new construction work
should be prepared based on the procedures, and the method of
constructing these facilities should be examined.

    e. Selecting demolition work method:
    (1) Selection of equipment and method to match the procedure and
the temporary facilities.
    (2) Selection of a demolition method to take account of the
features of the underground structures (large cross-sections and
reinforcing bars, etc.).

    f. Examining transport and disposal plan for leftover materials:
The transport and disposal plan for leftover materials should be
examined, taking into account road and traffic conditions in the
immediate vicinity.

    g. Examining safety and public hazard prevention measures.

3. Classification and application of demolition method

The demolition method must be selected with full awareness of the
features of the method, equipment, work conditions and scope of the
work.

3.1 Equipment and method for demolition
Although there are various methods of classifying the demolition
method, the classification by the principle, and the method, of
destruction shall be as follows:

407

a. Method using mechanical impact (example ... hand breaker, large breaker, etc.): Rapid, repeated impact is made by the tip of the chisel (making use of air pressure, hydraulic pressure, gas pressure, etc.) to break the concrete.

b. Method using hydraulic pressure (Example ... hydraulic crusher, jack, rock jack, etc.): Concrete is crushed with teeth, broken by pressing or expanding, using hydraulic machinery.

c. Method using grinding (Example ... cutter): Concrete is cut by a rapidly rotating diamond tipped blade.

d. Method using expanding pressure (Example ... static breaking agent): By mixing an expansive powder with water and injecting it into holes drilled into the concrete; in the resulting chemical reaction, concrete is broken through tensile force.

e. Method using explosives (Example ... concrete blaster, explosives): Holes drilled in concrete are filled with explosives, to break the concrete by the explosive force.

f. Method using heat (Example ... thermite process): Concrete and steel bars are burnt through by melting at a high temperature of 3,000-3,500°C, by burning iron or aluminum alloy in oxygen.

There is a wide variety of demolition equipment and methods, including those which have been put to practical use and those which are in the process of being researched and developed. Tables 1 and 2 show the classification and scope of application, etc. of the methods of demolition using equipment which has already been in regular use.

3.2 Selection of demolition method
The demolition method which must be selected to conform with the procedure mentioned in "How to proceed with demolition plan" must make full allowance for the construction period, workability, economy, safety, prevention of public hazards, etc.

Formerly, priority was given to the economy of the demolition equipment and method, along with the construction period. Now, emphasis has shifted to preventing public hazards, and keeping the noise, vibration, dust, etc. to within the allowable limits, in line with the environmental conditions in the area.

Demolishing underground structures involves larger cross-sections of members and heavier reinforcing bars compared with above-ground structures. In addition, demolition equipment, workability, etc. are also affected by the layout of the retaining wall, struts, and truck platform.

As shown in Table 1 and 2, each dismantling method has its pros and cons. Therefore two or more methods are used generally, so as to take sufficient advantage of each method. Table 3 shows examples of combination of various dismantling method. Fig. 1 shows the flow chart of determining dismantling method.

Table 1. Features of various dismantling machinery and methods and scope of application (1-1)

Classification of method	Machinery	Features of construction			Features of public hazards			Scope of application
		Crushing Ability	Secondary Crushing	Safety	Noise	Vibration	Dust	
Method using mechanical impact	Hand breaker	△	◎	○	△	○	○	Effective where heavy machinery cannot be used or in confined space used to supplement machinery and methods.
	Large breaker	◎	◎	◎	△	△	△	Structures can be dismantled more efficiently when this is used vertically rather than side ways. For mass concrete it is more efficient when used in combination with concrete breaker, etc.
	Hydraulic crusher	◎	◎	◎	○	△	△	Suitable for dismantling general building structures. Efficiency lowers when used for dismantling members with a large cross-section or mass concrete. Space for engagement is required around the members
Method using hydraulic pressure	Jack	○	○	○	○	△	△	Effective for dismantling RC slab, beams, etc. Dismantling of wall is possible if there is a reaction force. Strake may be insufficient where height of storey is large.
	Rock jack	○	○	◎	○ / Drilling △	◎ / Drilling	◎ / Drilling ○	Suitable for dismantling massive concrete structures without steel reinforcement or relatively few steel bars. For structures where the periphery is restricted a free surface must be provided in advance.
Method using grinding	Cutter	○	△	○	○	◎	◎	*Unlike other methods, cutter method cuts concrete together with steel bars and dismantles members with a large cross-section. A crane is required.

Table 1. Features of various dismantling machinery and methods and scope of application (1-2)

Classification of method	Machinery	Features of construction			Features of public hazards			Scope of application
		Crushing Ability	Secondary Crushing	Safety	Noise	Vibration	Dust	
Method using expanding pressure	Static breaking agent	○	○	○	◎ Drilling	◎ Drilling	◎ Drilling	Suitable for dismantling large concrete structures without reinforcing bar or relatively few reinforcing bars. It is possible to cause cracks in reinforced concrete.
	Concrete blaster	◎	○	△	△	△	△	Suitable for dismantling massive concrete structures. It is possible to use for reinforced concrete. Often used in urban districts in lieu of explosives.
Method using explosives	Explosives	◎	◎	△	△	△	△	Suitable for dismantling large concrete structures. Many restrictions on use in urban districts.
Method using flame	Thermite process	△	△	△	○	◎	○	Suitable for drilling concrete or cutting reinforcing bars by melting. Efficiency lowers when used for dismantling members.
Other auxiliary machinery	Core boring	—	—	—	○	○	◎	Used for drilling rock jack holes, filling holes of static breaking agent and charging holes with explosives. Sometimes used for cutting edges of large cross-section members.
	Rock drill	—	—	—	△	◎	○	

Legend for evaluation symbol

Symbol	Breaking ability	Secondary crushing	Safety	Noise	Vibration	Dust
◎	Larger than 20 m³/h	None	High	Lower than 65db (A)	Lower than 60db	None
○	0.5 - 2.0	Medium	Medium	66 - 80	61 - 65	A little
△	Smaller than 0.5 m³/h	Large	Low	Lower than 81	66 - 70	Caused

Note)
Noise level is the value at a distance of 30 m from the source vibration is the value at a distance of 20 m from the source.

Table 2. Application of dismantling machinery and methods, by member

Machinery and method	Member						Remarks
	Slab	Wall	Beam	Column	Founda-tion *	Pile	
Hand breaker	O	O	O	O	△	△	It is better to dismantle the column on the floor after it is pulled down. Efficiency lowers for dismantling of foundation and pile, as they are massive.
Large breaker	O	O	O	O	O	O	Can be used widely, but space is required for operating heavy equipment.
Hydraulic crusher	O	O	O	O	△	-	Use of this is not common for foundation and piles as they have a large cross-section. It is difficult to use where members contact directly with soil.
Jack	O	△	O	△	△	-	Examine how to take reaction force for wall. For columns, only pulling down can be done and crushing cannot be done. Insufficient ability for foundation.
Rock jack	-	-	△	△	O	O	Efficiency lowers for reinforced concrete.
Concrete blaster	-	-	△	△	O	O	Safety precautions must be taken in dismantling beams and columns. Effective for pile head treatment of in-situ pile.
Explosives	-	-	△	△	O	O	Use is restricted because of environmental conditions. Thorough examination and safety measures must be used.
Static breaking agent	-	-	△	△	O	△	Danger of collapsing and falling in the case of columns and beams. It takes time for dismantling.
Cutter	△	△	△	△	△	-	It is difficult to apply for foundations because of the large cross-sections and contact with soil.

* Foundations include mass concrete, such as underground beams, footing, etc.
  Legend for evaluation symbols O ... Wide applicability
  △ ... Applicable
  - ... Generally applicable.

Table 3. Examples of combination of various dismantling methods

Combined method	Essential points of combined method	Essential points
(1) Explosives + hydraulic crusher + large breaker	Large underground beams, foundations, etc. are crushed by blasting and slabs, beams, walls, columns, etc. are dismantled using a large breaker. If explosives are used, thorough care must be taken for safety, in consideration of enviroment conditions.	• Although the efficiency is high it is necessary to provide facilities to prevent noise and concrete from scattering. • Large instantaneous noise and vibration. • When breaking concrete into large pieces, it is better to out restraint reinforcing bars in advance.
(2) Hydraulic crusher + static breaking (static breaking agent and rock jack) + large breaker	After massive members, such as underground beams, pressure bearing slabs, foundations, etc. are broken by generating cracks or in large pieces, using static breaking agent, etc. they are crushed using the hydraulic crusher and large breaker, other members are dismantled directly using the hydraulic crusher, large breaker, etc.	• Although the efficiency lowers, safety is high. • Noise at the time of drilling holes and noise and vibration owing to use of large breaker. • Effective if restraint reinforcing bars are cut in advance.
(3) Hydraulic crusher + large breaker	Although beams, columns, etc. are dismantled using hydraulic crushers generally, and slabs, walls, underground beams, pressure bearing slabs, foundations, etc. are dismantled using a large breaker, what is to be dismantled is better determined in consideration of economy and measures against public hazards, etc.	• Although the efficiency is high measures against noise, vibration, dust, etc. become necessary. • One base machine will suffice if an attachment is replaced.

* Hand breaker was omitted as it is used to supplement each combination.
* There are many other combinations conceivable.

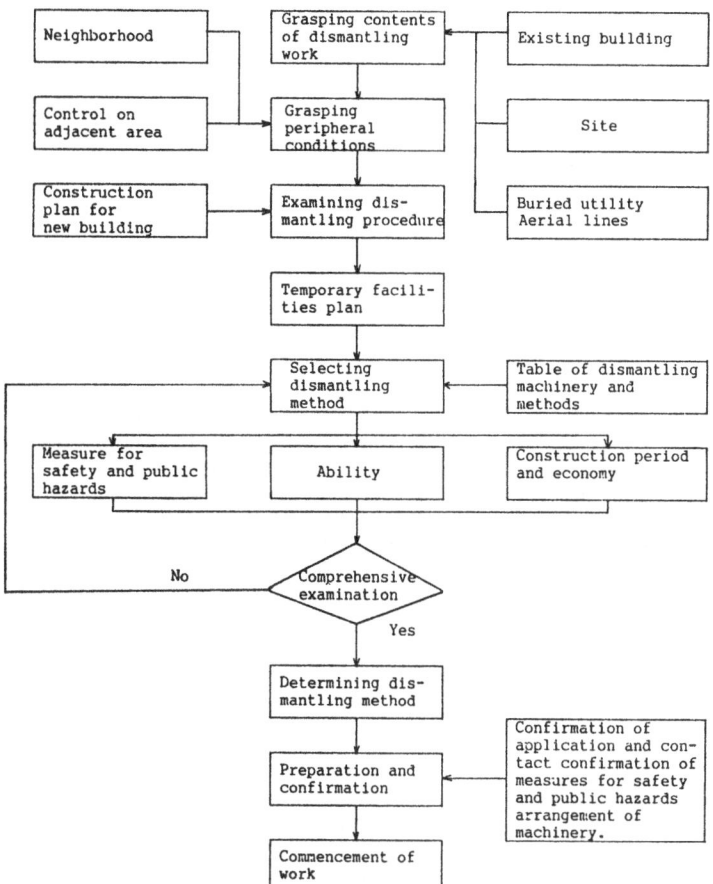

Fig. 1.   Flow chart of determining dismantling method

413

DEMOLITION METHOD FOR REINFORCED CONCRETE UNDERGROUND STRUCTURES IN
JAPAN (PART 2 DEMOLITION METHOD)

A. Enami   Nihon University
T. Kemi   Toda Construction Co., Ltd.
T. Suzue   Ohbayashi Corporation
J. Mase   Taisei Corporation

ABSTRACT
This report describes the methods for selecting optimum excavation and
retaining work methods as well as the methods for drawing up an
appropriate underground demolishing plan by clarifying the mutual
relationship between the existing underground structure to be
demolished and a structure to be newly constructed thereon. For the
demolition of piles, the methods for drawing out the existing piles as
well as breakage methods are explained. For the demolition of
basement floors, refer to Part 3, in which the demolition methods of
foundations, cast-in-place concrete piles, and mass concrete
structures such as caissons are also explained.
Key words: Reinforced concrete structure, Existing underground
structure, Existing pile, Demolition.

1. Total plan

Compared with demolishing structures above ground, demolishing below
ground floors and foundations, etc. has the following features:

(1) Generally, the demolishing work is executed simultaneously with
the new construction work. This may be to effectively use the land,
and to avoid the expense of dismantling existing underground
structures and piles independently.

(2) As the work involves excavation, the retaining wall is usually
required, and measures for prevention of settlement and deformation of
surrounding ground, etc. may be necessary.

(3) As the members to be dismantled, such as foundations, etc. have
larger cross-sections than those above ground, a so-called rational
demolition and transportation plan for massive concrete members
becomes necessary.
(4) The underground structures to be dismantled are in direct
contact with the soil at the perimeter and beneath the bottom members.
This restricts the demolishing method and removal of demolished
materials.

Consequently, demolishing underground structures requires
scrupulous planning. It must be seen as one overall job, integrated

414

with the new construction work, not merely considering the demolition work alone as in the case of the above ground portion.

Normally, demolition of the underground structure itself constitutes the critical path in the construction program. Hence, the total plan must provide for easier demolition work so as to assure the safety of the total underground work. Particularly, when the amount of demolition is large and there is a great deal of massive concrete structures to be dismantled, care must be exercised in preparing the total plan with the priority on demolition, or unexpected losses may result in the construction program.

This section describes the demolition of existing underground structures (massive concrete structures such as piles, basement floors, foundations, etc.), and outlines retaining wall work by trench excavation.

For simple demolition work for underground structures, simple retaining wall methods or slope cutting methods, used in advance and with sufficient backfilling are adequate.

However, when demolishing large underground structures, the method used must be determined after thoroughly examining the demolition method for massive concrete structures, such as aforementioned, and with regard to the pile and retaining wall work by trench excavation.

Fig. 2 shows the conceptional relationship between the new building and the existing underground structures to be demolished.

The numbering of explanations in the conceptional relationship drawing, Fig. 2, such as 1), 2), a, b, c, etc., matches those of the test that follows.

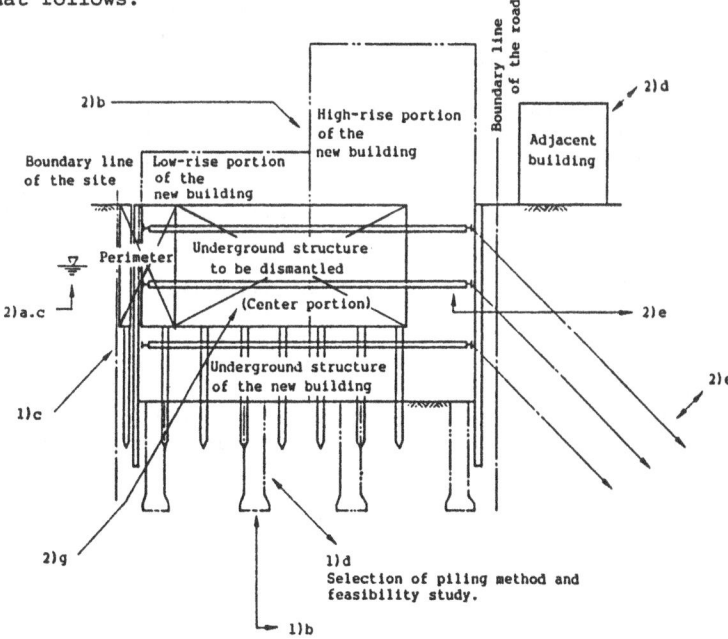

Fig. 2. Conceptual relationship between the new building and existing underground structures to be demolished

415

1) Relationship between the new building and existing basement floors

a. Before the design of the new building has been finalized, it is important that ample consideration be given to the relationship with the demolition work.

b. If the design of the new building has already been finalized, it is important to prepare the interposition drawings of plan and section of the new building and existing one, making a clear mutual relationship, and to establish a temporary retaining work plan and demolition plan.

c. When the existing underground external wall is close to the boundary line, it is difficult to build the new building to the boundary line of the site, unless an adjacent road can be used.

The usual approach is to leave the existing underground external wall as it is, and to build the new one inside the old one.

The existing underground external wall is sometimes used as the retaining wall, which can be economical.

d. Where the foundation of the new building is a pile foundation, a thorough study must be made in advance, as there may be design and construction restrictions owing to the existing building foundation being combined with existing piles. If the deep foundation method is used for the new building, there would not be too serious a problem as the excavation can generally be made from the reduced level after demolishing the existing underground structure, foundation, etc.

However, where the design uses the in-situ pile of the bored type Benoto pile, reverse circulation pile, etc., the level of equipment in operation becomes an important matter to study. Even though the operation is carried out using the existing 1st floor slab by reinforcing, it will be necessary to drill holes through the floor for new piles.

Where the position of the new pile, and that of the column center and pile of the existing building are interposed, the construction becomes difficult. Therefore, an advance study must be made to thoroughly examine the relative positions of the new pile, and column and pile of the existing building, from the viewpoint of construction. The method of construction, size of pile, position and work procedure must be considered.

Occasionally, it may be necessary to change the design of the piling method itself.

2) Relationship between the new building and existing excavation and retaining wall work

a. The appropriate excavation and retaining wall work method will be determined after the results of the geological survey and soil test have been reviewed. Where there is an existing underground structure, it will be necessary to investigate the retaining wall work method and backfilling condition, etc., at the time when the exsiting basement floors were constructed. In particular, where the retaining wall was embedded or special backfilling conditions exist, these factors will affect the determination of the method of construction of the new retaining wall.

b. When excavation is accompanied by the demolition of the underground structure, it is necessary to determine whether the whole

416

surface excavation or partial excavation should be carried out with
regard to the total construction program of the new building.
Especially, where there is a high-rise portion and a low-rise portion
in the new building the priority of the underground demolition must be
placed on the high-rise portion.

c. The selection of type of retaining wall must consider ease of
construction, in view of the relation to the existing underground
structure.

Where earth support of soldier beams and boarding is used, the
preceding demolition work, such as drilling, etc. can be minimized as
the position of column and beam of the old basement can be avoided by
adjusting the spacing of main piles.

For sheet piles or continuous underground walls, all existing
structures at the position of retaining wall must be demolished in
advance for separation, this entails problems such as method of
separation, temporary strut support, etc.

d. Where the demolition of the existing foundation concrete is to
be carried out close to the retaining wall, the vibration from the
work will affect the retaining wall for a lengthy period of time,
which in turn could aggravate deformation of surrounding work.
Therefore, measures against vibration and noise must be studied not
only for the peace of residents in the neighborhood but also to
minimize settlement of surrounding ground and adjacent buildings.

If possible, a high rigidity retaining wall executed using a
low-noise and low vibration method is preferable.

e. In the selection of the strut support method, the demolition
work may often become a bottleneck in the program. Thus, a large span
support method should be adopted, such as the concentrated strut
method or earth anchor method, so as to increase the efficiency of the
demolition work by securing a large open space.

It is also effective to use the plan in which the rigidity of the
retaining wall is enhanced and the upper and lower spacing of supports
is reduced to allow the demolition to proceed more easily.

f. It is important for supports to be laid out so as to make the
demolition of the basement and the excavation for the new building
easier and to maintain the total balance.

To improve the workability of the demolition, the rigidity of
supports may be increased and the structures to be demolished may
first be exposed from the ground so they can be readily worked on. In
many cases the posts for supports are provided in advance by partially
drilling holes in the existing underground structure.

g. The demolition of massive concrete structures, such as
foundation, etc. is normally accomplished using explosives to improve
efficiency, or crushed using a hydraulic crusher and large breaker,
etc.

The point is to secure a large space that allows the work to
proceed easily, promoting safe construction and making the removal of
waste more efficient.

h. As the retaining wall work involves earth and water the
condition of which varies with each project, it is necessary to be
able to cope with such situations case-by-case. There are many other
factors in demolishing and removing the existing underground
structures and piles that make standardization difficult.

417

2. Demolishing/removing piles

The existing piles may be wooden piles, RC piles, PC piles, steel piles, pedestal piles, in-situ piles (bored pile, Benoto pile, reverse circulation pile, deep foundation).

In considering demolition of these piles, the in-situ piles are treated in the same way as concrete structures, such as foundations, so mainly demolition and removal of wooden piles, RC pile, PC pile and pedestal piles are discussed here.

There are two methods of demolishing and for removing existing piles: One is to extract the piles by loosening the friction between the piles and the ground; the other is to crush or grind the piles directly in the ground. (See Fig. 3.).

In terms of the relationship to the retaining wall work method, there are cases where these existing piles interfere with the retaining wall (main pile, sheet pile column strip retaining wall, continuous underground wall) and posts (post for supports, inverted support center column) or permanent piles (in-situ pile, ready-made pile).

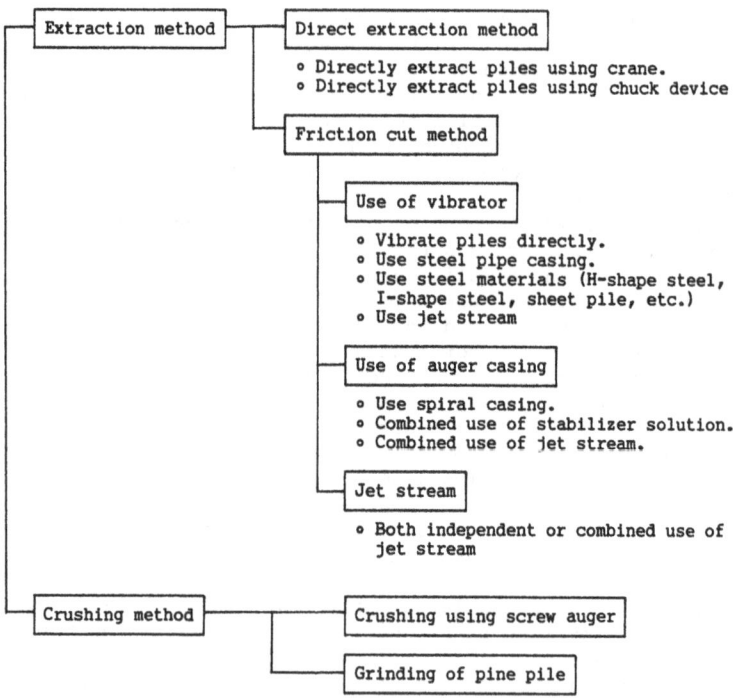

Fig. 3. Classification of demolishing and removal of existing piles

In general, the work should be done without demolishing and removing the existing piles. Even if demolition and removal become unavoidable, it is preferable to minimize the work, because the surrounding ground may be loosened due to extraction or demolition. Also, the ground should be stabilized by filling the voids after extraction or demolition with soil mortar, etc., however, the construction period becomes longer and costs increase.

1) Extraction method: There are two methods of extracting the existing piles i.e. the method in which the piles are directly extracted without friction being cut and the method in which the piles are extracted with friction being cut.

Where the piles are extracted directly, a special chucking device is installed on the head and the piles extracted using a crane, hydraulic jack, pulley, etc. Where the friction cut is used, adopt the method to serve the purpose, i.e. use of vibro hammer, auger casing or jet stream, etc.

Table 4 shows the method of extracting piles.

2) Cutting up by auger method

a. Cutting up using auger: This method is used for breaking up pine piles, RC piles and PC piles.

As shown in Fig. 4, the perimeter of the existing pile is excavated using a doughnut auger outer pipe. After driving the outer pipe to the tip of the pile, launch the screw auger inside to crush the pile. The outer pipe acts as a guide for the screw auger so that the pile does not escape during crushing.

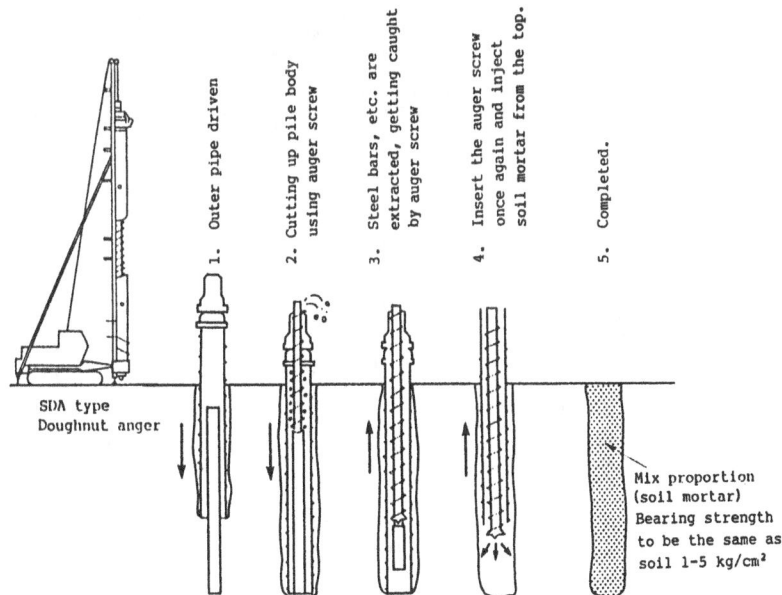

Fig. 4. Method of cutting up existing pile

419

Table 4. Table of extraction method of existing piles

Method	Direct extraction	Friction cut method	
	Direct extraction using chuck device	Vibro-casing method	Auger casing method
Summary of construction	A special chuck device is installed on the pile head and pile is extracted using crane, hydraulic jack, pulley, etc.	A casing larger in diameter than the pile (steel pipe) is driven by vibro hammer and pile is extracted by separating friction between pile and earth (bentonite solution may be injected).	A casing larger in diameter than the pile is driven by auger machine and pile is extracted by separating friction between pile and earth (over cut or bentonite solution, soil pile may be injected).
Summary drawing of construction	Wedge type / Chuck type / Cam type / Lever type	Vibro hammer / Casing steel pipe / Casing / Pile	Anger machine / Injection pipe / Casing
Ability	200-350 ϕmm $\ell = 5$-8 m	300-400 ϕmm $\ell = 10$-15 m	300-500 ϕmm $\ell = 15$-18 m
Efficiency	$\ell = 6$ m 15-18/day	$\ell = 15$m 8-12/day	$\ell = 12$ m 8-10/day
Safety	Δ	Δ	o
Public hazard	o	Δ	o

Method	Friction cut method	Friction cut method	
	Spiral casing method	Jet spray method	Vibro-jet method
Summary of construction	A special blade is installed on the perimeter of the casing with injection pipe running below. By injecting bentonite solution, friction is separated.	A jet pipe is inserted on the perimeter of the pile and the pile is extracted by separating friction between pile and earth by jet stream.	A jet pipe is installed on the casing and friction between pile and earth is separated by high pressure jet stream while driving the jet pipe using vibro hammer.
Summary drawing of construction	Spiral blade / Casing / Bentonite solution injection pipe / Overcut of blade tip / Pile / Overcut portion injection	Jet pipe / Jet	Jet pipe / Casing
Ability	Concrete pile 400-600 ϕmm $\ell = 15$-18 m	300-400 ϕmm $\ell = 10$-12 m	Larger than 500 ϕmm $\ell = 15$-18 m
Efficiency	$\ell = 12$ m 8-10/day	$\ell = 12$ m 5-8/day (8-12/day with high pressure)	$\ell = 15$ m 10-15/day
Safety	o	Δ	Δ
Public hazard	o	o	o

If the pile consists of two joined piles and is bent at the joint, it will be difficult to insert the outer pipe. For friction piles, the pile may slip downwards.

In recent years, a large bore rock auger has been used to drill the existing structures, so that existing piles of a large diameter can be crushed, though the efficiency lowers somewhat.

b. Method of cutting pine pile: A boring machine with a cutting core tube installed on the edge of the blade is used to cut the pine pile into a chip form.

The chips of wood produced by the blade are discharged outside from the tip of the core tube using the pressure of bentonite slurry. The special tube has a widening blade installed on the tip, which excavates the soil in the periphery of the pine pile. In the tube is a cutting blade which cuts the pine pile.

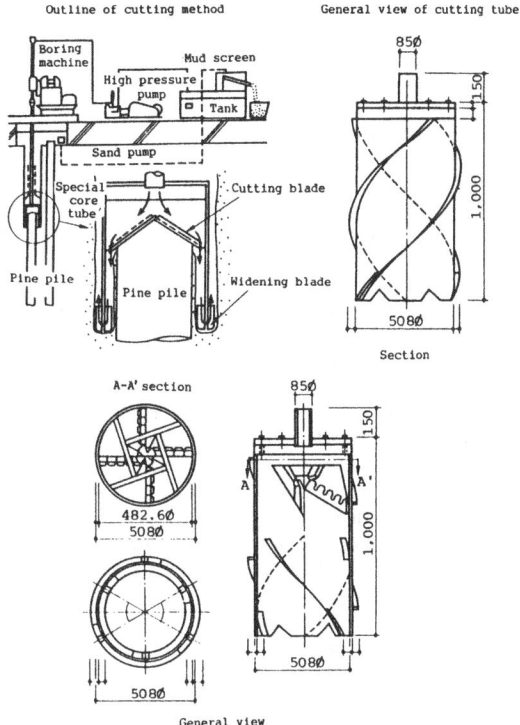

Fig. 5. Method of cutting pine pile[1]

3. Demolishing basement floors

As some examples are given in Part 3 of this series, this item will not be discussed here.

4. Demolition of foundation

The following shows the concept for the excavation, retaining wall and demolition work that accompanies the demolition of reinforced concrete, such as footings, in-situ piles, caissons, etc.:

1) As wide a space as possible must be secured for demolishing foundation concrete. For example, concentrated strut supports should be adopted to broaden the spacing of strut supports (horizontal space), and to broaden the spacing of each stage of strut supports by increasing the rigidity of the retaining wall (vertical space).

421

(2) By initiating trench excavation and by forming the operation platform for demolishing massive concrete structures, it may be possible to increase the efficiency of the work.

It is particularly necessary to try to separate the trench excavation and demolition operations, separating the carrying away of excess soil and waste according to the conditions of the dumping grounds.

The Example of Demolishing Underground Structures (IV) attached shows an example of excavating to GL-26 m while demolishing large concrete structures, such as caissons, etc.

Fig. 11 shows part of the outline.

The key points in planning demolition of footings are as follows:

(1) Where the demolition is carried out using large breakers, etc. noise presents problems as the operation continues for long hours.

Care must be taken in selecting equipment and the layout, also measures against public hazards should be studied.

(2) It is necessary to examine the use of explosives for the demolition of large concrete structures, especially where a large amount of demolition is involved, this method favors the work program. Although there is a preconception that explosives are dangerous, they are not so, if handled by skilled professionals following the technical standards prescribed by law.

(3) It is worth examining their use where there is only lower end reinforcement without the one of the upper and as the rock jack will work effectively.

(4) It becomes necessary to bore holes when explosives and rock jack are used. If the boring work is performed with the old 1st floor still there, in other words before demolition starts, it may help stifle noise caused by the operation.

(5) The demolition of caissons, etc. must be examined considering the relationship with the position of street supports at each stage for excavation and retaining wall work, and at what height and where the cutting should be done.

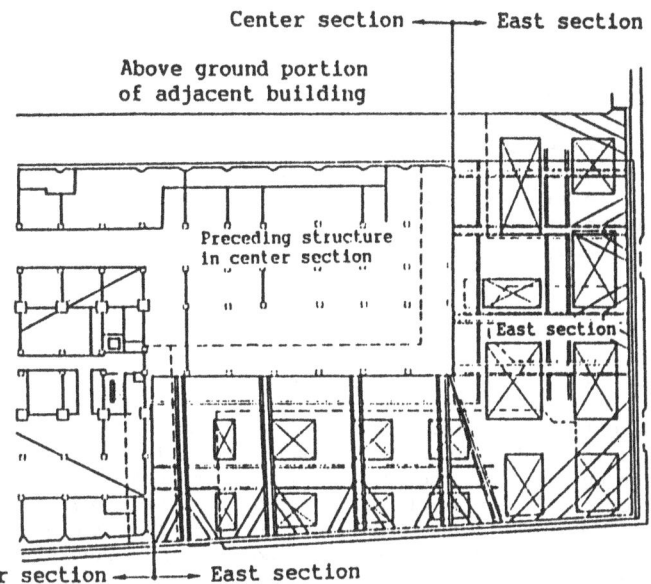

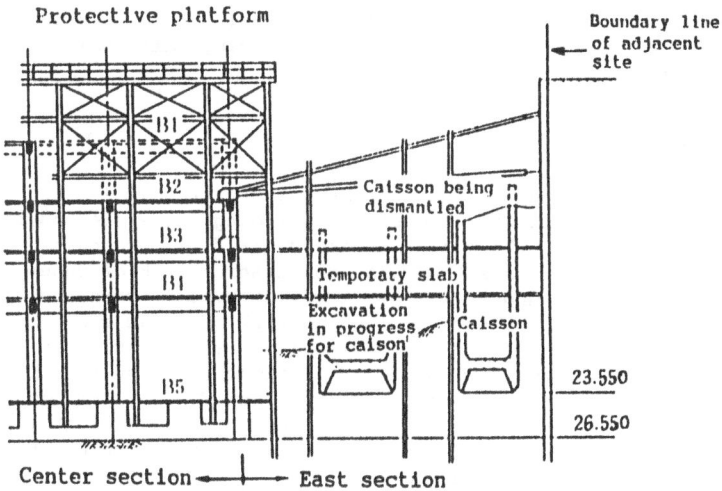

Fig. 6. Example of caisson foundation[2)]

DEMOLITION METHOD FOR REINFORCED CONCRETE UNDERGROUND STRUCTURES IN JAPAN (PART 3, PRACTICAL EXAMPLES OF DEMOLITION PROCEDURE)

A. ENAMI   Nihon University, Japan
I. SAWADA   Technology Development Division,
                  Taisei Corporation, Japan
S. KOSUGE   Construction Technology Department,
                  Shimizu Corporation, Japan
K. KATAOKA   Technical Research Institute,
                  Ohbayashi Corporation, Japan

Abstract
In many cases, buildings in urban areas are built to fully occupy the site in Japan, because of the acute shortage of land. Consequently, special methods must be devised to prevent settlement and deformation of peripheral ground and adjacent buildings. In particular, the utmost care must be taken in the retaining work and demolishing work. This report describes three practical examples of demolishing reinforced concrete underground structures in the Tokyo Metropolitan Area and the sequence of demolishing procedure and retaining work. The examples have been selected to represent three typical demolishing cases, and cover buildings with two, three and four underground floors.
Key words: Reinforced concrete structure, Existing underground structure, Existing pile, Demolition, Retaining wall, Existing underground external wall, Strut

1. Introduction

Japan is a small, mountainous country, with extremely limited space for its dense population. Therefore, building sites are generally tiny, and frequently the buildings are built fully occupying the narrow sites. It follows that when demolishing work of underground structures is to be undertaken, it involves special care and different considerations from the demolishing work of above ground structures. Retaining walls and shore supports are used as it is important to minimize settlement and deformation of surrounding ground, facilities of adjacent roads, and buildings. The basement to be demolished is considered an underground obstacle from the viewpoint of retaining work, as it hinders construction of the retaining walls and shore supports. Also, when shore supports are constructed, these will hinder the demolishing work of the underground structure. Thus, it is vital that retaining work to control settlement and deformation of surrounding ground and the procedure of demolishing work be examined carefully.

This report describes the sequence of retaining work and demolishing work to avoid settlement and deformation of surrounding ground for reinforced underground structures, ranging from two to four underground levels.

2. Practical example of demolishing existing two storey underground
   structure and constructing a new building with four underground
   floors (G-Building construction work)

This is an example of demolishing underground floors in Yaesu, Chuo
Ward, Tokyo. As there was an exisitng building with two underground
levels on part of the site, it was planned to keep the external wall
of the original underground floors intact, for protection of
surrounding roads and underground utilities. Since the new building
was to be a deep structure, i.e. four underground levels, the
retaining work was planned to use a method of preceding from the
ground floor and four-stage strut. As the demolishing work was
estimated to be extremely difficult when installing struts, a large
space, 10.4 m deep, was secured to make use of the slope after the
construction of the ground floor slab of the new building. The
demolishing work was executed efficiently.

    Fig. 1 is the superposition plan of the new and existing
buildings. Fig. 2 is the cross-sectional drawing at the time of
demolishing the original underground floors, and Fig. 3 the cross-
sectional drawing at the time the excavation was completed. The work
procedure of retaining work and demolishing work follows.

    1) The demolishing work started at the planned position for the
retaining H pile and temporary H column (temporary column to support
the ground floor of the new building). The work was carried out by
drilling with a giant breaker and hand breaker.

    2) To enable the heavy machinery work on the ground floor of the
building to be demolished, the ground floor was reinforced.

    3) Using the ground floor as a work platform, the retaining H pile
and temporary H column mentioned above were installed. The entire
depth of the retaining H pile was drilled using an auger, and was
filled with cement-bentonite.

    4) The gap between the retaining H pile and original underground
external wall was backfilled with crushed concrete and sand after
installing a horizontal lagging. Then a concrete slab was placed at
floor position of each floor to integrate the structure to secure
transmission of earth pressure.

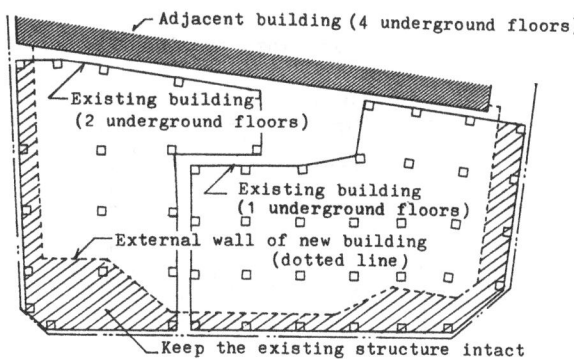

Fig. 1. Plan of the new building and existing buildings

425

5) A giant breaker was then set up on the ground floor and then this floor was demolished. Subsequently the columns of the B1 floor was demolished with primary excavation being done simultaneously.

6) The ground floor of the new building was constructed ahead of the others, to bear earth pressure.

7) The section where there was no structure to be demolished was excavated down to GL -10.4 m, leaving the temporary slope. Thus, a large space, 10.4 m deep was available for demolishing work. The retaining wall near the existing building would form a retaining wall integrating the retaining H pile and the existing underground external wall after the inner structures are demolished. However, it was felt possible that loosening and deformation of earth in the back could occur, so it was reinforced by chemical grouting in advance.

8) The giant breaker was lowered to the level of excavation, and the existing buildings were demolished. (See Fig. 2) Since a large space for demolishing operations unhindered by strut supports was obtained, relatively efficient demolishing work was made possible.

9) The first and second stage struts were installed.

10) Excavation work and supporting work of struts were repeated and excavated down to GL -22.2 m. (See Fig. 3)

This demolishing job aimed at increasing efficiency by using a combination of the ground floor preceding method and slope, and creating a large space in the ground. The noise generated from demolishing work was controlled from spreading by the ground floor slab, making the method advantageous in terms of noise pollution.

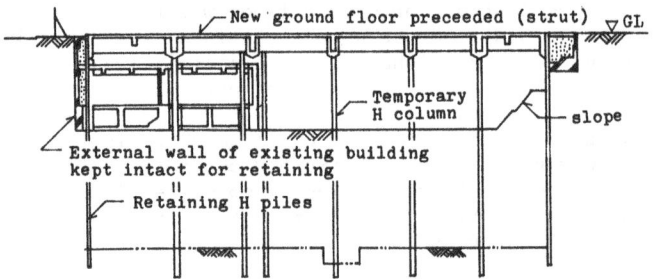

Fig. 2.   Cross-sectional drawing when demolishing the old building

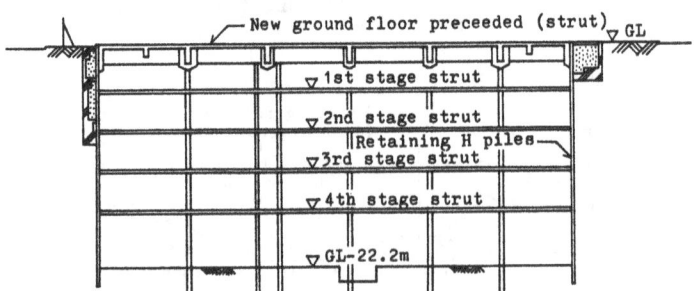

Fig. 3.   Cross-sectional drawing when the excavation was completed

426

3. Practical example of demolishing existing three storey underground
   structure and constructing a new building with three underground
   floors (Y-Building construction work)

This is an example of demolishing underground floors in Ginza, Chūō
Ward, Tokyo. The subject to be demolished was a three storey
basement. The external walls of existing building were close to the
boundary line of the site as in other examples. Since the new
building as to have three storeys underground and approximately the
same depth, the original underground external wall was used as a
retaining wall in its original state and the demolishing work was
executed to enclose the underground floors of the new building in it.
The retaining wall work was performed using concentratedly assembled
large span struts so as to aid the demolishing work by creating a
large space, for efficiency.

3.1 Concept of construction plan of underground floors
   1) As shown in Fig. 4, the existing underground structure was an
irregular three storey structure with some cast-in place concrete
piles. As the existing underground external wall was 250-300 mm from
the boundary line of the site, it was practically impossible to
install the retaining wall outside the existing external wall.
   2) In the planning stage, the designers measured the existing
underground structure, and the measurement drawing was used in
determining the position of the possible new external wall line.
   3) It was decided to utilize the original underground external
wall as the retaining wall for the general sections, and the external
wall line of the new building was to be inside of the columns of the
existing structure.
   4) The depth of the new building was to be shortened by about 80
cm from the bottom of the existing foundation, between which was to
be filled rubble concrete. The load of the new building was to be
transmitted to the ground through the existing foundation.

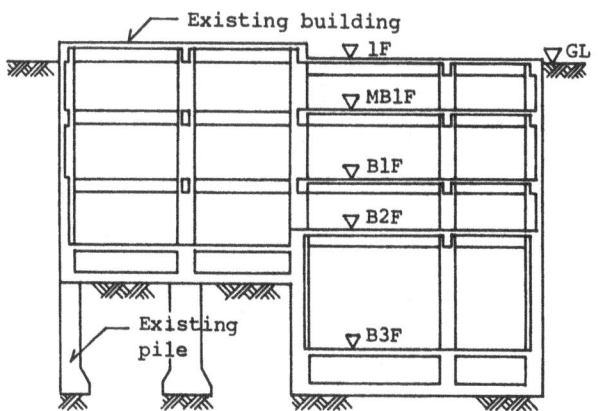

Fig. 4. Existing underground structure

427

5) The cast-in place concrete piles were to be demolished and a new underground structure built. The retaining wall of the portion was to be constructed by retaining H pile with vertical lagging.

6) The inner structure was demolished in sequence, and strut supports were installed on the existing underground external wall. A structural examination of the existing external wall was to be done to confirm safety.

3.2 Procedure of demolishing underground structure

1) Except for the beams, part of the ground floor was demolished. The positions where the retaining H piles and temporary H columns were to be driven were partly demolished by giant breaker and hand breaker, etc.

2) Using the ground floor as a work platform, the retaining H pile and temporary H column were driven using a vibratory pile driver. The embedded section was filled with cement-milk. At the deep portion, the B3 slab was demolished, and then the temporary H columns were installed with foot protection concrete.

3) The rest of ground floor was demolished as far as could be done while still maintaining the stability of the surrounding ground. The first stage strut supports were installed and preloading was introduced. The strut supports used were a combination of concentratedly assembled struts and large horizontal bracing, as shown in Fig. 5, so as to create a large space.

4) The B1 and B2 floors were demolished, the second stage strut supports were installed, and preloading was introduced.

5) Cast-in place concrete piles in the shallow portion and B3 section in the deep portion were demolished, the third stage strut supports were installed and preloading was introduced.

6) The remaining portions of the underground structure were then demolished, and excavation was completed. Fig. 6 shows the condition up to the time when the excavation was completed. Except for the use of the retaining H pile for part of the retaining wall, the existing underground external wall was fully utilized as the retaining wall, on which three stages of strut supports were installed.

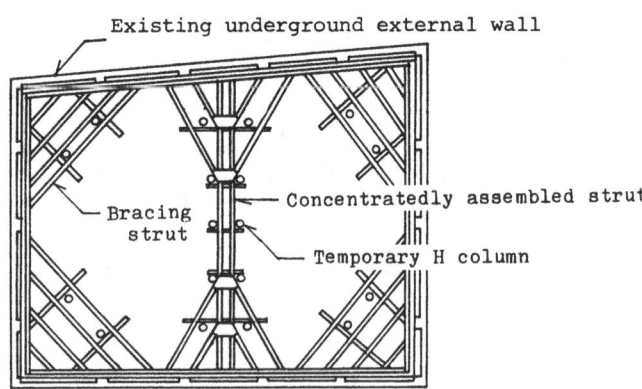

Fig. 5. Retaining by existing external wall and struts

7) Subsequently, the new building was constructed using ordinary procedure. After the foundation concrete was placed, the third stage strut supports were removed. The B2 floor concrete was placed and the second stage strut support were removed. Next, the B1 floor concrete was placed and the first stage strut supports removed. Finally, the ground floor concrete was placed. Thus the construction of the new underground structure was completed. Fig. 7 shows the condition at that time. It can be seen that the new underground structure was built so that it is completely enclosed in the external wall of the existing basement.

Because of the existing external wall was used as the retaining wall, the work caused practically no deformation of and settlement in surrounding ground during both the demolishing work and excavation work. However, when this method of construction is used, the area of the new underground structure is inevitably smaller. Although this may concern the Owner, it is considered unavoidable in Japan where there is so little land and there are many densely populated urban areas.

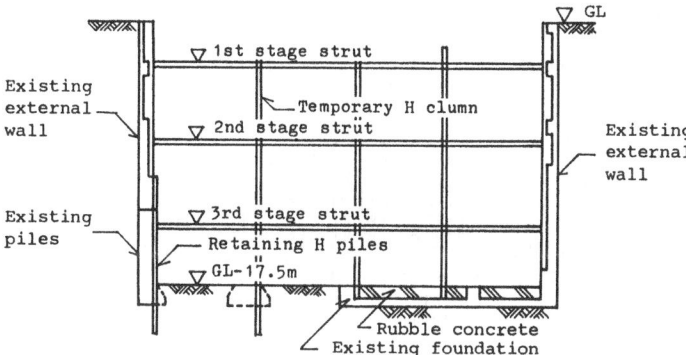

Fig. 6.   Sectional plan when demolishing was completed

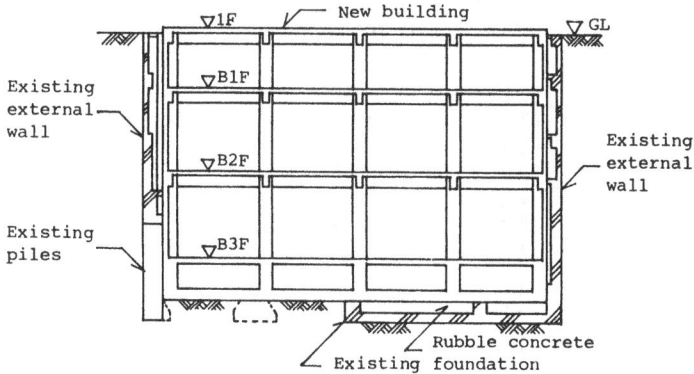

Fig. 7.   New building built enclosed in the existing wall

429

4. Practical example of demolishing existing four storey underground
   structure and constructing a new building with five underground
   floors (H-Building construction work)

This is an example of demolishing an underground structure in
Uchisaiwai-cho, Chiyoda Ward, Tokyo. The new underground structure
was extremely large, being 172 m long, 64 m wide, and 26.5 m deep.
Fig. 8 shows the original condition of the existing underground
structure on the site. The job was executed in three sections, i.e.
center section, west section and east section, and the construction
period was only 29 months from the commencement to the completion of
the building, which has 31 storeys above ground. The demolishing
work and excavation started in the center section where the high-rise
building was to be constructed, and the island method was used in the
west and east sections.

This report describes only the demolishing work and excavation
work of the west section. The west section had an existing four-
level basement structure. The new underground structure has four and
five storeys underground. To give priority to the schedule for this
section, an irregular demolishing method was adopted, in which the
four storey underground structure was to be demolished simultaneously

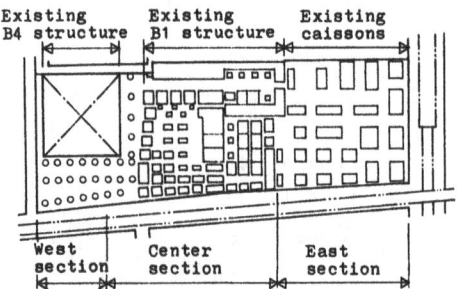

Fig. 8. Job site and existing underground structures

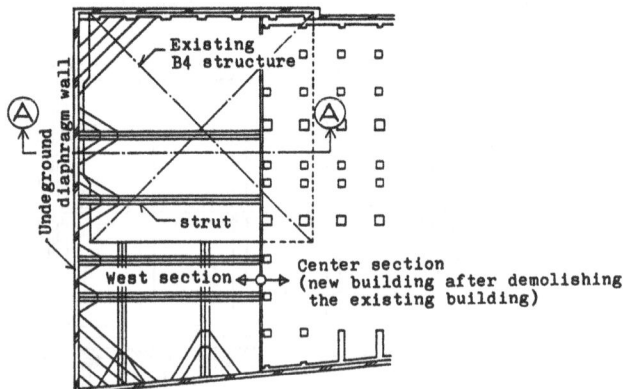

Fig. 9. Retaining plan in the west section

430

from above and below. The retaining wall of the structure was to be constructed using the underground diaphragm wall method and strut supports using the concentratedly assembled strut method and existing underground structure. (See Fig. 9). Demolishing procedure was as follows.

1) Fig. 10 shows the west section of the existing four storey underground structure. In the center section, the underground structure of the new building is shown. An underground diaphragm wall, 80 cm thick, was constructed in the periphery in advance.

2) If the demolishing work for the existing underground structure of the west section started after completing the new underground structure in the center section, the schedule would not be met. Therefore, to commence demolishing the existing underground structure in the west section earlier, the foundation of the center section was used to support the existing underground structure temporarily, the excavation was carried out by scooping out the soil under the existing structure, and the new foundation of five storey underground structure was built. At the upper level, the ground floor of the original building was demolished and the first stage strut supports were installed. Fig. 11 shows the condition at that time.

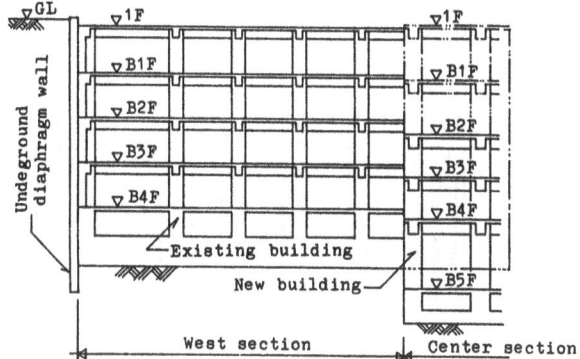

Fig. 10. Sectional design (A)-(A) in Fig. 9

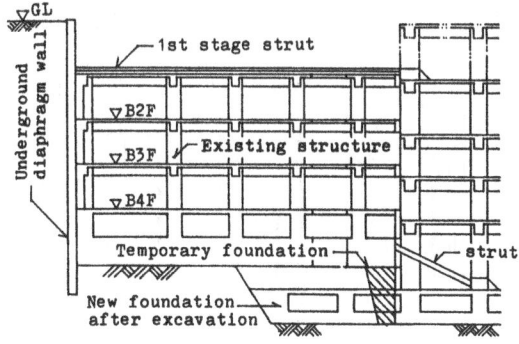

Fig. 11. Excavation under the existing building

431

3) The existing underground structure was supported by temporary H columns erected on the foundation of the new five storey structure and bored H piles. With these conditions, the B4 floor and foundation were demolished. At the upper level at this time, the B1 floor was demolished and the second stage strut supports were installed. Fig. 12 shows the condition at that time.

4) At the lower level, the B4 floor and B4 foundation were built. Concurrently, the B2 floor was demolished and the third stage strut supports were installed. Fig. 13 shows the condition at that time.

5) Then, the B3 floor was demolished, thus completing the demolishing work of the entire existing underground structure. Fig. 14 shows the condition at that time.

6) The procedure for subsequent work was the same as for ordinary construction; construction of the B3 floor, removal of the third stage strut supports, construction of the B2 floor, removal of the second stage strut supports, construction of the B1 floor, removal of the first stage strut supports, and construction of the ground floor.

Although the work proceeded in the above order, this is an example of the successful construction using the bold plan to simultaneously demolish the existing underground structure in soft ground from two

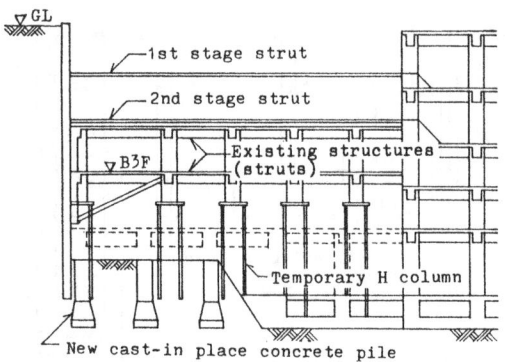

Fig. 12. Existing structures utilized as struts

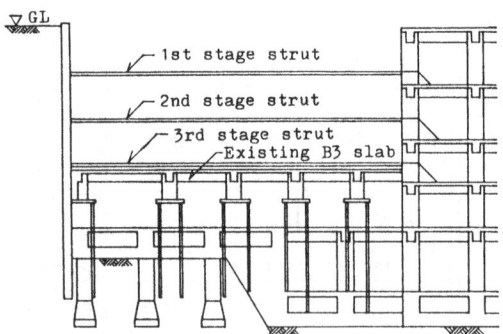

Fig. 13. Demolishing the existing B3 slab

432

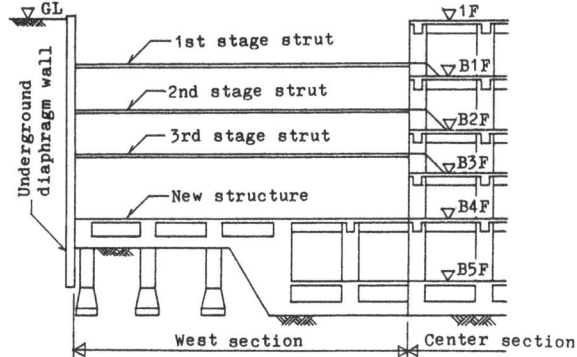

Fig. 14. Sectional drawing when the demolition was completed

directions, i.e. upper and lower. This was achieved through elaborate construction management that emphasized the secure transmission of earth pressure and the balance of force.

5. Conclusion

This report has introduced three examples of demolishing underground structures in Japan, focussing on the procedure of construction. Although the depths of the underground structures differ in these examples, various devices were used in all of them, taking their special design and construction into account. The essentials of demolishing underground structures, as illustrated by these examples, can be summarized as follows:

1) The condition of the existing underground structure must be surveyed in as much detail as possible, and the results must be reflected in the design of the new building.

2) When it is impossible to demolish the existing underground external wall, as it is too close to the borderline of the site, it may be serviceable as a retaining wall.

3) A study should be made to find a method of temporarily using the existing underground structure as the strut supports, to control settlement and deformation of the surrounding ground.

4) To make the demolishing work of the existing underground structure feasible, large span strut supports, large working space, etc., should be aimed for wherever possible.

5) The timing of demolishing of the existing underground structure and installing strut supports for the retaining wall should be reviewed carefully to enable secure and safe work.

References

RC Demolishing Method Committee, (1984), Guidelines for Demolishing Reinforced Concrete Underground Structures, Building Contractors' Society, Japan

BLASTING OF THE UNDERGROUND BEAM BY SLB

K. HASHIZUME and S. ASAKA
Explosives Division, Nippon Kayaku Co., Ltd.

Abstract

It is becoming a serious problem to demolish concrete
structures for the construction of new buildings with the
expansion of urban area. But in the urban area, it is
difficult to use explosives because of the ground vibration and blasting noise. On the other hand, the breakage
by the machinery has also problems such as short of hands
or its efficiency etc. We have developed a low-burning-
rate-explosive, which is called SLB(Safety and Low-sound-
level Breaker) in Japan and now widely applied to demolish
concrete structures. We will introduce SLB and a few
examples of its application. SLB has the burning rate of
40--60 m/sec. Therefore, the ground vibration and blasting noise are controlled to lower level and the flyrocks
can be controlled not to be observed. The powder factors
of SLB are remarkably changed by the blasting design and
the desired degree of the breakage. For an instrance, we
have the following figures to the underground beams.

(1) 240--300 g/m3 : cracks generated between the
                     shotholes.
(2) 300--360 g/m3 : cracks generated in all direction
                     from the shothole.
(3) 420--480 g/m3 : sufficient breakage effect, easily
                     treated by a hand-breaker.

Key Words: Explosive, Demolition, Blasting, Concrete
structure, Vibration, Noise, SLB

1. Introduction

Recently, there are a lot of works energetically under way,
such as breaking and removing concrete structure, boulders,
etc. in the urban area and factories, constructing and
expanding ground and roads. Blasting with high explosives
or works with mechanical rock breakers are accompanied by
great noise and vibration or scattering fragments that
often increase danger and public nuisance. The preventive
measures are one of the important problems to discuss

seriously. On the other hand, such works with public
nuisance must be under the strict application of regulations. Accidents in blasting are most frequently brought
on by scattering fragments.

Under these circumstances, SLB can be applied to reduce
ground vibration, blast noise and flyrocks. We can apply
SLB to demolish the following concrete structures
and rocks. Demolition of building structures, foundations,
beams, walls, columns, etc. Demolition of bridges and
piers. Demolition of equipment foundations. Demolition
of breakwaters and so on.

Among these applications, we will introduce, here, a
few examples which were applied to demolish reinforced
concrete structures.

2. Concrete breaker SLB

2.1 Properties of SLB
It is a commonly known fact that high explosives detonate
at a rate of 4,000--6,000 m/sec. to break and scatter
concrete or rocks by the shock wave. SLB, unlike the
conventional explosives and blasting powders, is mainly
composed of gas generating agents which are not dangerous
materials. SLB burns at a rate of 40--60 m/sec. under the
confined conditions, generating high-temperature gas. SLB
generates cracks around the shotholes and enables soft
fragmentation.

Properties of SLB are shown in Table 1. compared with
KIRI Dynamite.

Table 1. Comparison table of properties

Item	S L B	KIRI Dynamite
Specific gravity	0.8--0.9	1.3--1.4
Gas volume (cm3/g)	130--140	980--1,000
Heat of combustion (kcal/g)	0.7	0.95--1.0
Rate of burning		
open(cm/sec)	2--5	--
confined(m/sec)	40--60	5,500--6,000
5kg Drop hammer test(cm)	50--60	15--30
Stability test (70 C, 72hr)	no change	drop of stability

2.2 Construction and types of SLB
SLB consists of a cartridge and an igniting cap. It needs
one igniting cap for one cartridge to make SLB burn
perfectly. Fig. 1 shows the construction of SLB and
Table. 2 shows the types and sizes of the cartridge of SLB
and the igniting cap.

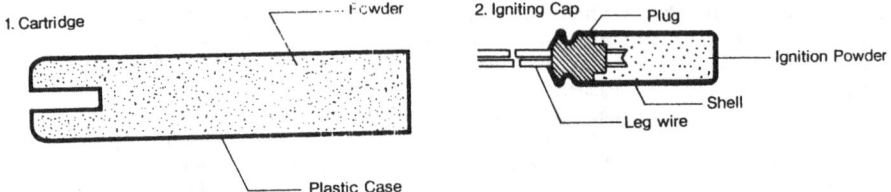

Fig. 1. Construction of SLB

Table2. Sizes of SLB
(1)Cartridge

Type	Quantity of Powder(g)	Outer Diameter(mm)	Length(mm)
P- 3	30	28	76
P- 6	60	28	134
P-18(B)	180	50	125
P-18(S)	180	30	338

(2)Igniting cap

Outer diameter (mm)	6.5
length of shell (mm)	25
length of legwire(m)	1.8
	3.0[P-18(B)]

2-3 Standard values of SLB consumption
SLB are changed in shotholes which are drilled to the
materials to be broken. The shotholes are tamped with
sands, quick cement, a rubber plug or special tamping
material "KAYATAMPER". Table 3. shows the burden and
the spacing of holes in the standard blasting design.

Table 3. Standard blasting design

Cartridge	Burden		Spacing of holes (cm)
	Plain concrete (cm)	Reinforced concrete (cm)	
SLB P-3	30--50	30--40	30--40
P-6	50--60	40--60	50--70
P-18	80--90	70--80	60--90

Table 4. shows the standard values of SLB consumption.
These values are shown for the powder factor and only for
a standard value. The powder factor varies not only from
the types of materials and structures but also from the
desired breakage.

Table 4. Standard values of SLB consumption

Materials	Powder factor g/m3	Remarks
Plain concrete	120--180	
Reinforced concrete	240--480	
Soft rock	120--240	2 free faces
Mediate rock	150--300	2 free faces
Hard rock	240--360	2 free faces
Most hard rock	300--420	2 free faces

3. Blasting of the underground beams

Demolition of the underground beam by SLB is described
here. The beams were demolished in connection with the
construction of new building. Concrete breaker, SLB, was
applied to demolish the beams for the following reasons.
It was hard to demolish beams by heavy machinery because
of the small area, and also it was the reason to have to
shorten the working period.

3.1 Environment
Working site was 5m from the pavement and 12 m from the
adjacent living building.

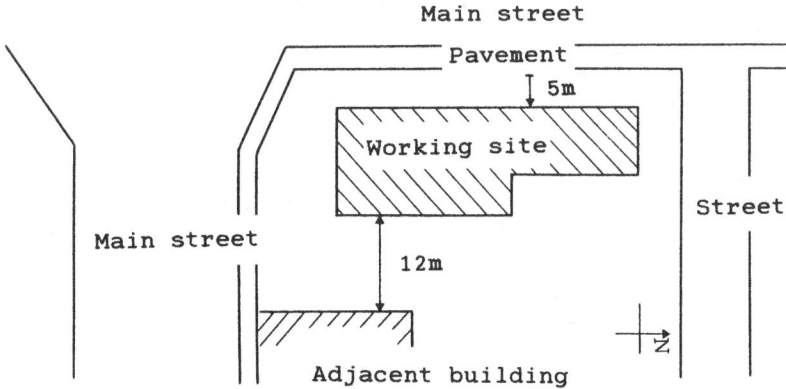

Fig. 2. Plan of working site

3.2 Test blasting

The numbers of SLB which can be ignited at one time depend on the kind of building structure, or space to the adjacent building. Required powder factor needed to demolish the underground beam depends on not only on the degree of reinforcement, thickness of the beam and the lateral soil but also on the degree of breakage etc.

The purpose of test blasting are the followings.

(1) To design the adequate blasting pattern: hole depth, burden, spacing and charge quantity.
(2) To obtain vibration data and to make charge quantity control chart.
(3) To select an adequate method to prevent flyrocks and blast sound.

After the test blasting, it was found that the following points were to be considered.

(1) The charge should be concentrated at the bottom of the hole because also the bottom concrete plate might expands.
(2) It is necessary to break out the beams sufficiently because the working area is small and heavy machineries can not be moved to the working area.
(3) After the measurement of the ground vibration, Fig. 5. was obtained.
(4) Protective covers were necessary to keep flyrocks off from the beams.

3.3 Blasting design

There were three types of the underground beam which were called Type A, B and C here. Considering the above four points, blasting designs were made according to a plan for each type of underground beam. The powder factor of SLB is 360--420 g/m3.

Generally speaking, blasting design for demolition by SLB is determined refering to the standard values of SLB consumption. First, the volume of the material to be broken is calculated, then depth of holes, spacing of holes and charge quantities are determined. The standard values of SLB consumption are shown in Table 4, for example, 120--180g/m3 for plain concrete and 240--480 g/m3 for reinforced concrete. The larger the concrete structure is, and the more the reinforcement is, the more the SLB consumption increases. The standard values of SLB consumption depends on not only on the sizes of concrete and the numbers of reinforcement but also on the desired degree of the breakage. The powder factor for the underground beams is obtained as follows:

(1) less than 180 g/m3 :   not sufficient breakage with
                           the blind shot or blown-out shot

(2) 240--300 g/m3 : cracks mainly generated between
                     the shotholes. assisted the
                     heavy machinery as an "Iron" etc.
(3) 300--360 g/m3 : cracks generated in all direction.
                     easily treated by a heavy machinery.
(4) 420--480 g/m3 : sufficient breakage effect, easily
                     treated by a hand-breaker.

Type A and B of the underground beams are designed to
treat the waste concrete by hand-breakers, and Type C is
by a machinery. Fig. 3. shows the blasting design of
Type A, B and C.

unit : cm

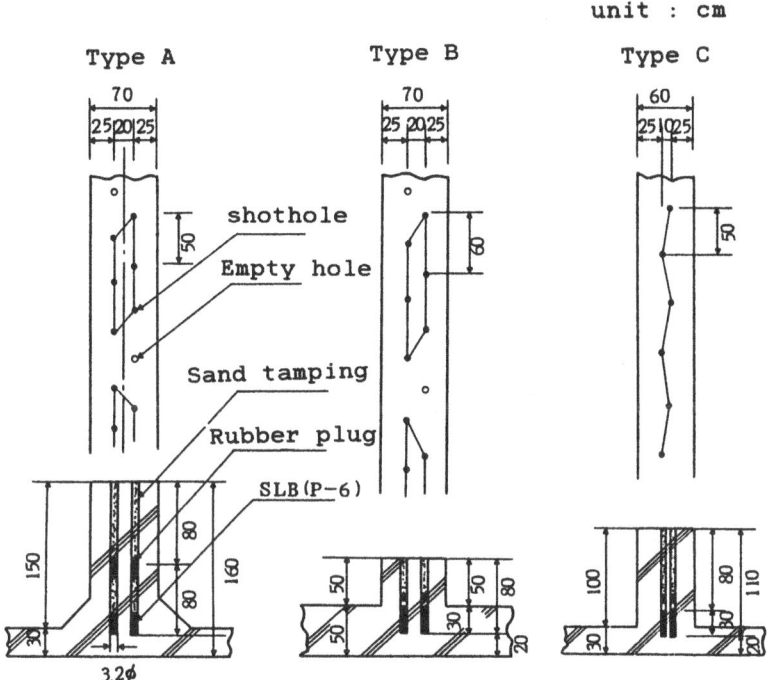

Fig. 3 Blasting design

3.4 Prevention of flyrock
It is very important to prevent flyrock, eapecially when a
adjacent building or a pave-ment is near the blasting site.
This serves as well as reducing the blast noise.
    As the result, flyrock is not observed but the blasting
sheet jumped a little at the shooting instance.

Table 4. Protective cover

Material		Size (m×m)	Weight (kg/piece)
Blasting mat	plastic	1.5×1.5	42
Blsting sheet	Nylon	4 × 6	15
Light blasting mat		1.8×3.6	25
Old Japanese mat		0.9×1.8	25

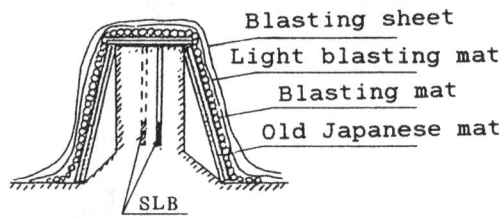

Blasting sheet
Light blasting mat
Blasting mat
Old Japanese mat

SLB

Fig. 4   Schema of Prevention

3.5 Ground vibration and blast noise
Vibration level meters (RION, Model VM-14E) and precision
sound level meters (RION, Model NA-60) were set on the
4th floor of the adjacent building to measure the blasting
vibration and blast noise. It was found that the blasting
vibraiton was shown by next expression.;
   V=0.26 W D
   where V (cm/sec) : particle velocity
         W (g) : weight of SLB
         D (m) : distance
Limited control value was determined to be less than
0.05 cm/sec. Fig. 5 shows the relationship between the
charge weight and the distance when the particle velocity
is 0.05 cm/sec.

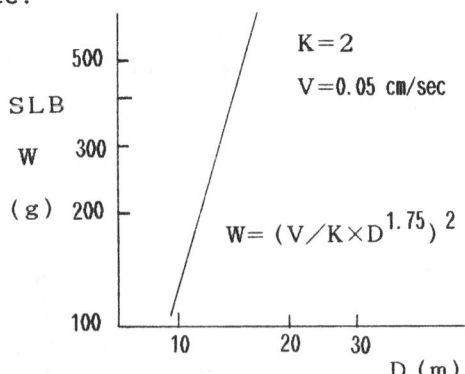

$K=2$

$V=0.05$ cm/sec

$W= (V/K×D^{1.75})^2$

Fig. 5.   Relationship between the charge weight and
          the distance

440

Blast noise was 67--82 dB(A), while traffic noise was 72--74 dB(A) and noise from the machinery was 67--82 dB(A).

4. Result of blasting by SLB

Table 5. shows the result of the breakage during the working period.

Table 5. Result of the working

	1st period	2nd period	Total
Volume of the underground beam (m3)	70	107	177
Number of holes	217	300	517
Quantity of SLB P-6 used (pieces)	518	673	1,191
Working day (days)	4.5	5	10
Powder factor (kg/m3)	444	377	404
Quantity per hole (g/hole)	120--180	120--180	120--180

For the 1st period, after the soil around the under-- ground beam was removed, the beams were crushed. The beams were crushed about 16m3 in a day and the powder factor was 444 g/m3. The degree of breakage was suffi- cient enough to be able to treat by a hand-breaker. For the 2nd period, the beams were crushed without removing the soil around the underground beam. The beams were crushed about 20m3 in a day and the powder factor was 377 g/m3. The degree of breakage was sufficient to assist the machinery.

Flyrock was not observed. The ground vibration was less than 0.05 cm/sec and the blast noise was controlled less than 82 dB(A).

5. Conclusion

Even in the urban area, blasting of the underground beam by SLB could be finished without any accidents. In addition to the adequate prevention of flyrock, to make the most careful plan and preparation are very important.

THE FEASABILITY STUDY AND CONTINUOUS CONTROL
OF THE DEMOLITION OF A LOCK WALL OF THE ZANDVLIET LOCK

F. HENDERIECKX
Belgian Building Research Institute

R. BOURGOIS
Royal Military School, Belgium

F. MORTELMANS
Catholic University of Leuven, Belgium

1. Introduction

Last year, the biggest lock in the world situated in Berendrecht, in the north of Antwerp, was imposingly set under water (dimensions : length 500 m, width 68 m and depth -13,5 m).
Although the greatest part of the civil works were achieved, one of the most sensational phases had still to be realized and started several months ago, i.e. the removal of one of the walls of the Zandvliet lock by demolition with explosives. The whole project consists in exploding 80.000 m^3 of reinforced concrete on a safety manner and recycling the rubble.

After many discussions, the contractor proposed a demolition technique by explosives, which was accepted by the client under the following conditions :
- the shipping to the Zandvliet lock should be disturbed as little as possible and remain constantly possible
- the demolition should avoid any damage to the industrial environment.

Therefore, an elaborated program of research and practical experments had preceded the final decision for the use of explosives. This program was ordered by the Building Research Institute of Belgium (W.T.C.B.) in collaboration with the Royal Military Academy (K.M.S.) and the building department of the Catholic University of Louvain (K.U.L.-R&D). The bridge office-section of Liège (M.O.W.) also participated in this realization.

KEYWORDS :
Feasability study - Control - Demolition - Blasting operation - Explosives - Lock wall - Quay wall - Method of measurement - Risk analysis - Explosion - Blasting vibration - Environmental protection.

2. General description of the project

When the decision of the Belgian government was made to build up the Berendrecht lock near the Zandvliet one for economic reasons, this enormous project was undertaken (in 1983).

Figure one gives a survey of the different parts :
- the Berendrecht lock
- the access to the river "De Schelde"
- the demolition of the south quay wall of the Zandvliet lock
- the construction of the pier
- three steel lift bridges
- the lock doors and the mechanical equipment
- the dredging operations.

The quai wall that should be demolished is composed of two parts :
- a high founded quay wall of concrete, 10 m high and 2,5 m thick.
  This construction with a volume of 9600 m^3 is built on metallic caisson piles with a length of 20,75 m and a weight of 32 tons.
  Special concrete beams (height 5 m, width 1 m) were used as strengthening between the two walls (see fig. 2)
- a deep founded quay wall with a total length of 484 m composed of 22 parts of 22 m. The total height reaches 25 m and the foundations has a width of 18,5 m. The total volume of each part is 3100 m^3 reinforced concrete (see fig. 3)
  An explosive charge of 4500 kg distributed into 200 holes is used for the demolition.

In the first part of the research program, the most effective time interval between two initiations was fixed at 50 milliseconds. It means that one row of holes situated in one cross section detonates simultaneously; the following row explodes 50 ms later.

The public works, buildings or other elements are :
1) the doors of the existing Zandvliet lock
2) the duct under the same lock containing industrial pipes
3) the radar building for the shipping of the river
4) the industrial plant of BASF
5) the refining company BRC
6) the nuclear power plant of Doel
7) the power pylon of EBES
8) the navigation on the river.

In common consultation with DOLSO and the industrial companies, the different measuring positions were defined for the installation of the transducers. According to the necessity some positions were changed or new ones were chosen.

A good radio-communication network was used between the central point (for ignition of the explosives) and the different measuring points situated at distances of 500 m to 3 km.

3. The effect of the demolition technique on the environment

Although the demolition technique by explosives have been used for several years in Belgium, it was necessary to make an evaluation of the risks consequent upon the use of such a big quantity of explosives.

The contractor had indeed planned the demolition of the lock by 22 m long pieces with a volume of 3100 m^3 reinforced concrete, using for the detonation of each piece some 4500 kg of explosives distributed in holes and following a well-defined shoot plan. A literature study did not give any information about such an application and the consequences for the industrial environment. Moreover there exists no specific standard in Belgium which could give innformation on the acceptable effects on people or constructions caused by the shocks and vibrations due to any demolition work. There-fore, references are often made to the German standard DIN 4150 or to the French standard AFTES which gives well-defined values for acceptable vibration levels expressed in terms of velocity or acceleration, taking into account the type of the building and the age of it.

In table I we give some values for maximal accepted velocity vibrations measured on the foundations of a construction according to well-defined rules, when a source of vibration is working in the vicinity, for example when building activities such as soil compacting, sheet piling and blasting are going on.

DIN 4150 Project 1987 – Limits for velocity mm/s			
Type of Construction	Foundation		Ceiling of highest level
	10 Hz	10–50 Hz	50–100 Hz	all frequencies
Industrial	20	20–40	40–50	40
Residential	5	5–15	15–20	15
Specially sensitive buildings	3	3–8	8–10	8

More specific information can be found in the literature. Concerning the human reactions to vibrations we can say that man, who feels some vibrations coming from sources which are difficult to identify, for instance blasting, registers quite disturbing phenomena, especially if there has been no warning beforehand. On the other hand as human beings are much more sensitive to vibrations, the limits of the maximum acceptable levels are much lower than the limits imposed for the damage to buildings.

In the specific case of the environment of the Berendrecht lock, the scientific committee has taken after several meetings the following decisions :
a) a well-defined research programme will be executed on the site allowing to come to a definite conclusion about the feasibility of the demolition by explosives;
b) the different industrial companies have defined their most sensitive constructions and have indicated the maximal acceptable levels for these buildings.

	Power Pylon	Turbine	Commercial Building		Nuclear Power Plant	Radar tower
			Floor	Top		
Value	2 m/s^2	2 m/s^2	$2,5\text{m/s}^2$	30mm/s	$0,2 \text{ m/s}^2$	no

c) before the explosion, some buildings are to be evacuated : the radar tower and the BASF commercial building;
d) the demolition has to take place when few administrative people are present; it means that the explosion has to take place either on saturdays, either in the evening or in the night;
e) a good radio-communication must be established between every control point and the shoot master with the rader tower as well.

4. Measuring techniques for shocks and vibrations

The explosion of the reinforced deep founded concrete quay wall produces transient vibrations which propagate in the surroundings and make the construction vibrate.
There are a great number of terms used to describe the vibration phenomena. The characterization can be expressed in the field of time or in the field of frequency. The quantity to be measured is the acceleration, the velocity or the displacement; the acceptable limits for the damage to buildings are expressed in this same way. In the last few years, great progress were made in the analysis of dynamic phenomena such as shocks and transient vibrations. High speed analog to digital converters allow us to use computers for the transformation of time signals into frequency components. Signals can also be treated, filtered and recomposed in the field of time.

445

This fast Fourier transformation and Inverse Fourier transformation connected to a computer and a plotter give us all information required.

In practice, two techniques for the measurements of the shock phenomena and the structure were used.

The signal is detected by the transducer usually an accelerometer, sometimes a strain-gauge or hydrophone (for the dynamic pressure in water). This signal is conditioned by a charge amplifier, provided with a low frequency band pass-filter. This signal is fed to an analog frequency modulated tape-recorder.

It was possible to make a direct visualization using modern digital ascilloscopes with a memory or fast Fourier analyzers. Our conclusion after some practical measurements was that the phenomena are best stored on a magnetic tape-recorder and later on analysed in laboratory. The signals have often to be processed to eliminate non-relevant information such as DC components or noise. The Fourier analyzers we used, allowed us to obtain information in the fields of time and frequency and to receive when using integration techniques the velocity or displacement signals. The block diagram gives a survey of the two registration techniques.

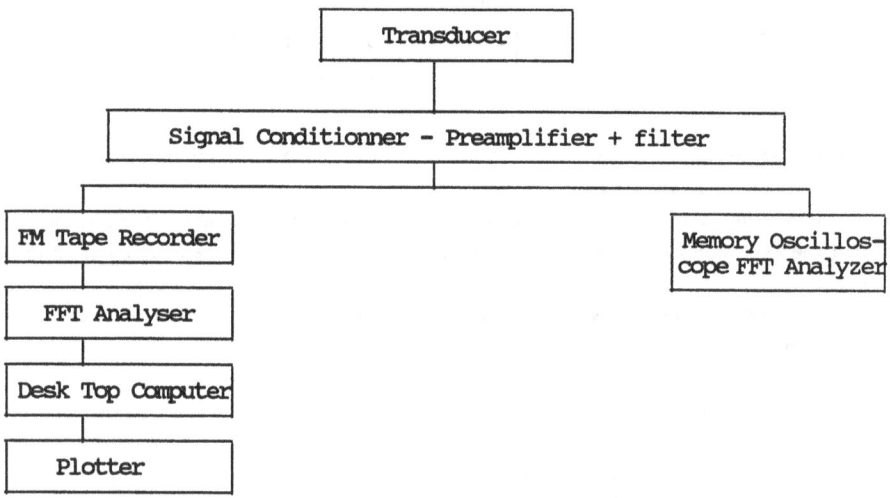

5. Short description of the feasibility study of the demolition project

As we had to overcome many practical problems, we developed a research program in collaboration with the contractor DOLSO and the different laboratories. This program consisted in the following items :

a) calibration and choice of the adequate measuring equipment using low charges to explode in the soil and in water;

b) explosion of free charges of 27-55-110 and 220 kg at two depths in the soil (12 and 22 m);

c) explosion of free charges in water with growing weight of 1, 2,5, 8, 10 and 15 kg;

d) explosion of a charge of 8 kg in a concrete block to simulate the real situation.

These experiments gave us a better understanding of the vibration phenomena. On the one hand, it was clear from the literature that a great number of parameters could have some influence such as the composition of the ground layers, the homogeneity of the soil, the degree of the water saturation, the river and the reflections on the quay wall, the distance to the buildings and so on. On the other hand, different wave forms are excited, each of them propagating with different wave velocities. It was also comprehensive that the first images on the FFT screen were very complex and that some special analysis techniques had to be applied to obtain signals which were ready to be used.

The first study brought a good information on the vibration phenomena such as the amplitude and frequency content and the duration of the signals.

At the same time the analysis methods could be optimized and a first tentative was made to find out if the formulas given in the literature were appropriate.

During this period we resolved a number of practical problems of coordination between the different measuring teams.

6. Enlargement of the research project to the demolition plan

After a long examination of the obtained results with the contractor, the industrial companies and the research team, we adapted the experimental project.

As the demolition plan foresaw the demolition of the quay wall of 4500 m^3 concrete in one time, the total charges of explosives needed was very high. Therefore we proposed to divide the program into the following steps :

- study of the vibration phenomena in the different buildings when free charges of growing weight from 25 to 250 kg explode in water;
- determination of the relationship between a free charge explosion in water and the same charge (of 8 kg) explosion in a concrete block;
- optimization of the time delay detonators to obtain a minimum vibration level at the measuring points situated in the critical buildings;
- determination of the relationship between a series of charges exploding in deep water and in a concrete block;
- study of the reduction influence of an air bubble screen surrounding the explosion point on the vibration level in the river to the sluice doors.

After the realization of this research program and the analysis of all the results, we obtained a lot of practical informations which helped us to make the final decision to agree the definite demolition plan from the contractor.

The main conclusions were the following :
- the air bubble screen has a very high efficiency between 90 to 98 % on the reduction of the amplitude vibrations in the water near the sluice door. The effect on the vibration level measured in the radar tower or the BASF buildings is negligible. This means that the vibration energy is preferably transmitted through the soil layers;
- the best time delay for the explosion of the different charges is 50 ms, as can be deduced from the results given in the following table.

Relation 10 x 8 kg / 1 x 8 kg	Effect of the time delay Charges exploded in the concrete block				
Time delay	Power Pylon EBES	BASF		Radar tower	Nuclear Power Plant (Doel)
		Turbine	Building		
12,5 ms	2,06	1,36	1,5	2	1
25 ms	2,03	1,19	1,5	2,18	1,4
50 ms	1,2	0,71	(< 1,5)	0,94	-

- The reduction coefficient for an explosion of the same charge of 25 kg in a concrete block or free in water is situated between 0,4 and 0,6 as a function of the different measuring points. The results are represented in the table below.

Charge in concrete / Charge in water	Reduction coefficient for the charges				
	Power Pylon EBES	BASF		Radar tower	Nuclear Power Plant (Doel)
		Turbine	Building		
8 kg	0,33	0,5	0,27	-	0,63
25 kg	0,51	0,4	0,5	0,45	0,6

- the following table gives the results of the explosion free in water of different charges at a distance of 800 m from the radar tower.
  This building is one of the most critical points because it is situated in the linear direction of the quay wall which will be demolished and at a distance of about 350 m to the nearest point.

Relationship between charge and maximum velocity	
Charge (kg)	Velocity (mm/s)
80	< 10
116	< 10
136	22
158	24
176	19
200	26
218	28
248	24
LIMIT	30

7. Results of the real domolition of the quay wall

The analysis of all the results obtained during the experimental study has allowed to conclude that the proposed demolition plan from the contractor Herbosch Kiere should be applied for a final test by demolishing the first part of the quay wall. Some 3.200 m^3 of concrete should be broken in debris smaller than 1 m^3 by a total charge of 4.557 kg. Each explosion of 217 kg occurs with a time delay of 50 ms; the total duration of the explosion comes approximately to one second.

The first explosion was executed on 18 August 1987 at 10 pm. The results showed that no limit in the building measuring points was reached and that the definite demolition program could start at a rythm of one explosion every fourtien days from 1 September 1987 (see figures 4 to 7).

8. Conclusions

- To optimize the blasting work, it is necessary to make some kind of risk analysis in order to find out, first, what sized vibrations the environment will accept and secondly what is the maximum charge which can be blasted at a certain distance without exceeding the vibration limits. Particular attention must be paid to equipments which are sensitive to vibrations such as electronic equipment, computers,...
- It is necessary to establish the relation between the vibration level, the distance and the weight of the charge. Some formulas from the literature could be used but a test program with small charges has to be achieved to determine the constants in these formulas. Once the total charge is known, the time delay interval between two explosions must be optimized.
- When important demolition works take place in some critical environments, a study is required to find out the parameters necessary for the demolition plan.

449

- The collaboration between the different research teams, the contractor and the client has always been quite successful. Regular contacts especially in the preparation period were therefore necessary.
- It seems very important to inform all people living in the surroundings about the explosions, so that they do not worry when feeling the vibrations which are not at all dangerous for the structure of the building.

9. References

(1) Vibrations generated by traffic and building construction activities
   by Roger HOLMBERG
   Swedish Council for Building Research, 1984.
(2) Vibrations of Soil and Foundations
   by RICHARD, HALL and WOODS
(3) Blasting vibration and structural damage
   by Dr. G.S. SEN, 1981
(4) A manual for the production of blast and fragment loading on structure.
   by U.S. Department of Energy, 1980.
(5) Erschütterungen im Bauwesen.
   DIN 4150, 1983, Deutsche Normen
(6) Effects of vibrations on man.
   ISO 2631.
(7) AFTES : Normes françaises concernant les vibrations.

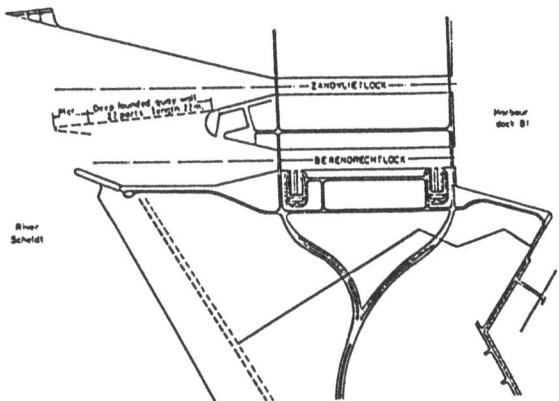

Fig. 1 - Survey of the Zandvliet and Berendrecht locks

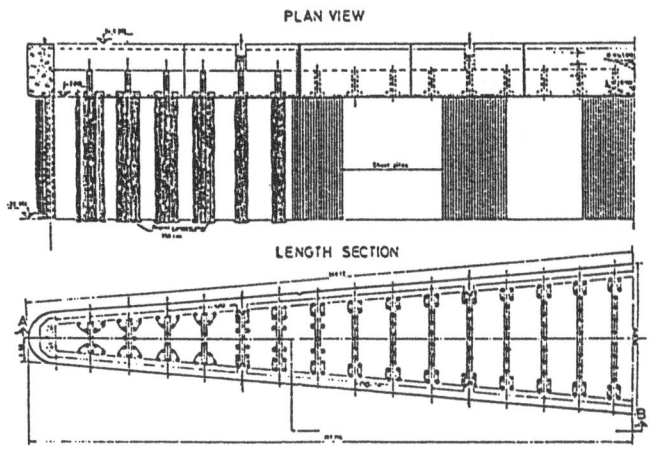

Fig. 2 - Zandvlietlock Pier

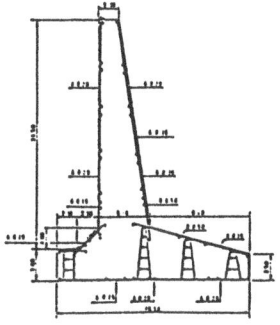

Fig. 3 - Zandvlietlock Quay-wall

Fig. 4 - Start of the demolition

Fig. 5 - Explosion after 0,75 s.

Fig. 6 - Projection of concrete

Fig. 7 - Evacuation of gases
after the explosion

APPLICATON OF NON-EXPLOSIVE DEMOLITION AGENT FOR REMOVING AN UPPER
PORTION OF A CAST-IN-PLACE CONCRETE PILE

M. SATO, S. YAMADA, H. KONDO and I. MATSUI
PRODUCTS RESEARCH LABO., SUMITOMO CEMENT CO., LTD.

Abstract
Application of the non-explosive demolition agent to a cast-in-place
concrete pile to remove its top has eliminated the environmental
pollution and industrial problems such as noise, vibration and
dusting, which were associated with the conventional chipping with a
breaker or the like.
 In the PPC method described hereunder, where the concrete is cast
after installing a tube filled with CAB, a special non-explosive
demolition agent, to the steel reinforcement cage of the pile top,
CAB develops its expansive pressure to demolish the surplus concrete
head while the concrete pile is being cured. This method has an
advantage of easy removal of pile head in a short period.
 PPC method is grouped into a slurry system and a briquette system.
The briquette system is further divided into a bulk type and a
capsule type.
 CAB is characterized by the development of expansive pressure that
matches the hardening rate of concrete. For the purpose of the
slurry method, CAB is designed to have a proper slurry fluidity
before charging and hardening property after charging. Since the
briquette produces a larger expansive pressure than the slurry system
and the field workability is good, this system is expected to become
more widely adopted.
Key Word : Non-explosive demolition agent, Cast-in-place concrete
           pile, Pile head removal

1. Introduction

The head of a cast-in-place concrete pile is brittle because the soil
excavated during drilling and the bentonite muddy water used for
stabilizing the hole wall are included therein. For this reason, a
surplus concrete head is provided above the required height for the
sake of safety. The surplus concrete head is traditionally removed
by using a breaker, causing environmental pollution like noise,
vibration and dusting as well as other industrial damages. The use
of non-explosive demolition agent has been solving these problems.
Especially, in case of the PPC method, which was developed by the
authers et al, a tube filled with CAB, a special non-explosive
demolition agent, is installed to the steel reinforcement cage at the

boundary of surplus concrete head and pile body before concrete casting. Therefore, the expansive pressure of CAB develops at the time of concrete hardening, and the surplus concrete is demolished before the pile head is removed. The method has a remarkable advantage, for the surplus concrete head can be demolished and removed easily in a short time.

The PPC method owes its superiority mostly to the characteristics of CAB. This report describes the outline of PPC method and the characteristics of special non-explosive demolition agent, CAB, centering in the development of expansive pressure.

2. Outline of PPC method

The work procedure to remove the pile head by PPC method is illustrated in a flow chart of Fig. 1.

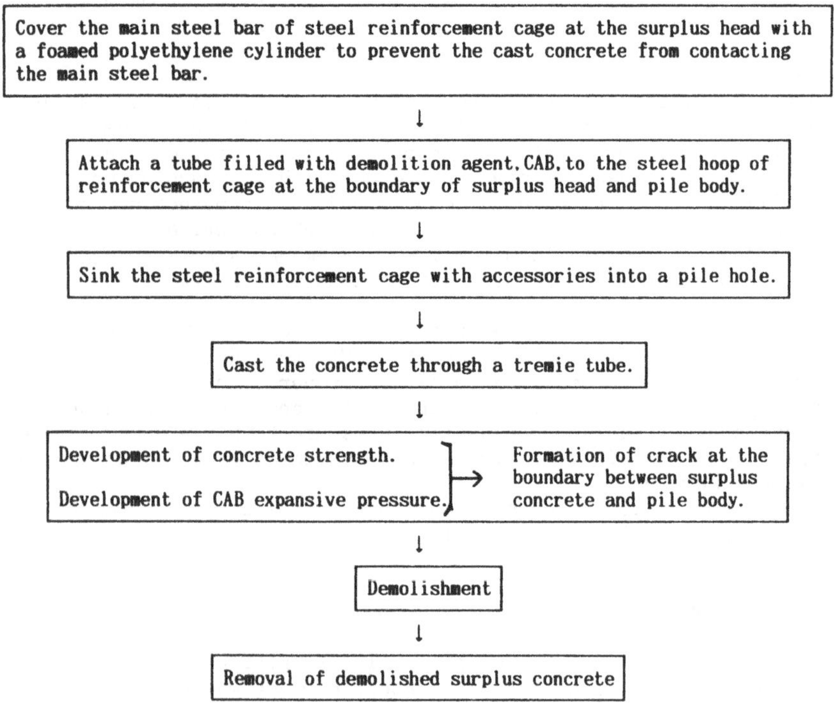

Cover the main steel bar of steel reinforcement cage at the surplus head with a foamed polyethylene cylinder to prevent the cast concrete from contacting the main steel bar.

↓

Attach a tube filled with demolition agent, CAB, to the steel hoop of reinforcement cage at the boundary of surplus head and pile body.

↓

Sink the steel reinforcement cage with accessories into a pile hole.

↓

Cast the concrete through a tremie tube.

↓

Development of concrete strength.

Development of CAB expansive pressure. → Formation of crack at the boundary between surplus concrete and pile body.

↓

Demolishment

↓

Removal of demolished surplus concrete

Fig. 1 Work procedure to remove the pile head by PPC method

The PPC method is classified as shown in Fig. 2, according to the type, filling method and handling of CAB.

The slurry method is a method where the CAB slurry, prepared by mixing the powder of demolition agent CAB with water at a water/CAB ratio of 30%, is filled in a pipe attached in advance to the steel hoop of reinforcement cage.

On the other hand, the briquette system consists of the bulk type and the capsule type. The bulk type is a method where a pipe filled with the granule of CAB at the job site, followed by filling water, and sealed, is attached to the steel reinforcement cage, or alternatively, a pipe attached to the steel hoop of reinforcement cage in advance is filled with CAB and water.

The capsule type is a method where a capsule made by a slightly winding and flattening spiral sheath pipe of several tens centimeter length, is filled with the granule of CAB and by sealing the both ends with nets, is attached to the steel reinforcement cage. In this case, the expansion of CAB takes place by absorbing the water in the muddy water after the steel reinforcement cage is sunk in the pile hole.

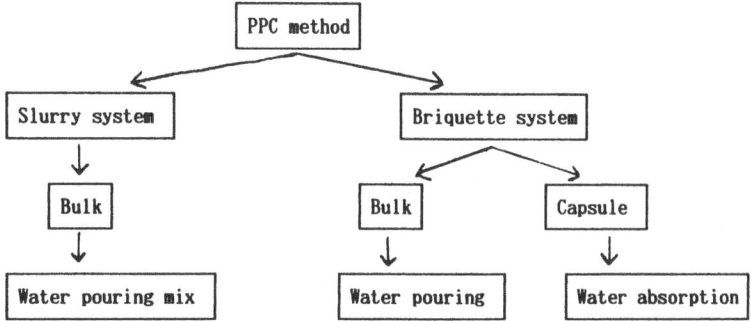

Fig. 2 Classification of PPC method

3. Characteristics of demolition agent, CAB

CAB basically consists of hard calcined lime as an element to generate the expansion. It also contains inorganic and organic materials such as calcium fluoroaluminate and cane sugar for the purposes of retarding CaO hydration reaction, and providing fluidity and hardening property of slurry.

As an indication of hydration reactivity of CAB, Fig. 3 shows a chronological development of heat of hydration measured by conduction calorimeter, in comparison with an ordinary demolition agent. The heat of hydration mostly consists of the heat of hydration of CaO. Fig. 3 shows that the heat of hydration generation of CaO in CAB is remarkably retarded in comparison with the ordinary demolition agent. Accordingly, in case of ordinary demolition agent, the expansive pressure due to CaO hydration reaction develops before the concrete hardens enough, and the control of crack formation becomes difficult. However, CAB has an advantage that the control of crack formation can be done effectively because CaO hydration expansion takes place simultaneously with the hardening of concrete.

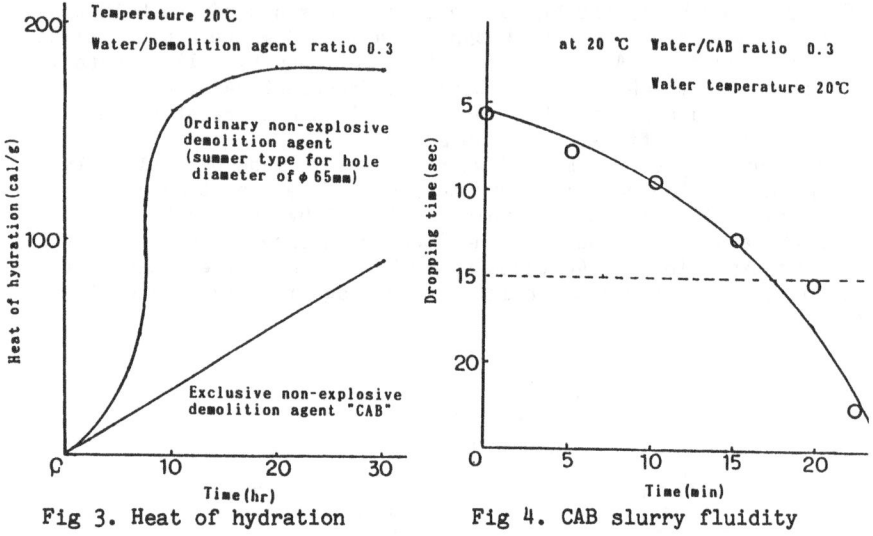

Fig 3. Heat of hydration   Fig 4. CAB slurry fluidity

Fig. 4 shows the chronological change of CAB slurry fluidity obtained
by J-funnel dropping time test according to the specified test of
Japan Highway Public Corporation. As an indication of smoothness of
slurry pouring workability, J-funnel dropping time needs to be less
than 15 seconds. Fig. 4 tells that a pouring work time of about 20
minutes can be assured if kneaded with 20°C water. CAB slurry starts
setting and hardening after its fluidity is lost.

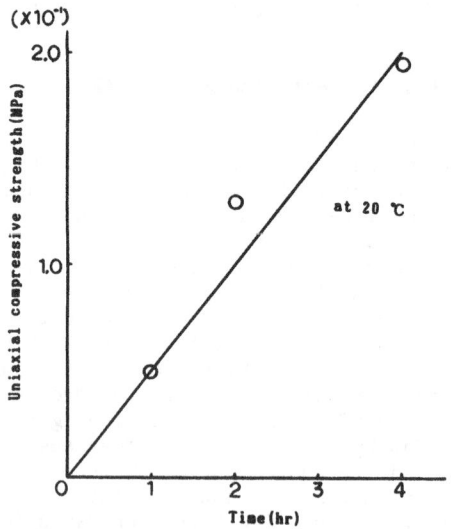

Fig. 5 Strength after hardening of CAB slurry

Fig. 5 refers to the examination of strength after hardening of CAB slurry, showing the uniaxial compressive strength of a ⌀50 x 100mm cylindrical specimen measured with a strain control type strength tester. Fig. 5 illustrates that CAB slurry reaches the uniaxial compressive strength of about 0.05MPa in one hour, and that the strength continues increasing. In other words, in one hour after pouring into a pipe, CAB slurry hardens to form a combined body with the pipe. Thus, if a steel reinforcement cage attached with a CAB filled pipe is sunk in mud water of pile hole, CAB does not leak into the mud water. If the tremie pipe hits the pipe during concrete casting, the pipe can stand well against the breakage.

Fig. 6 shows the chronological change of expansive pressure in case of CAB slurry and 1-5 mm dry briquette of CAB poured with water. The temperature was set at 35°C based on the experiential temperature measurement at the surplus head of cast-in-place concrete pile. The expansion pressure was measured by the revised method of the Association of Non-Explosive Demolition Agent. From Fig. 6, we can tell that both CAB slurry and briquette develop an expansive pressure of over 30 MPa, enough to break brittle bodies such as rock and concrete, after 48 hours.

In comparison between slurry and briquette, the briquette develops larger expansive pressure, supposedly because of the consolidation effect due to briquetting of CAB particles under high pressure and the water/CAB ratio of about 24%, which is smaller than the slurry's ratio of 30%.

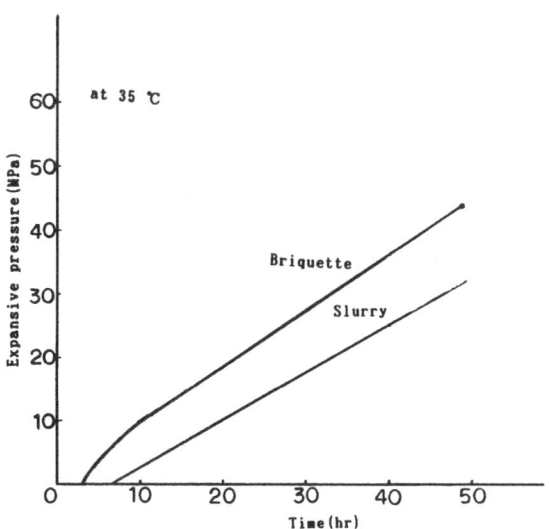

Fig. 6 Expansive pressure development of CAB

457

4. Experimental demolition of concrete specimen

4.1 Experimental method
(1) Measurement of internal temperature of concrete and the temperature of filled CAB
A concrete specimen of 600 mm cube was prepared, which contained at its center a circular spiral sheath of 40 mm diameter filled with CAB, as shown in Fig. 7 and was cured in a 35°C constant room chamber to approximate to a comparatively massive surplus concrete. An ordinary concrete of JIS A 5308 Ready Mixed Concrete Specification was used, with a nominal strength of 24 Mpa and slump of 120 mm. The internal temperature of concrete was measured with a self-recording thermometer by installing thermocouples at 3 locations. A.B.D. in Fig. 7. The temperature of CAB was measured with a self-recording thermometer by installing a thermocouple at the location C in the spiral sheath pipe.
   (2) Measurement of demolition time
The crack formation time was measured with a self-recording strain meter by attaching strain gauges at 4 locations, E.F.G.H. on the concrete surface of Fig. 7.

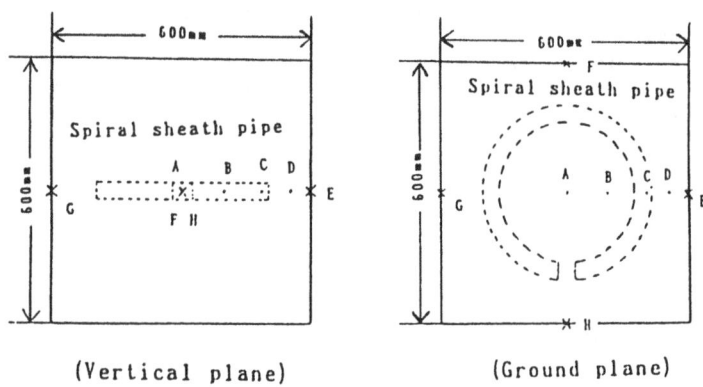

(Vertical plane)   (Ground plane)

Fig. 7 Concrete specimen containing spiral sheath pipe

   (3) Measurement of concrete compressive strength
The concrete of same kind and quality as for the concrete specimen body was cast into a mold of ⌀100 x 200 mm, and was cured, without removing from the mold, in a steam curing vessel under the temperature control to simulate the temperature increase pattern of the internal concrete temperature measured under (1). The average of 3 locations, A,B,D, of (1) was used as the internal concrete temperature. The compressive strength was first measured at 6 hours after starting the curing, followed by measurements at 6-8 hours intervals.
   (4) Measurement of expansive pressure
According to the Revised Method of the Association of Non-Explosive Demolition Agent, CAB slurry was filled into a steel tube with a strain gauge attached and sunk in a constant temperature water vessel, and the expansive pressure was calculated from the strain

458

measured with the self-recording strain meter. CAB slurry was cured
by controlling the water temperature in accordance with the recorded
temperature at the location C of CAB in the sheath pipe measured
under (1).

4.2 Experiment result

The experiment results are summarized in Fig. 8, where the internal
temperature of concrete is shown in an average value.

The hysteresis of internal temperature of surplus concrete of the
cast-in-place pile is approximate to the conventional measurement.
The CAB temperature is also close to the average internal temperature
of concrete. Cracks occurred 17 hours after concrete casting, when
the concrete strength was 6MPa, and the expansion pressure 16MPa.
From the above result, cracks are deemed to be formed at the surplus
concrete of PPC method at a relatively young age of the half a day to
one day. It is indicated that the expansive pressure continues
increasing after generation of cracks, which is considered to be
effectively functioning to enlarge the crack width. Photo 1 shows
the complete breakage of concrete after two days.

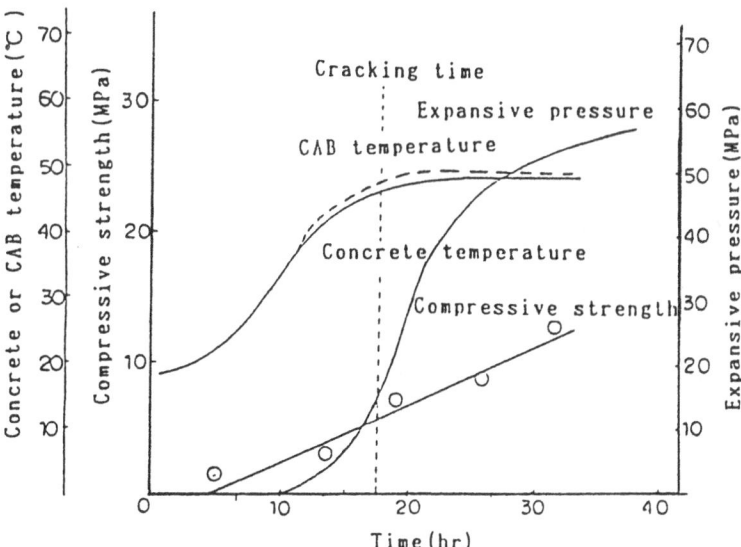

Fig. 8 Result of experimental demolition of concrete specimen

Photo 1 Breakage condition of concrete specimen

5. Examples of site application

Application of various types of PPC methods and the breakage conditions of surplus concrete are illustrated below by the pictures.
  (1) Example of application of slurry system PPC method
Photo 2 shows the slurry of 30% water CAB ratio being filled into a spiral sheath circular pipe attached to a hoop steel bar of steel reinforcement cage at the boundary between the surplus and the pile body. Although the measurement of CAB and water and mixing work to make slurry is troublesome, the charge is easy.

Photo 2 Slurry system   Photo 3 Briquette system bulk type

  (2) Example of briquette system bulk type PPC method
Photo 3 shows how CAB briquette is being filled into a spiral sheath

circular pipe before it is attached to a steel reinforcement cage.
After pouring water and sealing the inlet with tape, it is attached
to the steel cage. This method does not require the troublesome
measurement and mixing associated with the slurry system, and has an
advantage that the briquette can be filled at any time and location
before installing the pipe to the steel cage.

(3) Example of briquette system capsule type
Photo 4 shows the capsule being attached to the hoop steel bar of the
reinforcement cage at the boundary between the surplus head and the
pile body.

Photo 4 Cupsule attaching

(4) Example of breaking condition of the surplus concrete
Photo 5 shows the demolition at the boundary between the surplus
concrete and the pile body. The surplus concrete head is lifted by a
crane for disposal. Photo 6 shows the demolition condition when
additional CAB filled circular pipes were installed in the surplus
concrete at two locations besides the boundary. Photo 5 and photo 6
show large break and small break, respectively.

Photo 5 Large break   Photo 6 Small break

6. Conclusion

It was known from an experimental demolition test of a concrete simulated to the surplus concrete that cracks take place at a young age of half a day to one day and that the expansive pressure continues to increase thereafter to enlarge the crack width and produce enough demolition effect. PPC method started from slurry system, and has developed to briquette system bulk type and capsule type. The work has become easier with higher accuracy of demolition. In construction of buildings in the urban district, a large number of cast-in-place concrete piles are widely used as a representative method to avoid the environmental construction pollution such as noise and vibration. Under these circumstances, the importance of this method to employ the non-explosive demolition agent will be recognized more and more to solve the environmental problems and labor accidents associated with removal of surplus head of cast-in-place concrete piles.

Reference
The Association of Non-Explosive Demolition Agent, "Measuring method of expansive pressure of Non-Explosive Demolition Agent Explanation of it" (1984).

DECOMMISSIONING PROGRAM OF JAPAN POWER DEMONSTRATION REACTOR

T. HOSHI, M. TANAKA and M. KAWASAKI
Department of JPDR, Japan Atomic Energy Research Institute

Abstract
Japan Atomic Energy Research Institute (JAERI) is conducting the
Japan Power Demonstration Reactor (JPDR) Decommissioning Program
which was initiated in 1981. JAERI developed necessary decommissioning technologies since 1981 till 1986 and started the physical dismantlement of JPDR in 1986. All buildings and facilities of JPDR
will be dismantled except for the administrative building, warehouse
and machine shop. Highly activated components such as the reactor
internals, the reactor pressure vessel (RPV), and the biological
shield concrete will be carefully dismantled using remotely operated
tools to prevent radiation exposure of workers. The dismantlement
will be completed by 1992.
    Outline of the JPDR Decommissioning Program is described in this
paper.
Key words : Reactor decommissioning, Decommissioning technologies,
Dismantlement radioactive waste, JPDR Decommissioning Program.

1. Introduction

The "Long-Term Program for Development and Utilization of Nuclear
Energy" published by Japan Atomic Energy Commission in 1982 pre-
scribes that reactor decommissioning technologies for a commercial
power reactor should be developed and demonstrated through the JPDR
Decommissioning Program of the JAERI. In accordance with this philosophy, JAERI initiated the JPDR Decommissioning Program under contract from the Science and Technology Agency (STA). JAERI developed
decommissioning technologies from 1981 through 1986 and is conducting the physical dismantlement of JPDR using these technologies from
1986 through 1992. During this dismantlement, a wide variety of
data is being collected to be used for commercial nuclear power
reactor dismantlement in the future.

2. Outline of JPDR

The JPDR is a BWR type power demonstration reactor. It first generated electricity in October 1963. The power was initially 45MWt
(JPDR-1) and was increased to 90 MWt in 1972 for enhancement of

463

neutron irradiation capability (JPDR-II). During the operation of
JPDR-I and JPDR-II, a large quantity of data was aquired for nuclear
power reactor operation and operators for follow-on commercial nuclear power reactors were trained. Table. 1 shows the major specifications, operation history and radioactive inventory of JPDR.

Table 1   Major Specifications, Operation History and Radioactive
          Inventory of JPDR

Specifications	
Type of reactor	BWR
Thermal power	90 MWt (45 MWt initially)
Pressure vessel	
material	ASTM-A302-56 Gr B
inner diameter	2.1 m
height	8.1 m
thickness	7 cm
Biological shield	
material	reinforced concrete
thickness	1.5 to 3 m
Reactor enclosure	
inner diameter	15 m
height	38 m
Operation history	
Operation time	17,000 hours
Output of electricity	1.4×10^6 KWH
Radioactive inventory	4,100 Ci (as of December, 1986)

3. Dismantling activities

3.1 Procedures

All buildings and facilities of JPDR will be dismantled except for
the administrative building, warehouse and machine shop. The JPDR
site will then be renovated and landscaped. The area to be dismantled is shown in Fig. 1.
     Radioactive components and equipment are removed as early as pos-
sible to reduce worker exposure. The containment vessel and building
walls are utilized as a confinement boundary for radioactivity during
dismantlement of radioactive components and equipment. Pre-decontam-
ination and post-decontamination of components, buildings, etc. are
conducted to reduce worker exposure and radioactive waste volume, if
necessary.

464

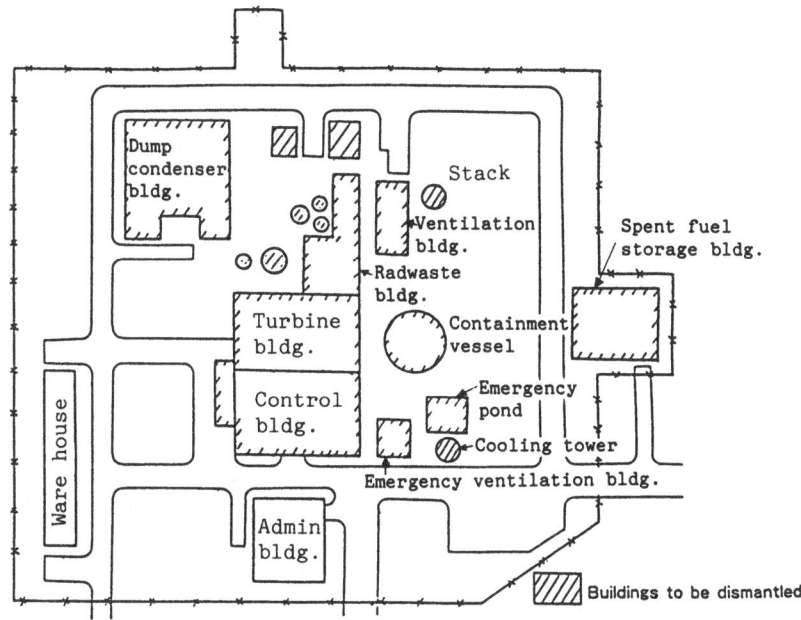

Fig. 1   Plan View of JPDR Facility

Exisisting utility systems, electricity supply systems, water supply and drainage systems, ventilation systems, etc. are utilized during the dismantlement. Therefore, the dismantlement of these systems is scheduled carefully. Fig. 2 shows the JPDR dismantling schedule together with the techniques to be used. The main dismantling work will start with removal of the reactor internals, followed by removal of the RPV wall, connected pipes and biological shield concrete. The dismantling procedure of containment vessel is shown in Fig. 3.

The total manpower and worker exposure of the JPDR dismantlement are estimated at 73,000 man-days and approximate 100 man-rem respectively. The waste from the JPDR dismantlement is estimated at 27,800 tons consisting of 2,200 tons of metallic waste and 25,600 tons concrete waste. Radioactive waste will be 4,100 tons.

465

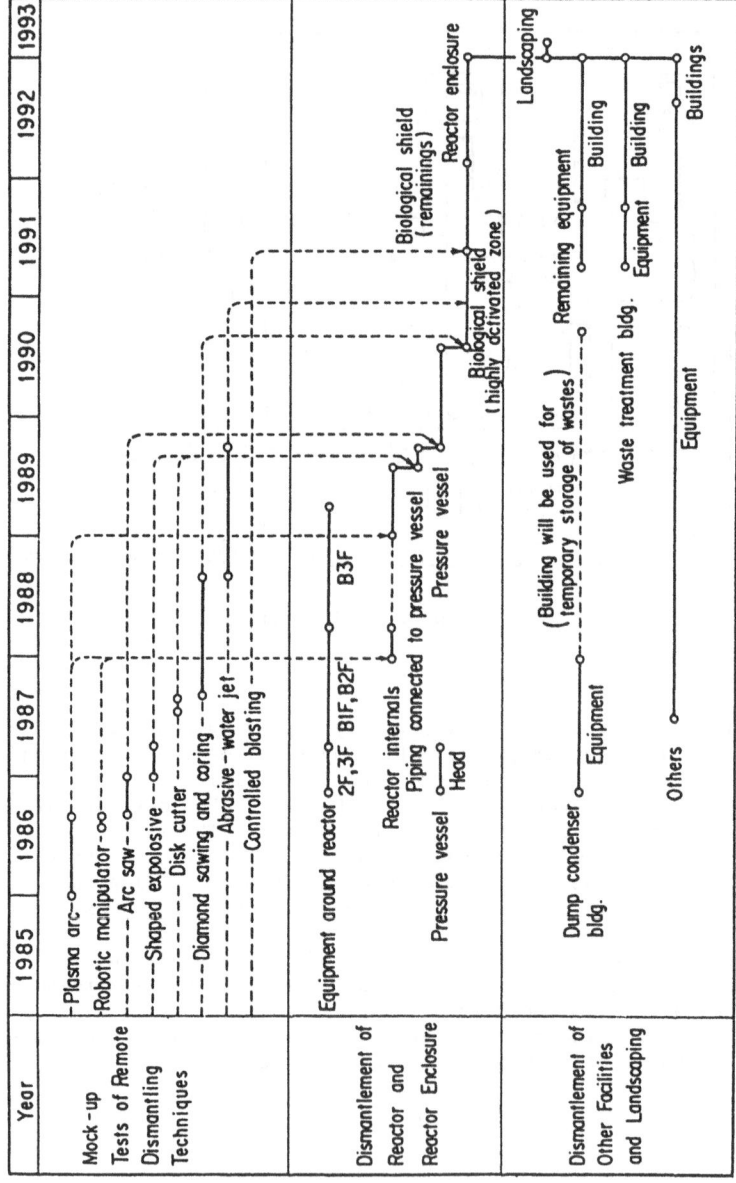

Fig. 2 JPDR Dismantling Schedule

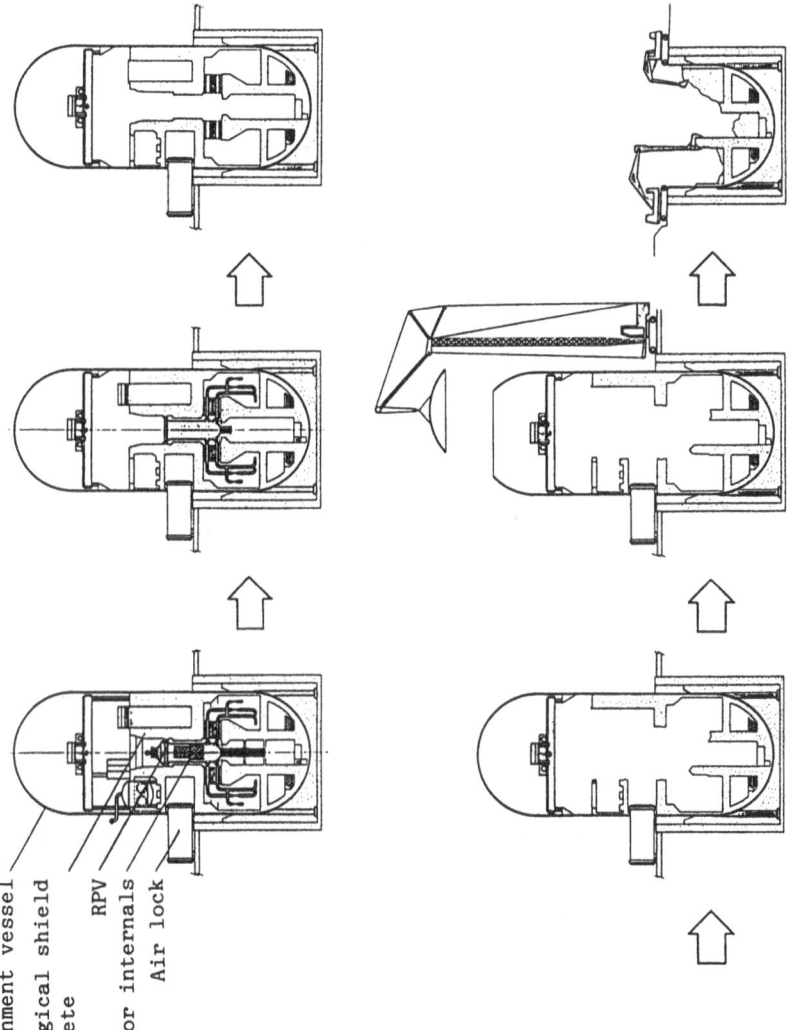

Containment vessel
Biological shield
concrete
RPV
Reactor internals
Air lock

Fig. 3   Dismantling Procedure of Containment Vessel

467

3.2 Dismantling of metal structures

The metal structures with high radioactivity such as the reactor internals, the RPV wall and pipes connected to the RPV will be dismantled using remote cutting systems to reduce worker exposure. Disassembly techniques applied to these metal structures are shown in Table 2. The other pipes and components are dismantled using conventional techniques such as gas cutting, band saw, etc.

Table 2   Disassembly Techniques Applied to Metal Structures with High Radioactivity

Object	Technique	Example of performance
Pressure vessel	Arc saw	Cut carbon steel 250mm (in water)
Reactor internals	Plasma arc	Cut stainless steel 130mm (in water)
Piping connected to pressure vessel	Rotary disk knife	Cut stainless steel 12inch, Sch 80
	Shaped explosive	Cut carbon steel 26inch, Sch 80

3.3 Dismantling of biological shield

The biological shield of JPDR is a reinforced concrete with inner diameter of about 3 m and thickness of 1.5 to 3 m. A steel plate with thickness of 13 mm is lined inside biological shield.

The radioactivity distribution in the biological shield is shown in Fig. 4. The radioactivity decreases from inside towards outside. The radiation dose rate inside the biological shield is estimated to be about 250 mR/h even after the RPV is removed. But it is expected that radiation dose rate will be less than 5 mR/h, if highly activated inner portion of 40 cm in thickness is removed.

Therefore, it is decided to dismantle of the 40 cm of biological shield concrete in depth by using remote cutting machines such as the abrasive-water jet and the diamond-saw and coring machine. The remaining potion will be demolished by the controlled blasting since workers can access there. Fig. 5 shows the diamond-saw and coring system for dismantling of the biological shield.

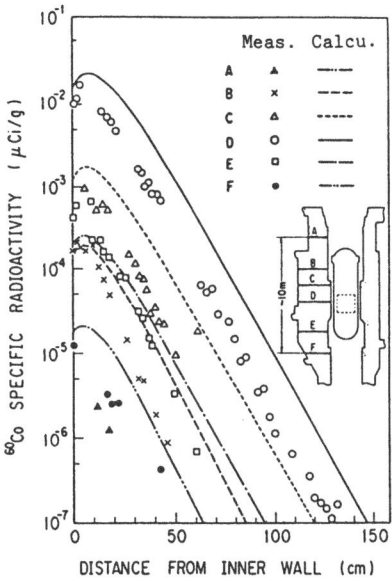

Fig. 4   The Radioactive Distribution in the
Biological shield

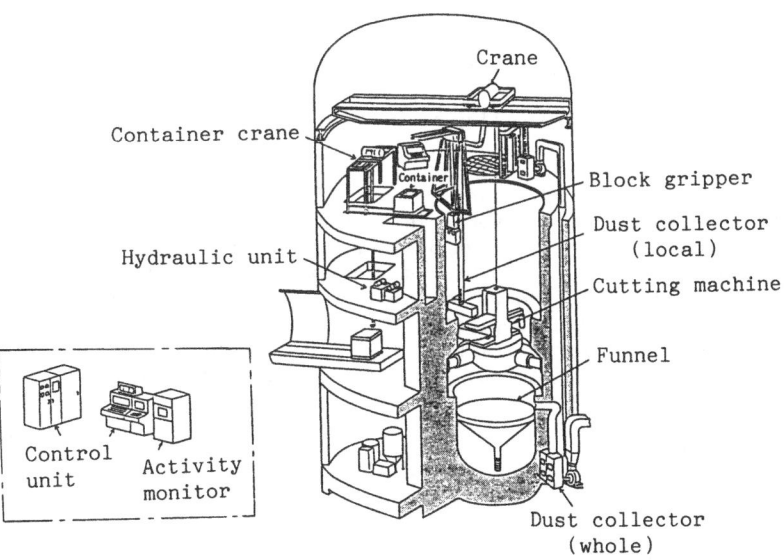

Fig. 5   Diamond Sawing and Coring System

469

3.4 Dismantling of buildings

Some areas of concrete floors are radioactively contaminated. A typical measured result of contamination on the floor is shown in Fig. 6. Contaminated concrete should be treated as radioactive waste, therefore, the contaminated concrete will be removed carefully by a scabbling technique, a micro-wave technique, etc. After removal of all contamination, buildings will be demolished by conventional techniques.

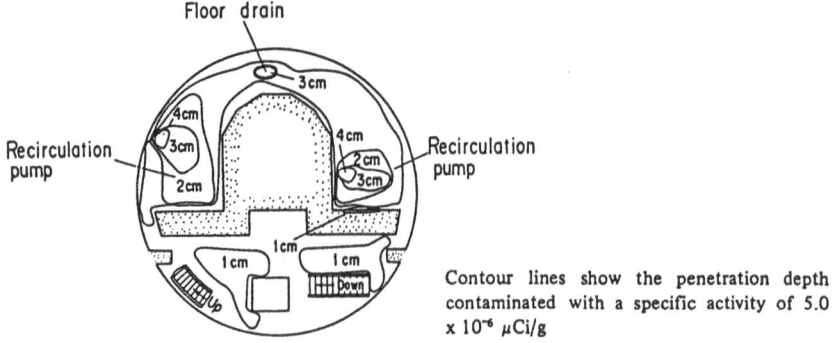

Contour lines show the penetration depth contaminated with a specific activity of 5.0 x 10^{-6} μCi/g

Fig. 6 Contour Map of Radioactive Contamination on B2 Floor of Containment Vessel

4. Progress of the dismantling

The progress of the dismantling is as follows.

1986.12 - 1987.3
 -Dismantlement of RPV head
 -Dismantlement of components in containment vessel
  (emergency condenser, high pressure poison tank, etc.)
 -Dismantlement of components in the dump condenser building
  (dump condenser, feed water pump, condensate water pump, sea
  water circulation pump, etc.)
 -Refurbishment of control building
  (reconstruction of access room to controlled area and set-up of
  monitors)

1987.3 - 1987.12
 -Dismantlement of components in containment vessel
  (forced circuration system, steam and feed water system, etc.)
 -Dismantlement of components in the dump condenser building
  (condensate demineraizer, feed water control pannel, etc.)
 -Dismantlement of components in the emergency ventilation
  building (motor control center, A.C generator, emergency
  ventilation system, etc.)

Because of the low radioactivity, these components were cut manually using conventional cutting tools, such as oxygen torch and band-saw.

Almost components in the dump condenser building were taken out so that the dump condenser building could be used as both a temporary storage yard of wastes and a place for decontamination of removed components.

The control building has been partially refurbished for effective management of worker access to the radiation controlled area at the peak time of dismantling activities. It is estimated that the JPDR dismantling will require about 150 men/day of workers at the peak.

All components of the emergency ventilation building have been removed so that the control device of diamond sawing and coring machines for biological shield concrete dismantlement can be installed in it.

The total worker exposure between Dec. 1986 and Sept. 1987 was 0.25 man-rem because the radiation dose rate of dismantled components were very low. Even on the surface of the forced circuration system, the dose rate was below 10 mR/h. Film badges and alarm pocket dosimeters are used for measurement of worker exposure during these works.

As of September 30, 1987, the total manpower used in these activities was about 14,000 man-days. Table 3 shows the solid wastes of JPDR dismantlement between December 1986 and September 1987.

Table 3   Total Amount of Solid Radioactive Waste
(Dec. 1986 ~ Sept. 1987)

Material	Weight (ton)	Number of Containers				
		1 m³ Container	200 Liter Drum	Package (ton)	Carton	Filter
Metal	290	70	420	130	—	—
Concrete	60	—	230	20	—	—
Consequential Waste	20	—	—	—	6450	230
Total	370	70	650	150	6450	230

6. Concluding remarks

The JPDR dismantling Program is on schedule. In January 1988, the dismantlement of core internals will be initiated and followed by the RPV wall dismantlement. During the JPDR dismantlement, many technologies developed by JAERI for reactor dismantlement are being used, and a wide varaiety of data is being collected to be used for future commercial power reactor dismantlement.

References

Ishikawa, M. and Kikuyama, T. (1985) Decommissioning plan and present status of technical development in JPDR. <u>Proceedings of Interna-tional Nuclear Reactor Decommissioning Planning Conference,</u> Bethesda, 450-468.

Ashida, S., et al., (1987) Development of cutting tools for JPDR core internals. <u>Proceedings of 1987 International Decommissioning Symposium,</u> Pittsburgh, VI48-VI60.

Konno, T., et al., (1987) Abrasive water jet cutting technique for biological shield concrete dismantling. <u>Proceedings of 1987 Inter-national Decommissioning Symposium,</u> Pittsburgh, IV270-IV284.

DEMOLITION TECHNIQUE FOR REACTOR BIOLOGICAL SHIELD CONCRETE USING DIAMOND SAWING AND CORING

S. YANAGIHARA and H. GOHDA

K. KOHYAMA, M. TOKOMOTO
and H. ZAITA

Department of JPDR
Japan Atomic Energy Research Institute
Department of Engineering
Shimizu Corporation

Abstract
A diamond sawing and coring technique has been developed to dismantle the biological shield concrete of Japan Power Demonstration Reactor (JPDR) in Japan Atomic Energy Research Institute(JAERI).
  Preliminary cutting tests were conducted using a prototype machine and the concrete structure simulating the JPDR biological shield. Based on the data obtained, dismantling cutting system was designed for safe and efficient dismantlement of the JPDR biological shield concrete. The system includes cutting machine with diamond sawing and coring units, concrete handing machine, water reprocessing equipment and local ventilation, etc. Through the study, it was confirmed that the diamond sawing and coring technique was applicable to the JPDR decommissioning.
Key words: Sawing, Coring, Biological shield, Reinforced concrete, Abrasives, Dismantling

1. Introduction

It was desired to develop a new dismantling technique to remove a biological shield concrete surrounding a reactor pressure vessel safely and efficiently, because the biological shield is composed of massive and heavily reinforced concrete and is highly activated by neutron irradiation during the reactor operation. A diamond sawing and coring technique has been developed for dismantling the JPDR biological shield concrete together with other techniques such as abrasive water jet and controlled blasting in the JPDR decommissioning program. The study was mainly concentrated on development of cutting ability and designing the dismantling system for the JPDR biological shield concrete. Prior to applying the diamond sawing and coring technique to the actual dismantlement of the JPDR biological shield, the prototype machine equipped with a saw blade and a core bit was made, then cutting tests were conducted using the prototype machine and the concrete structure simulating the JPDR biological shield to obtain the data such as cutting characteristics and work efficiency. Based on the data, the dismantling system was designed for the actual dismantlement of the JPDR biological shield. This paper describes the results of the cutting tests and dismantling system to be applied to the JPDR decommissioning.

2. Outline of the JPDR

The JPDR is a BWR demonstration reactor. It first generated electricity in Japan, October 1963. The power was initially 45 MWt(JPDR-I), then it was increased to 90 MWt(JPDR-II) for enhancement of neutron capability.

The biological shield of the JPDR is composed of ordinary concrete of 300 kgf/cm^2 in compressive strength and steel bars of 29 mm in diameter. The inner surface of the biological shield is lined with 13 mm thick carbon steel plate. Cooling pipes and ion chamber guide tubes are contained in the concrete structure. Figure 1 shows a cross-section of the JPDR biological shield together with activity levels and dismantling techniques to be applied to the actual dismantlement.

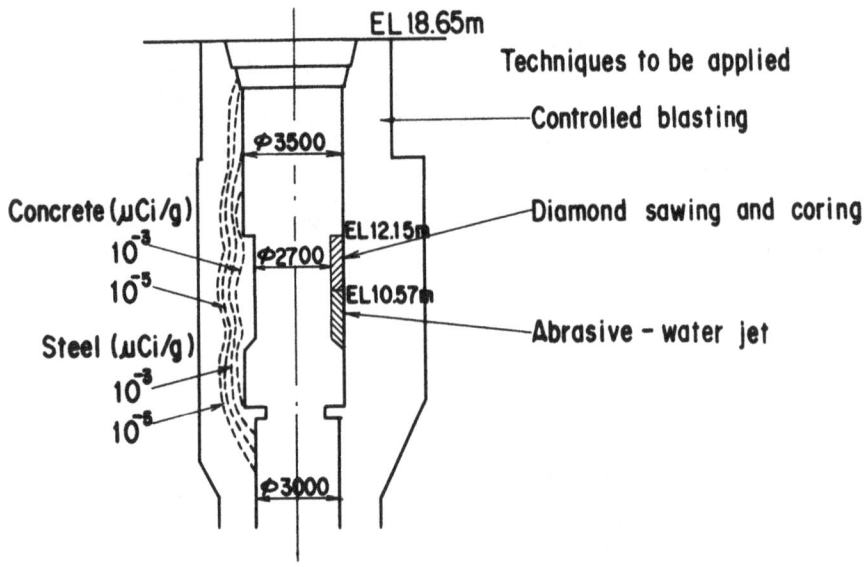

Fig. 1 JPDR biological shield

3. Diamond sawing and coring

A diamond sawing and coring uses a saw blade and a core bit to cut concrete structure. They are tipped with cutter edge molded out of metal power and diamond abrasive. It assures continuous cutting ability for relatively long operation because the fresh surface of the cutter edge appears due to being sharpened by the friction between cutter edge and concrete. The saw blade and core bit are cooled by water during the operation.

The diamond sawing and coring has a capability to remove a

concrete block together with reinforced steel bars from previously
planned position. It is also easy of automatized, and can be handled by remote operation even in a limited space. Therefore, being
used with remote handling machine and water reprocessing device, it
must satisfy the following requirement necessary for reactor
decommissioning.

- prevent spreading of radioactive material
- minimize worker exposure
- minimize radioactive waste volume

Figure 2 illustrates the typical cutting pattern for dismantling a
concrete structure by the diamond sawing and coring technique.

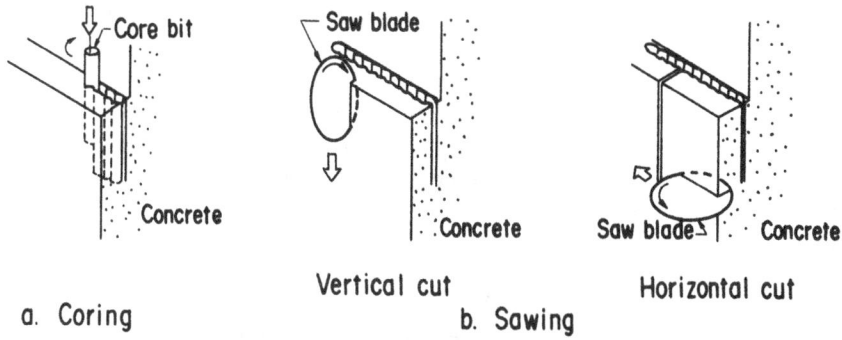

a. Coring   b. Sawing

Fig. 2 Cutting patterns for sawing and coring

4. Cutting tests by prototype machine

4.1 Prototype machine

The prototype machine was equipped with a saw blade and a core bit.
The saw blade was 1077 mm in diameter and 5.5 mm in width. The core
bit was 156 mm in diameter and 4 mm in width. A special 15.0 hp.
high-frequency motor was used for driving the saw blade and another
7.5 ph. for driving the core bit. Cutting tests were conducted under
constant current supply condition to the motor as shown in Table 1.

Table 1 Test condition

Items	Sawing	Coring
Current supply (A)	30	8
Cutter rotation(rpm)	788	436

4.2 Test results

4.2.1 Cutting speed

It is possible to evaluate cutting ability by measuring cutting speed under various cutting conditions. Reinforcing steel-concrete ratios and cutting patterns were changed to measure the cutting speed under various test conditions with constant current supply. Figure 3 shows the measured cutting speed under various test conditions. As for the sawing, the vertical cutting is faster than the horizontal cutting. This might be caused by the following reasons; The friction between the saw blade and concrete surface is larger under the horizontal cutting than the vertical cutting, since the cooling water is supplied effectively to the cutter edge under the vertical cutting rather than the horizontal cutting. It is also obvious from the figure that the cutting speed decreased with increasing concrete-steel ratio.

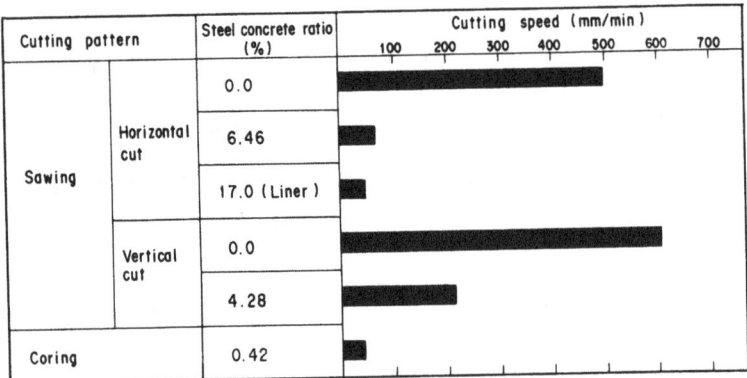

Fig. 3 Cutting speed under various test conditions

4.2.2 Cooling water flow rate

As described before, cooling water is supplied to the cutter edge to remove the friction heat as well as to prevent the spread of radioactive airborne particles. However, the water used for cooling must be drained as radioactive liquid waste after completion of the dismantlement of the concrete structure. It is, therefore, desirable to minimize the use of the cooling water. The cutting speed was measured changing the flow rate of the cooling water to evaluate the minimum amount of the cooling water necessary for the sufficient cutting. Figure 4 shows an example of the test results. The cutting speed decreased to 75% of that under the flow rate enough for cooling when the water flow rate became less than one litter per minute. Based on the data obtained, the practical flow rate for cooling was evaluated to be 4 and 3 litter/min for sawing and coring, respectively.

476

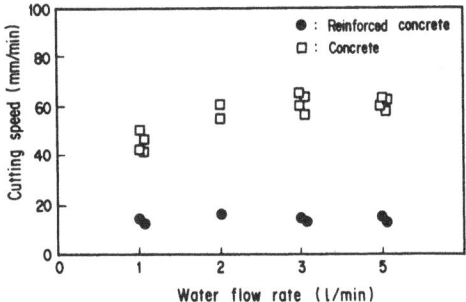

Fig. 4 Measured cutting speed vs. water flow rate at horizontal
sawing

4.2.3 Cutter wear rate

Since the cutter edge of the saw blade and the core bit is worn by
the friction between the cutter and concrete, the saw blade and core
bit must be changed when the thickness of the cutter edge decreased
not to be usable due to wearing. It is, therefore, useful to evaluate the wear rate for estimating the necessary number of the saw
blade and core bit to be used in the dismantling activity. The wear
rate of the cutter edge was measured at sawing and coring under various steel-concrete ratio as shown in Fig. 5. The cutting area is
estimated to be approximately 11 m^2 and 45 m^2 for sawing and coring,
respectively in the dismantlement of the JPDR biological shield. The
average reinforcing steel-concrete ratio in the biological shield is
also estimated to be approximately 2.42% excluding the lining region
and 1.0% for sawing and coring, respectively. Based on the measured
cutter wear rate and cutting area, it is estimated that two cutters
will be necessary for both sawing and coring during the dismantlement of the JPDR biological shield concrete.

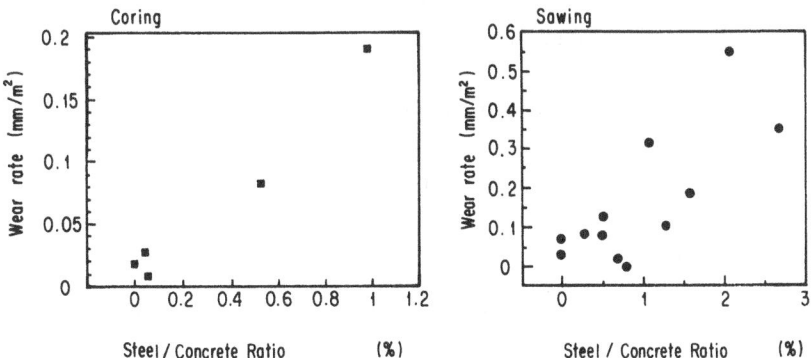

Fig. 5 Measured cutter wear rate vs. steel/concrete ratio

4.2.4 Working time

The working time of each activity during the cutting tests was measured to evaluate the work efficiency for the dismantlement of the
JPDR biological shield concrete. Figure 6 shows the measured working
time of each activity. The activity for exchanging the cutter units
took the half of the whole dismantling work, which was relatively
long duration in comparison with the dismantling activities such as
sawing and coring. As shown in the figure, the coring and the sawing
took 30% and 10% to the total duration of the works, respectively.
Since the working time of coring and sawing depends exactly on the
machine characteristics, the working time will be saved by shortening
the cutter exchange time.

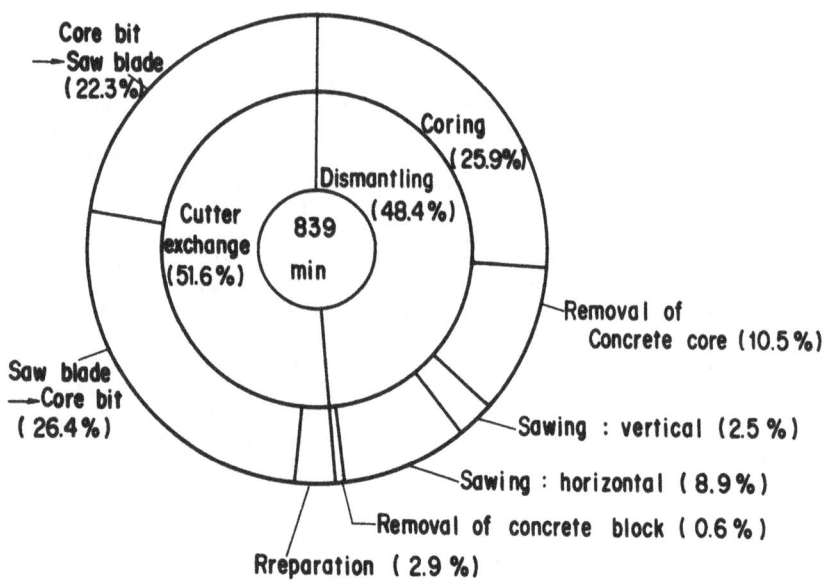

Fig. 6 Measured working time

5. Dismantling system for the JPDR biological shield concrete

5.1 System design

Based on the results obtained in the preliminary tests, the dismantling system including the cutting machine and auxiliary equipments
was designed for the actual JPDR dismantlement. Figure 7 illustrates the cutting machine for the actual dismantlement of the JPDR
biological shield concrete. It is composed of five main parts as
shown in the figure. These are sawing and coring units, actuators

for moving the sawing and coring units, a frame for the actuators, blade and core bit grinding units and outriggers for positioning the frame. The sawing and coring units are installed in the cutting machine facing in opposite directions. This allows the exchange of the units for sawing or coring by rotating the units. This is done remotely without a worker's direct assistance. The cutting machine has mobility in three directions, that is, axial, radial and circumferential all while confined in a cylindrical shaped space.

In addition to the cutting machine, the dismantling system is composed of the concrete block handling machine, a local ventilation equipment, water reprocessing equipment, monitoring instruments, etc. Figure 8 shows a schematic of the dismantling system. The role of the each equipment is as follows; Being once separated, concrete pieces are handled by one of two clamps, one for prismatic and another for columnar shapes. The clamps have the ability to hold a 1200 kg prismatic concrete or a 50 kg columnar concrete even if the power supply fails. The local ventilation system, composed of a hood and dust collector, collects mist and dust close to the cutting machine. The local ventilation system reduces the load on the existing ventilation system as well as keeping the working area clean. The air drawn into the hood is exhausted through the dust collector to an existing duct of the building ventilation system.

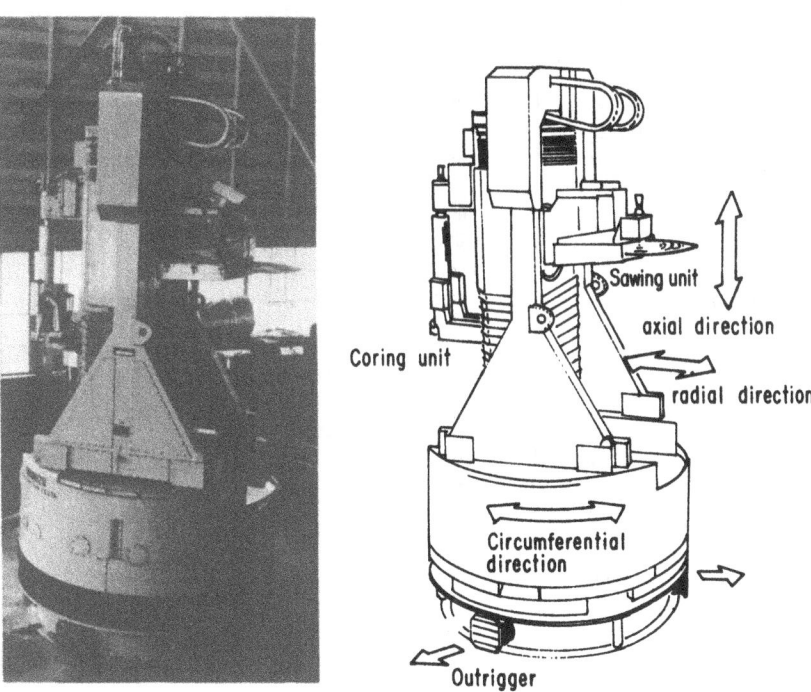

Fig. 7 Cutting machine for dismantlement of the JPDR biological shield concrete

479

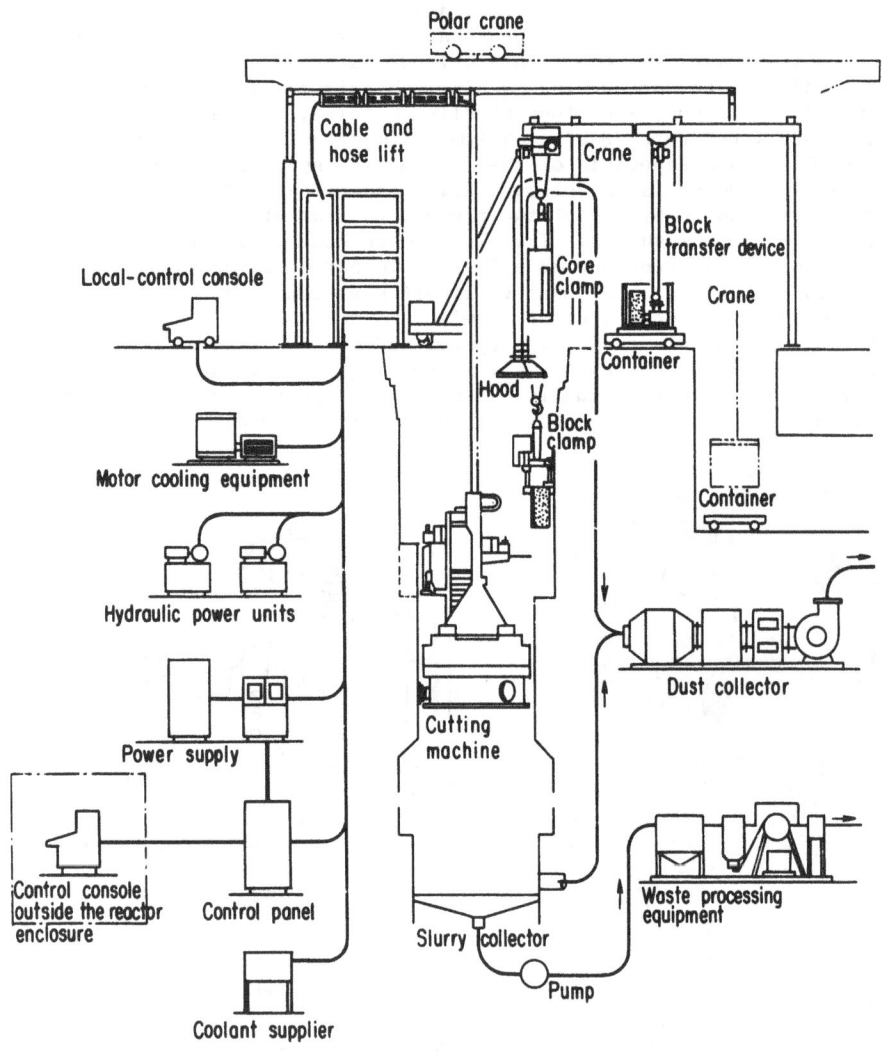

Fig. 8 Schematic of the dismantling system
for the JPDR biological shield

5.2 Dismantling Procedure

The JPDR biological shield concrete will be dismantled after removing the reactor pressure vessel and all pipes and components from the cavity. Preparation work will be conducted to prevent the spread of contamination during the dismantlement of the biological shield concrete before the dismantling activity. The dismantling procedure is summarized as follows.

- Openings in the biological shield will be closed by attaching steel plates to cover the openings and by filling with grout in the pipes.
- A slurry collector will be installed in the bottom of the cavity.
- The cutting machine, suspended from the polar crane will be lowered into the cavity surrouned by the biological shield. It will be then rigidly positiononed against the wall with its outriggers.
- The concrete structure will be cut at circumferential direction by swing.
- A columnar portion will be separated from the concrete structure by coring.
- The separated concrete core will be removed and placed in a container.
- The concrete structure will be cut along with the vertical direction by sawing.
- The prismatic concrete piece will be removed and placed in a container.
- The radiation level of the container surface will be monitored.
- The containers will be shipped out of the reactor building.

These activities will be repeated. Approximately 15 tons of concrete will be dismantled by sawing and coring during the 3 month operation.

6. Conclusion

The diamond sawing and coring technique has been developed to dismantle the JPDR biological shield concrete safely and efficiently in the JPDR decommissioning program. Various cutting tests were conducted using the prototype machine equipped with a saw blade and a core bit to evaluate the cutting characteristics and work efficiency. Through the study, it was confirmed that the diamond sawing and coring was applicable to dismantling the JPDR biological shield. Based on the data obtained, the dismantling system was designed for JPDR decommissioning. This system will be applied to the actual dismantlement of the JPDR biological shield in 1990.

References

ISHIKAWA, M. and KIKUYAMA, T., (1985) Decommissioning plan and present status of technical development in JPDR, Proceedings of International Nuclear Reactor Decommissioning Planning Conference, Bethesda, 450-468

ISHIKAWA, M., et al., (1986) Decommissioning program of the Japan Power Demonstration Reactor by JAERI, Proceedings of International Low-, Intermediate-, and High-Level Waste Management and Decontamination and Decommissioning Meeting, Niagara Falls

ISHIKAWA, M., et al., (1987) Present status of JPDR decommissioning program, Proceedings of 1987 International Decommissioning Symposium, III18-III30

ISHIKAWA, M., et al., (1987) Reactor decommissioning in Japan - philosophy and first program-, IAEA-CN-48/152

RESEARCH AND DEVELOPMENT OF A MACHINE FOR THE REMOVING SYSTEM OF BIOLOGICAL SHIELD WALL OF NUCLEAR REACTOR

R. FUKUZAWA, N. KONDO and O. ASAKAWA   Nuclear Power Division
                                                 Toda Construction
T. HIRAGA   Institute of Construction Technology
                                                 Toda Construction

Abstract
The removing system of heavily reinforced biological shield wall using newly developed diamond saws (disc saws) for decommissioning of nuclear facilities is presented.
    Cutting experiment of full model scale has been executed, in which the cutting speed was controlled automatically and remotely. This proves that the heavily reinforced concrete section with D51 mm in diameter deformed bars, can be cut in the depth of 100 mm speedily, safely and continuosly with precision.
Key Words : Decommissioning, Dismantling method, Diamond saw, Disc saw, Biological shield wall, Remote control.

1. Introduction

Today in Japan, 35 nuclear power plant units are in operation, and the total amount of capacity, 27,880MWe, supply 30% of total electric demand.
    However in near future, the decommissioning after life exhausted is inevitable, so the R and D on its technique would be definitely required, because the biological shield wall surrounding the reactor is not only highly activated by neutron irradiation but heavily reinforced. For this purpose "a diamond sawing and coring technique" could be thought to be superior. Nevertheless in this technique, there are some difficulties in using conventional disc saw, which have cemented diamond tips (cutting edge) which are available in market. So it was newly required to develop not only the remote control system which avoids more exposure, but the particularly specific disc saws (blades) which met the demand.
    Here the result of R and D on its purpose is presented.

2. Dismantling system

2.1 Outline of system

The fundamental composition of dismantling system is shown in Fig.1. For the purpose of safe dismantling with less exposure, all devices that compose the system are remotely surveyed and controlled. The

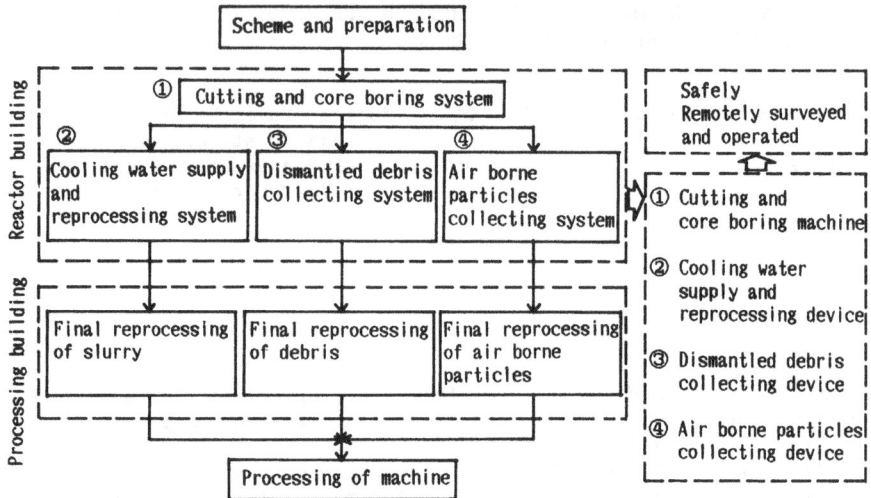

Fig. 1 Fundamental concept of dismantling system

Fig. 2 Cutting and core boring machine

Fig. 3 Conventional cutting machine

saw cutting and core boring by remote operating are controlled responding to the fluctuation of amount of steel (reinforcement) at cutting cross sections.

The cooling water for cutting and boring are collected as much as possible at the spot, and collected water are used again after reprocessing of slurry. The air borne particles are avoided to spread into the interior of the reactor building by being collected as possibly in cavity.

2.2 Saw cutting and core boring

The conventional cutting machines (Fig. 2 and 3) for ordinary RC structures has been improved to a specific one for nuclear facilities which is smaller and lighter to adapt themselves to the biological shield wall (Fig. 4). In dismantling the biological shield wall, firstly lap line boring of the previously defined outer line of the block is done, then vertical and horizontal cut proceed. The cylindrical and block shapes debris after boring and cutting as shown in Fig. 5 are finally collected with "collecting device".

484

3. Cutting experiment

3.1 Specimens

The configurations of specimens are shown in Fig. 6. The reinforcement of the specimens were determined conforming to actual shield wall (Table 1). In reinforcing, C and D specimens represented the shield walls for BWR and PWR respectively, A and B for JPDR (Japan Power Demonstration Reactor) of JAERI (Japan Atomic Energy Research Institute). The compressive strength of concrete was 346kg/cm^2.

3.2 Diamond blades

In order to cut heavily reinforced concrete, the tip abrasive for blades are compound of diamond and boron nitride which have high thermal resistance for metal. Cobalt based soft metal bond for abrasive is adapted, because which makes self dressing of tips easily.

In consideration of experiments for D51 bars and plate liners, the best diamond blade was selected as shown on Table 2. The result for plate liners is shown in Fig. 7.

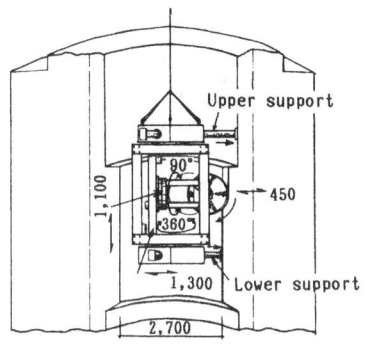

Fig. 4 Cutting machine

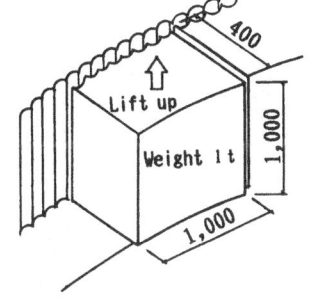

Fig. 5 Cutting and core boring

Table 1 Specimens

Specimen	A	B	C	D
Reinforce-ment	D 29	D 29	D 38	D 51
Steel pipe		$\phi 34 \times 2.2$		
Pitch	@150	@75	@200	@200

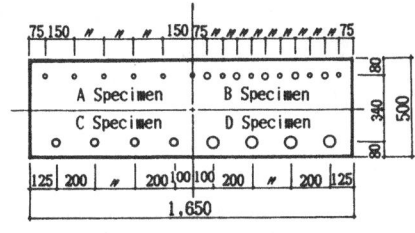

Fig. 6 Configurations of specimens

485

Table 2 Blade tips

Sort of blade		C 3 3	C 6 0
Metal bond		Cobalt based	Cobalt based
Concentration of abrasive (%)	diamond	3 0	6 0
	boron nitride	3 0	0
	total	6 0	6 0

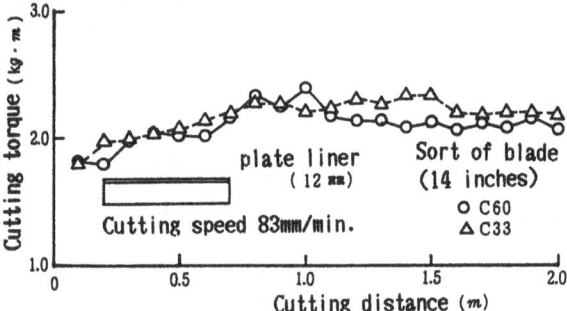

Fig. 7 Cutting distance and cutting torque

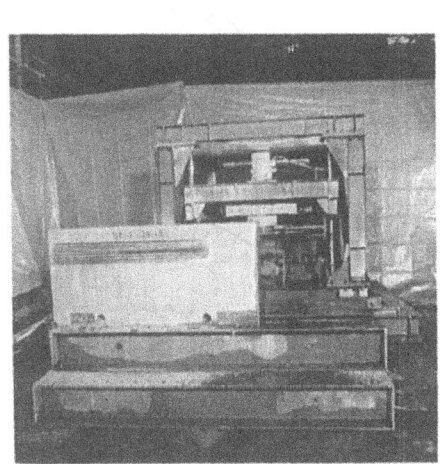

Fig. 8 Cutting machine

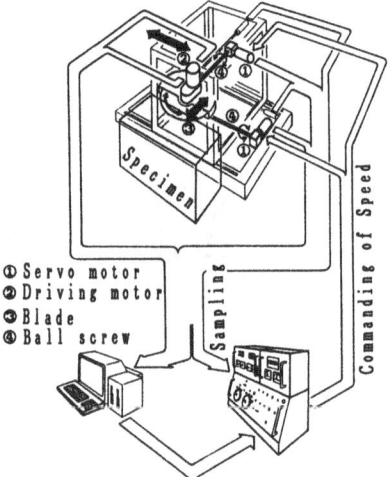

Fig. 9 Cutting machine

3.3 Cutting machine

The composition of cutting machine is shown in Fig. 8 and 9.
The cutting machine mocked up the actual one. The large size blade
of 42 inches in diameter (1,000 RPM) was used to enable cutting of
max. 400 mm in depth. The driving motor is 40 HP of high frequency,
and for traverse and horizontal movements DC servo motor.

486

3.4 Cutting method

900 mm length of horizontal cutting at 300 mm in depth were executed, by way of "multi layers cutting".

The cutting speed which responds to the fluctuation of cutting load due to the degree of reinforcement, is controlled automatically under the constant current. This is called "constant current cutting". In this experiment, the scheduled current were 40 and 50 A respectively. Additionally the "constant speed cutting" and "manual cutting" were also executed for reference. In constant current cutting, command speed increment is determined by current value in cutting, in the formula described below.

$$\Delta V_k = K_P(I_{k-1} - I_k) + K_I(I_S - I_k) + K_D(2I_{K-1} - I_{k-2} - I_k)$$

$$\Delta V \ : \ \text{Speed increment}$$
$$Is \ : \ \text{Constant current}$$
$$Ik \ : \ \text{Measured current}$$
$$K_P, K_I, K_D \ : \ \text{Control parameter}$$

The constant parameters in the formula are determined by the Table 3 using the result of unit step response as shown in Fig. 10 and 11. And PID control is adapted as the result of preliminary tests.

Table 3 Control parameters (PID)

K_P	K_I	K_D
$\dfrac{1.2}{R(L+\Delta T)} - \dfrac{1}{2}K_I$	$\dfrac{0.6\,\Delta T}{R\left(L+\frac{\Delta T}{2}\right)^2}$	$\dfrac{0.6}{R\Delta T}$

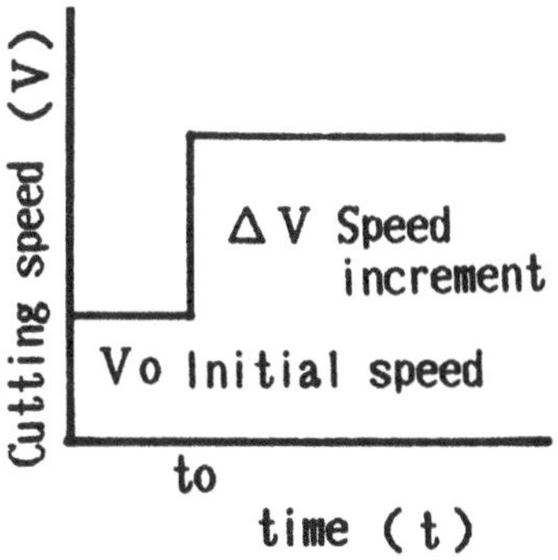

Fig. 10 Step response 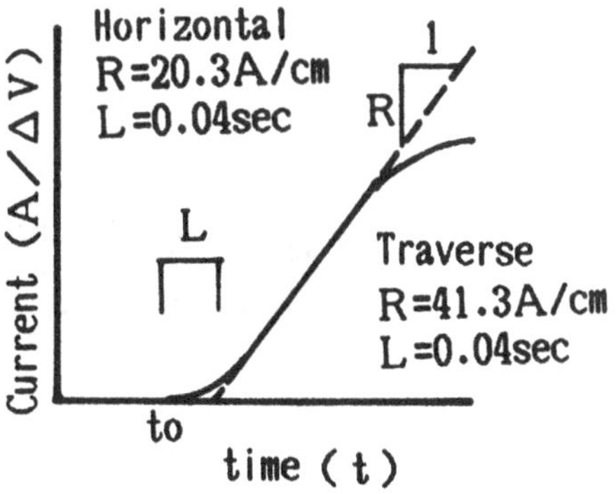　　　Fig. 11 Response current

3.5 Remote control cutting

The flow chart of automatic remote control is shown in Fig. 12. The control was executed by the control computer program. Prior to cutting, number of layers, cutting length and depth, scheduled

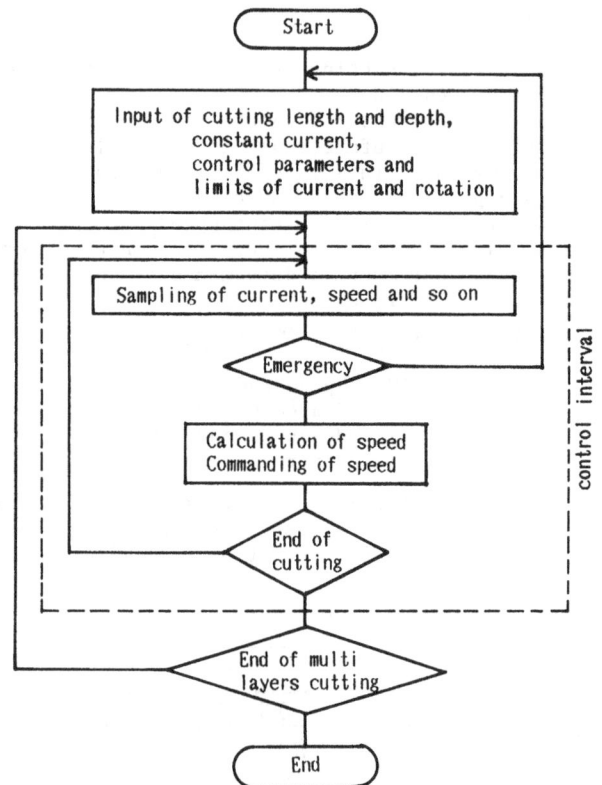

Fig. 12 Flow chart of automatic remote control

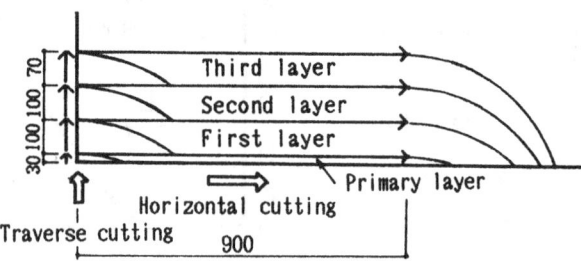

Fig. 13 Multi layers cutting

current, control parameters and limits of current and rotation were
input (shown in Fig. 13). At every step (control time interval,
40∿100 msec), current, speed, revolving number of blade and position
were measured, and the speed of next step was determined in the
computer. In an emergency it automatically stopped.

488

4. Results and discussions

4.1 Cutting efficiency

In the cutting by way of "constant current control" with a high power motor and a actual size blade, the range of current fluctuation to setting current was small and, cutting of heavily reinforced concrete with D51 was done speedily in stability as shown in Fig. 14. Cutting section is shown in Fig. 15. The average speed of cutting for specimens with D29, D29 and Ø34, D38 and D51 which were executed one after another respectively twice up to the length of 900 mm is shown in Fig. 16. The speeds for each specimens were 126 mm/min. for D51, 308 mm/min. for D38, 392 mm/min. for D29 and Ø34 and 386 mm/min. for D29 respectively. Both kinds of blades cut 7,200 mm in total horizontal length and, 572 cm^2 (30-D29, 14-Ø34, 11-D38, and 11-D51), respectively. The cutting speed was stable for a virgin blade but, began decaying with the number of cut section increase. The cutting speed with "C33" blade was stable till cutting of D29 and Ø34, but after cutting of D38, it began decaying. The cutting speed of "C60" decayed after cutting of D29 and Ø34. "C33" blade maintained better efficiency in duration than "C60".

In horizontal cutting, the relation between speed and reinforcement is shown in Fig. 17. The speeds for every specimens were 138 mm/min. for D51, 220 mm/min. for D38 and 316 mm/min. for D29, and the cutting speeds showed the same tendency. For smaller size bars than D38, the cutting speed for the second cutting was almost same as the first. In comparison of cutting speeds of each bars for one layer, for up to the specimens with D38, the speeds for the first and last bars tend to be lower than other bars.

The relation between the ratio of total section and reinforcement and cutting speed is shown in Fig. 18. The speed comparisons of horizontal and traverse cutting is shown in Fig. 19. In this Figure, traverse speed was approximately 50% of horizontal one. The speed comparison in constant current cutting and constant speed one is shown in Fig. 20. The reason why the constant current cutting was fastest of all, was that the current for non-reinforced concrete part was so large that the cutting speed was fast. The constant current cutting was 3 times faster than constant speed one. In constant current cutting, current was 40 A, but in the case of 50 A, better

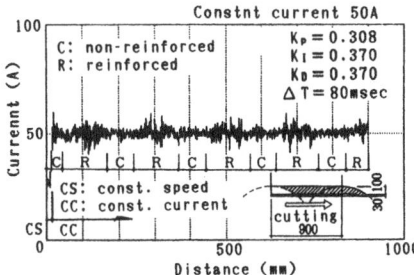

Fig. 14 Constant current cutting

Fig. 15 Cutting section (D51)

489

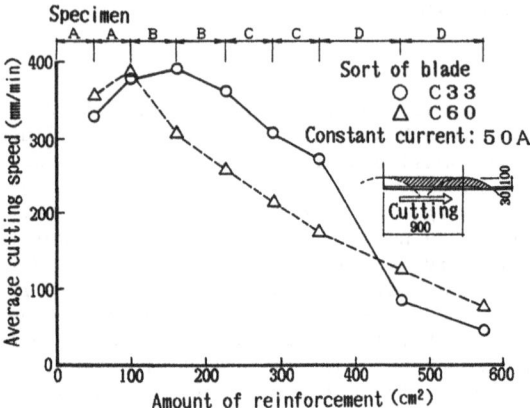

Fig. 16 Amount of reinforcement and average cutting speed

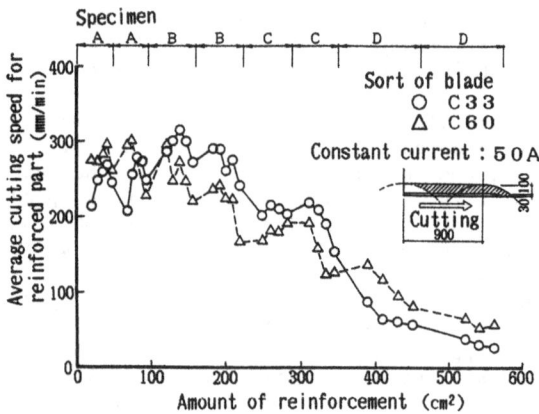

Fig. 17 Amount of reinforcement and average cutting speed
        for reinforced part

result was expected.

The main reasons of high cutting efficiency obtained are presumed
to be as below.

1 : The abrasive of cutting edge is the compound of diamond and
    boron nitride that has more thermal-resistance features than
    diamond for steel cutting.
2 : The softer metal bond which maintains abrasive so as to enable
    self dressing of cutting edge to be speedy.
3 : The cutter driving motor is 3 ∿ 12 times as powerful as the
    former one.
4 : High efficiency blade depresses heat generation, prevailing
    from melting and sticking.

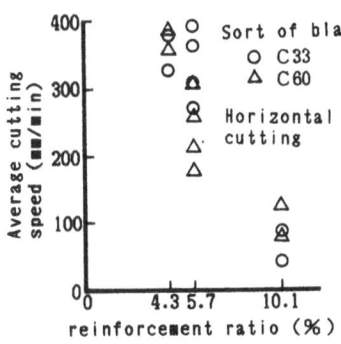

Fig. 18 Reinforcement Ratio and
average cutting speed

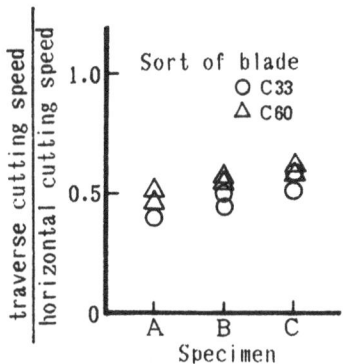

Fig. 19 Average speed for traverse
and horizontal cutting

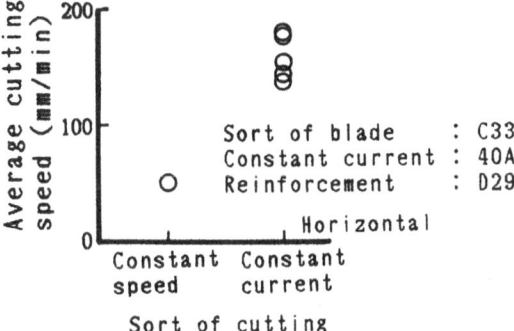

Fig. 20 Average speed for constant
current and constant speed

4.2 Remote control

In cutting, the shift from the early constant speed control to the
speed control responding to the load under constant current and the
shift from non-reinforced part to reinforced one were smoothly
switched as shown in Fig. 21 and 23. Especially, sudden increase of
current which was worried to be at the boundary between
non-reinforced and reinforced area did not occur (Fig. 24).

The varying of control interval (40∿100 msec) which was a
barometer of response speed did not afford bad influence on current
increment, range of current fluctuation and off-set (Fig. 21 and 22).
This demonstrates that the current intervals in the test are quite
adequate for heavily reinforced concrete. It is assumed that the
reason why there is no occurence of bad current increment is that
large number of cutting tips touch concrete because of large diamter
of the saw.

491

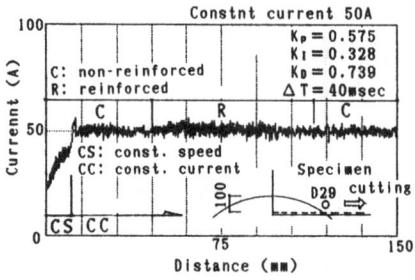

Fig. 21 Constant current cutting

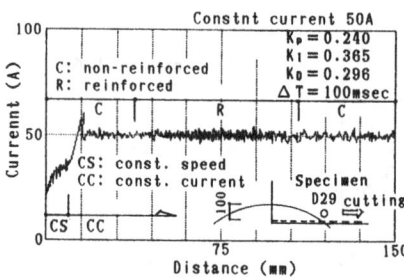

Fig. 22 Constant current cutting

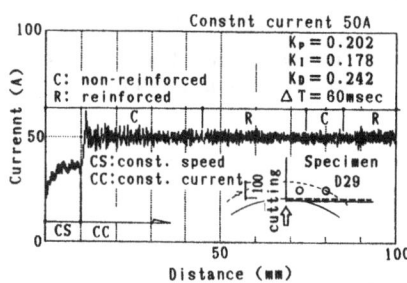

Fig. 23 Constant current cutting

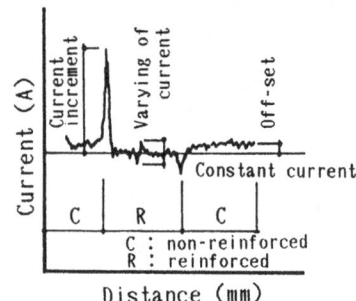

Fig. 24 Constant current cutting

5. Conclusion

By using actual size blades, the remote control cutting under
constant current in which speed was controlled responding to the load
due to the amount of steel reinforcement, successfully demonstrated
safe and precise cutting on schedule.

References

Hiraga, T. (1983) A study on the cutting technology of reinforced
    concrete member and its application in construction sites, Toda
    Tech. Res. Report.
Asakawa, O. et al. (1986) A study on dismantling method of biological
    shield wall by cutter (Part 1∿3), Summ. Tech. Papers Annual Meet.
    AIJ. 497-502.
Kondo, N. et al. (1987) A study on dismantling method of biological
    shield wall by cutter (Part 4∿5), Summ. Tech. Papers Annual Meet.
    AIJ. 311-314.
Takahashi, Y., Chan, C.S. and Auslander, D.M. (1970) Parameter tuning
    of linear DDC algorithms, ASME Paper, 70-WA/Aut-16.

ONE PIECE REACTOR REMOVAL SYSTEM

M.WATANABE, S.TANIGUCHI and K.SUGIHARA
Nuclear Power Department, Shimizu Corporation

Abstract
The Japan Research Reactor 3 (JRR3), a research reactor of Japan
Atomic Energy Research Institute (JAERI), has been utilized for the
production of radioisotope as well as for the development of reactor
technologies. In order to match recent versatile and highly
advanced researches, however, JAERI decided to remove the old
existing reactor and the equipment and pipings attached to it and to
install a new reactor at the same spot. As a removal method for the
reactor, the one piece removal, to remove the radio-activated core
materials as one block enclosed in the biological shield concrete
walls, has been developed for minimizing the worker's exposure to
radiation. The one piece removal of reactor, the world-first attempt
of its kind is the total engineering forming a system all based on
conventional technologies.
Key words: One piece removal, Reactor, Temporary structure, Oil jack,
Load equalizing, Roller, Radiation control.

1. Introduction

The Japan Research Reactor 3 (JRR3) is one of the research reactor of
Japan Atomic Energy Research Institute (JAERI). It reached the
criticality in 1962 as the first domestic reactor and it had been
used for various experiments and studies as well as for radioisotope
production.
    In order to meet recent versatile and highly advanced research
demands, JAERI decided that JRR3 should be reconstructed and upgraded
through replacing the existing reactor with a new reactor.
    General requirements for removal of nuclear facilities are mini-
mization of worker's exposure, reduction of radioactive waste, saving
of cost and time, etc. Besides above, in the JRR3 case, the removal
required planning to avoid damage on existing reactor building for
the purpose of its reuse.
    How to handle the life-expiring reactors has been under intensive
study in various countries in the world. In Japan, JAERI implemented
various full scale reactor disassembly tests in collaboration with
several companies.
    Through those tests JAERI and Shimizu Corporation had developed a
method to remove a reactor as one block without disassembling it,
"One Piece Reactor Removal". The one piece removal met the requirements mentioned above and JAERI selected it as the most effective
technique to remove the reactor of JRR3.

The preparation work started in February 1986 and the reactor removal was successfully completed in November of the same year. This paper describes planning, execution and results of the reactor removal of JRR3.

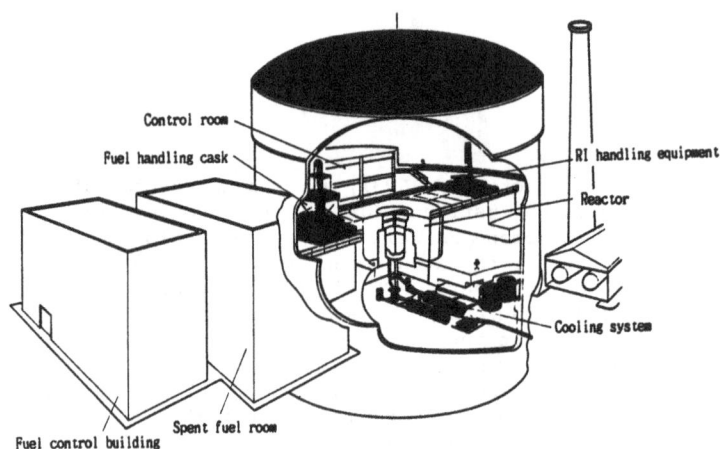

Fig. 1.   General view of JRR3
(before the reactor removal)

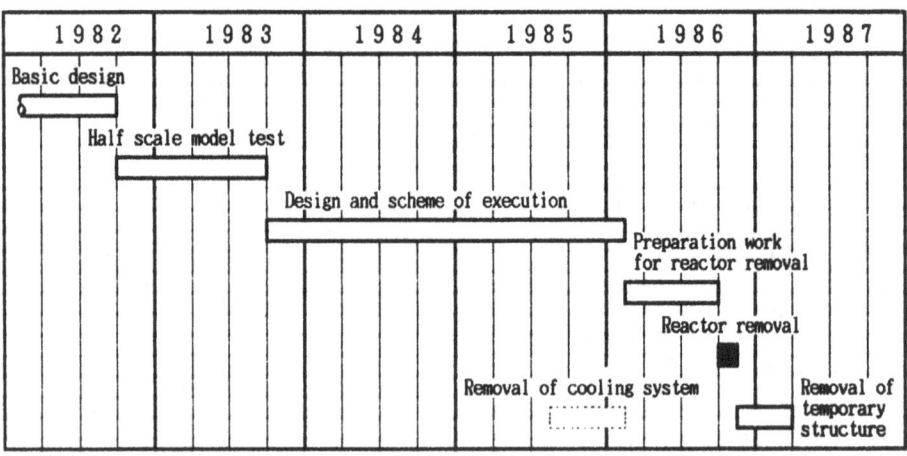

Fig. 2.   JRR-3 One Piece Reactor removal schedule